Recent Advances in Secondary Processing of Pharmaceutical Powders

Recent Advances in Secondary Processing of Pharmaceutical Powders

Editor

Colin Hare

MDPI

Basel • Beijing • Wuhan • Barcelona • Belgrade • Novi Sad • Cluj • Manchester

Editor
Colin Hare
School of Engineering
Newcastle University
Newcastle upon Tyne
United Kingdom

Editorial Office
MDPI
St. Alban-Anlage 66
4052 Basel, Switzerland

This is a reprint of articles from the Special Issue published online in the open access journal *Pharmaceutics* (ISSN 1999-4923) (available at: www.mdpi.com/journal/pharmaceutics/special_issues/pharm_powders).

For citation purposes, cite each article independently as indicated on the article page online and as indicated below:

Lastname, A.A.; Lastname, B.B. Article Title. *Journal Name* **Year**, *Volume Number*, Page Range.

ISBN 978-3-7258-0678-2 (Hbk)
ISBN 978-3-7258-0677-5 (PDF)
doi.org/10.3390/books978-3-7258-0677-5

Contents

Preface

Pharmaceuticals are becoming increasingly important in modern society. In most cases, active pharmaceutical ingredients (APIs) are formed through crystallisation but need to be combined with excipients to form a tablet, or with carriers to form an oral suspension. The formation of tablets requires several secondary processing steps, such as milling of the API and mixing with excipients (often involving granulation) before finally being tabletted. Due to the differences in physical and mechanical properties that individual APIs and excipients exhibit, these secondary processing steps are often fraught with challenges. This Special Issue comprises 11 research articles and 1 review which outline the recent advances in secondary processing of pharmaceutical powders, and aims to enhance our understanding of powder behaviour and help us tackle these challenges. Through careful experimentation and the application of sophisticated computational techniques, the authors of these papers demonstrate advances in our understanding of many of the processing steps for pharmaceutical products. With this Special Issue, I hope to engage a broad readership in order to spark new ideas to further advance our knowledge of powder processing and to optimise production.

Colin Hare
Editor

pharmaceutics

MDPI

Article

Experimental Elucidation of Templated Crystallization and Secondary Processing of Peptides

Vivek Verma [1,*], Isha Bade [1], Vikram Karde [1] and Jerry Y. Y. Heng [1,2,*]

1 Department of Chemical Engineering, Imperial College London, London SW7 2AZ, UK; isha.bade16@imperial.ac.uk (I.B.); v.karde@imperial.ac.uk (V.K.)
2 Institute for Molecular Science and Engineering, Imperial College London, London SW7 2AZ, UK
* Correspondence: v.verma@imperial.ac.uk (V.V.); jerry.heng@imperial.ac.uk (J.Y.Y.H.)

Abstract: The crystallization of peptides offers a sustainable and inexpensive alternative to the purification process. In this study, diglycine was crystallised in porous silica, showing the porous templates' positive yet discriminating effect. The diglycine induction time was reduced by five-fold and three-fold upon crystallising in the presence of silica with pore sizes of 6 nm and 10 nm, respectively. The diglycine induction time had a direct relationship with the silica pore size. The stable form (α-form) of diglycine was crystallised in the presence of porous silica, with the diglycine crystals obtained associated with the silica particles. Further, we studied the mechanical properties of diglycine tablets for their tabletability, compactability, and compressibility. The mechanical properties of the diglycine tablets were similar to those of pure MCC, even with the presence of diglycine crystals in the tablets. The diffusion studies of the tablets using the dialysis membrane presented an extended release of diglycine through the dialysis membrane, confirming that the peptide crystal can be used for oral formulation. Hence, the crystallization of peptides preserved their mechanical and pharmacological properties. More data on different peptides can help us produce oral formulation peptides faster than usual.

Keywords: pharmaceutical manufacturing; peptides; crystallization; tableting; powder mixing

Citation: Verma, V.; Bade, I.; Karde, V.; Heng, J.Y.Y. Experimental Elucidation of Templated Crystallization and Secondary Processing of Peptides. *Pharmaceutics* **2023**, *15*, 1288. https://doi.org/10.3390/pharmaceutics15041288

Academic Editor: Anne Marie Healy

Received: 31 March 2023
Revised: 15 April 2023
Accepted: 18 April 2023
Published: 20 April 2023

1. Introduction

Peptides are a category of biomolecules usually with a molecular weight of 500–5000 Da. They closely mimic natural pathways and can exhibit increased potency and high selectivity [1,2]. The development of novel synthesis strategies to produce peptides with modulated pharmacokinetic properties and target-specificity has resulted in more accessible pharmaceutical-grade peptides [3]. Since the synthesis of the first therapeutic peptide, human insulin (a peptide with 51 amino acids), in 1921, more than 80 peptide-based therapeutics have been approved by different drug agencies and have been launched on the market, while more than 500 peptides are in pre-clinical development and 150 in clinical trials [1,4]. As a result of the increased investment and research efforts in the field of peptides, the maturing of peptide synthesis technology, the success of biologics, and pressure on the pharmaceutical industry to maintain approval rates for new drugs, we anticipate peptide therapeutics to continue growing and expanding. Thus, the development of peptide drugs is the most emergent topic in pharmaceutical research.

Regardless of whether a peptide is formed synthetically or expressed using recombinant technology, the purification and/or isolation of peptides are often the bottlenecks of the manufacturing process [5]. High-performance liquid chromatography (HPLC) is the most common technique used for purification and ion exchange. However, the large amount of aqueous/organic waste generated from the purification process makes it an unsustainable process. Alternatively, precipitation can be used to purify and isolate peptides, but precipitation can result in amorphous solids or significant solvent-adsorbed volumes.

Therefore, crystallization is often viewed as an environmentally friendly and economically advantageous substitute to chromatographic separations since it uses much less solvent [6]. In addition, crystallization offers us the opportunity to improve particle attributes, such as crystal shape, size, and form. However, the flexible nature of the peptide molecules poses a challenge to their crystallization, therefore requiring researchers to improve the peptide crystallization process [7,8].

The presence of templates (either soft or hard) has been shown to control the crystallization of proteins, such as insulin [9] and lysozymes [10,11], but sometimes it also decelerates the crystallization process by delaying the induction time [12]. Link and Heng have demonstrated the influence of dissolved amino acids acting as soft templates for insulin crystallization. The intermolecular interactions between insulin and the l-arginine molecule in their study led to insulin stabilization, resulting in an improvement in the crystal occurrence compared to that of l-glycine [9]. Similarly, Li et al. confirmed that the presence of silica particles facilitated the crystallization of lysozymes with an improved induction time and crystal yield [11]. Furthermore, the presence of glass beads accelerated the nucleation of glycine, diglycine, and triglycine. This was potentially due to the enhanced number of probable non-covalent interactions between the hydrogen bond donor of glycine and its homopeptides and the hydrogen bond acceptor of the glass beads. This was further corroborated with the molecular dynamics simulations, which confirmed the enhanced residence time of triglycine on the surface of the glass beads due to an increased number of hydrogen bond interactions [13,14].

Although crystallization could potentially improve the purification and isolation of peptides, the use of peptide crystals in drug products is far from a reality. In addition to the advantage of target specificity and selectivity, the unfavourable physicochemical properties of peptides, such as large molecular weight, inactivation by gastric pH values, hydrophilicity, low intestinal permeability, and susceptibility to digestive enzymes, work against successful oral peptide delivery [15]. Therefore, the most widely used route of peptide administration is the parenteral route, which is generally inconvenient, painful, and requires medical knowledge. Although, many drugs other than peptides, such as proteins, monoclonal antibodies, small molecule drugs, etc., are administered parenterally, it is estimated that ~5% of the population strongly prefers other routes of drug administration [16]. Therefore, the oral delivery route is considered as the preferred route of administration as it is non-invasive and has a high level of patient compliance. Additionally, it provides a chance to extend the patent life of expiring injectable peptides with novel formulations.

The oral delivery of peptides is influenced by food and water intake [17]. Further, the inherent physical and chemical properties of peptides, such as molecular size, proteolytic stability, hydrophilicity, and ionic charge, also influence their oral absorption [18]. Although oral peptides and proteins face barriers to delivery, the number of formulations that have been launched or are in clinical trials is steadily increasing. There are multiple strategies for improving the oral delivery of peptides, including permeation enhancers (PEs), multiparticle systems, targeted particles, nanotechnology, enzyme inhibitors, colonic delivery methods, and modifications to the peptides themselves (cyclisation or the use of non-natural amino acids) [19–23]. Since the publication of the first paper reporting the use of alcohol to improve the oral absorption of insulin in 1923 [24], other researchers have also attempted to develop peptide oral formulations. This led to the approval of Sandimmune®, the first oral formulation of a cyclic peptide, cyclosporin A, by the FDA in 1990, followed by the approval of Neoral® developed by Novartis, which was an updated formulation of Sandimmune® [22]. Later, oral semaglutide (Rybelsus® in 2019) and oral octreotide (Mycapassa® in 2020) were approved by the FDA. These formulations demonstrated that the oral delivery of peptides is feasible if the peptides and formulations are optimized for the routes of administration. Crystallization offers peptide stability, therefore diversifying the potential formulations for various routes of administration, including oral dosage. Additionally, it offers the controlled release of the peptide from the polymeric encapsulation, leading to an improvement in the half-life of peptides [25]. Peptide crystals further allow for high-concentration doses without increasing the viscosity of the suspension as well as the ability to reach a concentration greater than 200 mg/mL in the final dosage form, leading to a higher bioavailability and lower dosage requirement [25]. Hence, peptide crystallization offers great applicability for oral doses of the current peptide molecules on the market.

At present, oral formulations are available for several hormones, including insulin, vasopressin, somatostatin, calcitonin, parathyroid hormone (PTH), uroguanylin, thyroid hormone-releasing hormone, and GLP1. These treatments can be classified into two categories based on their intended action within the body—those that require oral absorption and those that require retention in the gastrointestinal tract [26,27]. All of these formulations use one of the above-mentioned strategies; more specifically, well-known permeation enhancers such as sodium caprate (C10; also known as decanoic acid) and sodium N-[8-(2-hydroxybenzoyl)amino] caprylate (SNAC; also known as salcaprozate sodium) improve the transcellular and paracellular permeation [20]. However, there has not yet been a report of an oral formulation of the peptide using just the peptide crystals on their own. This is potentially due to the fact that no peptide drug is small enough to be compliant with Lipinski's rule of five (MW < 500 g/mol, H-bond donors < 5 and H-bond acceptors < 10, LogP < 5, rotatable bonds < 10, and total polar surface area < 140 Å^2) for predicting good absorption and permeation [28]. Surprisingly, the peptides that have been identified as having good oral activity and that could be potential candidates for oral formulations, in addition to the one in the clinical trial, are mostly cyclic [29]. This suggests that further research is needed to explore more cyclic peptides with good oral activity.

This work focuses on improving the crystallization of peptides using the templated crystallization approach, already established by the authors [9,10,13,14]. The templates interact with the crystallising solute through functional group complementarity, allowing the sequestration of solute molecules on the surface of the templates with h-bond interactions for a long enough time to achieve a fully grown crystal. This was recently complemented with a molecular dynamics simulation, as mentioned earlier in the section [14]. In this work, diglycine was crystallised in the presence of different pore sizes of silica particles to examine the effect of template pore size on the crystallization time and rate. Diglycine is made by combining the simplest amino acid, i.e., glycine, with itself though a peptide bond, making it the smallest known peptide, which behaves similarly to a small molecule due to the lack of degrees of freedom, defined unit cell, and well-defined intermolecular contacts. The crystallised peptide was later blended with a widely used pharmaceutical excipient, microcrystalline cellulose (MCC) [30], to obtain an oral formulation. The formulation was later compressed to form tablets, which were studied to obtain the tabletability, compressibility, and compactibility for the peptide tablets. This is the first study reporting the tablet properties of a peptide formulation. The tablets were later studied to obtain the permeability rate of different formulations.

2. Materials and Methods

2.1. Materials

Diglycine (Digly, >99% by titration) and microcrystalline cellulose (MCC, Avicel® PH-101, 50 μm particle size) were supplied by Sigma-Aldrich and used as received. Deionized (DI) water was supplied by a Sartorius Arium® Advance (Göttingen, Germany). Silica particles (40–63 μm) of different pore sizes (6 nm, 10 nm, 30 nm, and 50 nm) were purchased from Element and used as received.

2.2. Isothermal Colling Crystallization Experiments

Mettler-Toledo EasyMax 102 was used for the cooling crystallization experiments, allowing for precise control over experimental variables, such as reactor temperature, stirring rate, and heating and cooling rates. A diglycine solution (concentration of 284.28 mg/mL, total volume of 40 mL, and saturation temperature of 40 °C) was prepared in DI water, as per the diglycine solubility data published previously by our group [31]. The saturated diglycine solution was heated to 5 °C above the saturation temperature for complete dissolution of diglycine. This solution was cooled to 32.7 °C to induce crystallization at relative supersaturations of 1.20, which is in the metastable zone width limit, enabling us to capture the effect of templates on heterogenous nucleation of diglycine. The change in concentration upon nucleation of diglycine in the absence and presence of porous silica particles ((10% w/w loading) was captured using Mettler-Toledo ReactIR 15 system, an in situ Fourier transform infrared (FTIR) probe. Each experiment was carried out at least twice to ensure the reproducibility of the results. The induction time was accessed from the desupersaturation curves using the tangent method that was previously used by the authors [14].

2.3. *Dynamic Light Scattering (DLS)*

DLS was carried out to measure the hydrodynamic radii of diglycine in water. A dilute solution of glycine was prepared, filtered using a disposable syringe filter with a pore size of 0.2 μm, and analysed using Zetasizer μV (Malvern, UK). Samples with a polydispersity index lower than 0.10 were used for the size measurements.

2.4. *Solid State Characterization of Isolated Solids*

2.4.1. Powder X-ray Diffraction (PXRD)

PXRD was recorded for the diglycine porous silica composites isolated after the isothermal cooling crystallization experiments using a PANalytical Empyrean diffractometer (Malvern Panalytical, Malvern, UK) with a Cu radiation source (λ = 1.541 nm) at 40 mA and 30 kV. Scans were performed between 5 and 35° 2θ at a scan rate of 0.013° 2θ/min.

2.4.2. Scanning Electron Microscopy (SEM)

The as-received and isolated samples from the crystallization experiments were analysed for the surface morphology and crystal habit using SEM (Zeiss LEO Gemini 1525, Cambridge, UK) at a working distance (WD) of 5–8 mm and a voltage of 5 kV.

2.4.3. True Density Measurement

The Micromeritics® Accupyc II helium gas displacement pycnometer was used to measure true density of the samples at ambient temperature (25 °C) based on USP 699 standard procedure. True density measurements were perfomed using the 1 cm^3 sample cell and 20 purge cycles.

2.5. *Tabletting of the Diglycine–Silica–MCC Composite*

The diglycine–silica–MCC composite tablets were compressed using a Gamlen R-series compaction simulator (Gamlen Tableting Limited, Beckenham, UK). Composite tablets of 100 mg (22.5 mg diglycine, 2.5 mg silica, and 75 mg MCC) were prepared using a cylindrical 6 mm punch die at a punch speed of 10 mm/min. All the samples were compacted at compaction pressures of 1–3 kN with a step increase of 0.5 kN to thoroughly study the compaction behaviour of diglycine–silica–MCC composite powder. MCC was also compressed at the same compaction pressure as the composite and acted as the positive control. The tensile strength of the tablets was measured using the EZ50 equipment (Lloyd Instruments, Bognor Regis, UK) equipped with a load cell on 10 N and an operating speed of 0.1 mm/min. The tabletability, compactability, and compressibility of the powder were evaluated to assess its mechanical properties.

2.5.1. Tablet Porosity

Equation (1) was used to calculate tablet porosity.

$$Tablet\ porosity = 1 - \frac{\rho_{app}}{\rho_{true}} \tag{1}$$

The ρ_{true} is the true density (g/cm^3) calculated using the helium pycnometer mentioned in Section 2.4.3, while ρ_{app} is the apparent density calculated by dividing tablet weight by its volume.

2.5.2. Tabletability

The tabletability of a material is represented by the relationship between its tensile strength and the compaction pressure applied. Tabletability is expressed by a linear relation between tensile strength and compaction force, as given by Newton et al. [32] and presented in Equation (2).

$$\sigma_t = C_p P + b \tag{2}$$

where P is the compaction pressure, C_p is the tabletability parameter, and b is a constant.

2.5.3. Compactability

An exponential relation between the tensile strength and porosity expresses the compactability of a composite. Ryshkewitch–Duckworth proposed a mathematical equation to understand compactability, as shown in Equation (3) [33].

$$\sigma_t = \sigma_{t0} e^{(-bP)} \tag{3}$$

where tablet tensile strength (MPa) is represented by σ_t and tablet tensile strength at zero porosity (MPa) by σ_{t0}. The tablet porosity is given by P, while b is an empirical constant representing bonding capacity, in which stronger bonding between primary particles is expressed by higher b value [34].

2.5.4. Compressibility

Compressibility profile shows the change in the tablet porosity with the increasing compaction pressure. A tablet with low porosity is associated with the capping problem and can result in slow tablet dissolution. Compressibility can be assessed by change in tablet porosity with compaction pressure as expressed by Heckle's model according to Equation (4).

$$-\ln \varepsilon = \ln\left(\frac{1}{1-D}\right) = kP + A \tag{4}$$

The compressibility of a powder can be indicated by the values of the Heckle coefficient (k) and its reciprocal giving yield pressure (P_y). In the Heckle's model, tablet density, relative tablet density, compression pressure, and intercept are represented by ε, D, P, and A, respectively.

2.6. *In Vitro Diffusion Studies*

The diffusion profile of diglycine forming the composite tablets was studied using a dialysis tube as a substitute for the intestinal membrane [35]. Diffusion tests were completed using the Pur-A-Lyzer™ Maxi Dialysis Kit-Maxi 6000 (Sigma-Aldrich Co., LLC, London, UK) with a molecular weight cut-off of 6–8 kDa and volume capacity of 0.1–3 mL. The diffusion was performed in 100 mL water in a Duran flask, at 37 °C in a water bath, and stirred at 100 rpm. The dialysis tubes were soaked in the diffusion media for 30 min before adding the tablet. The dialysis tube contained 2 mL water and the composite tablet, while the flask contained 98 mL water, making a total volume of 100 mL. All the diffusion experiments were repeated at least three times on different days to exclude human error. Samples of 1 mL aliquots were withdrawn from the diffusion flask at fixed times of 1, 3, 5, 10, 15, 30, 45, 60, 90, 120, and 1440 min, and were analysed by UV spectrophotometry at 215 nm. A calibration curve was previously made at this wavelength with an R2 value of 0.99 and was used for the determination of concentration.

3. Results and Discussion

The induction time of the diglycine crystallised in the absence and presence of the porous silica templates is presented in Figure 1. The silica templates have a positive yet discriminating effect in that they reduce the time required for crystallising diglycine at a supersaturation of 1.20. The significant reduction in the induction time (the time required to observe a significant change in the solution concentration from the isothermal holding time at the crystallization temperature) [30] of diglycine was observed when it was crystallised in the presence of silica with a pore size of 6 nm.

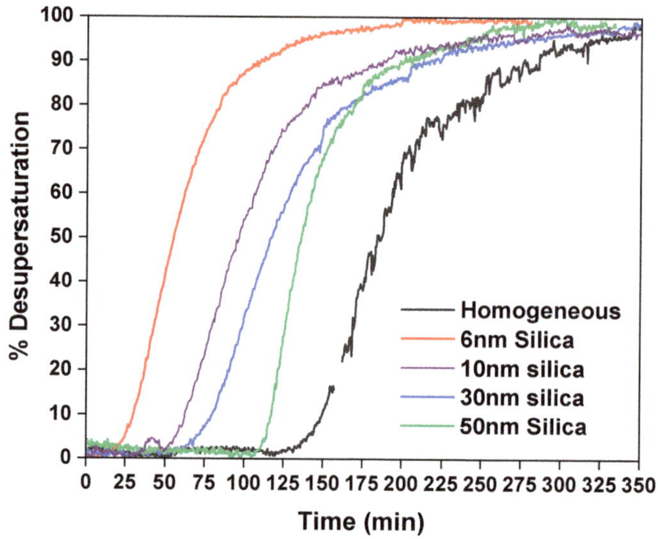

Figure 1. Comparison of % desupersaturation curves of diglycine in the absence and presence of porous silica at S = 1.20; volume = 40 mL; T_{sat} = 40 °C; T_{cry} = 32.7 °C.

The reduction in the induction time in the presence of porous silica has a direct relation with the pore size. The smallest silica with a pore size of 6 nm exhibited a five-fold reduction in the induction time compared to that of the homogeneous nucleation, as presented in Table 1. This was followed by silica with pore sizes of 10 nm, 30 nm, and 50 nm, which had 3-fold, 2-fold, and 1.5-fold reductions, respectively.

Table 1. Average induction time of diglycine in the absence and presence of porous silica at S = 1.20; volume = 40 mL; T_{sat} = 40 °C; T_{cry} = 32.7 °C.

Pore Diameter	Induction Time (min)
Homogeneous	145 ± 7
6 nm pore	29 ± 5
10 nm pore	43 ± 17
30 nm pore	72 ± 3
50 nm pore	115 ± 7

This direct relation of pore size to the reduction in the induction time is potentially due to the diglycine cluster size matching of pore sizes. DLS data suggest a hydrodynamic radius of 0.94 nm for diglycine in water, while a unit cell of diglycine contains four diglycine molecules [13], bringing the diglycine cluster size in the range of silica with a 6 nm pore size. A pore size (length) of 6 nm is sufficient to sequester a critical size cluster of diglycine, creating local supersaturation and resulting in the nucleation of diglycine in the pores. Similar results were obtained by Shah et al. [36] when different-molecular-weight proteins were crystallised using the templates with engineering pores with an optimum size similar to their hydrodynamic radius. Further, an antibody named anti-CD20 was crystallised using the specifically designed porous silica template with a pore size in the range of the molecular diameter of the antibody [37,38]. Silica with a 10 nm pore size also influenced the diglycine induction time; 6 nm and 10 nm pores are significantly not different, and hence are also able to crystallise diglycine and reduce the induction time. Due to the large pore size of the other pores' silica, the diglycine induction time was not affected much by the other pores' silica compared to the silica with a pore size of 6 nm.

Figure 2 presents a solid-state analysis of the isolated composite solids from the desupersaturation experiments in the presence of different-pore-size silicas. The PXRD graphs in Figure 2A show the crystallization of diglycine stable form (α-form) in the presence of porous silica. There is a preferred orientation observed for the peaks at (100) and (20–2) when crystallised in the presence of porous silica. This is potentially due to the non-interaction between the functional groups (-COO, carboxyl group) present on these surfaces with the silica hydroxyls. Figure 2B presents the SEM micrographs of the isolated solids along with the bare silica and diglycine. The diglycine crystals obtained in the presence of porous silica seem to be associated with the silica particles, as observed in the SEM micrographs, suggesting the crystallization of diglycine either on the silica surface or the pores. The particle size of the diglycine crystals obtained is in the range of 20–100 μm.

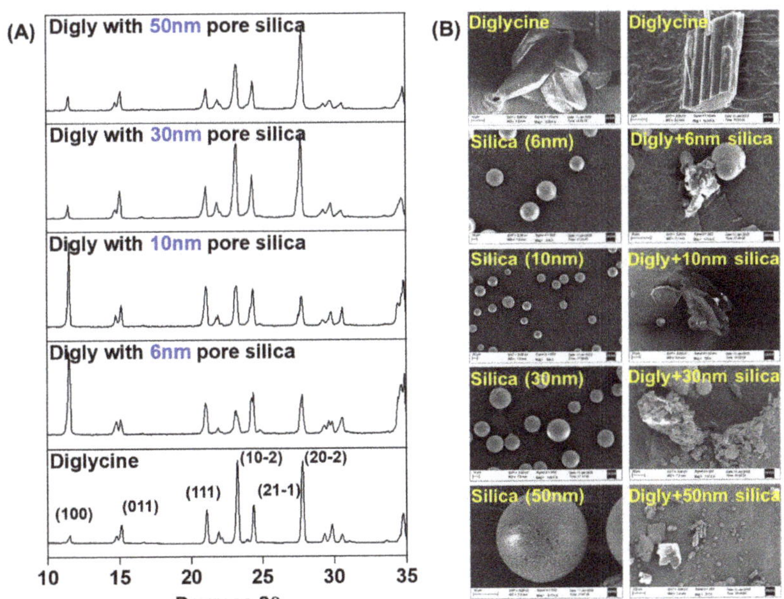

Figure 2. (**A**) Powder X-ray diffraction spectra of the diglycine–silica composite solids isolated upon complete desupersaturation of diglycine in the presence of porous silica at S = 1.20, along with the diglycine patterns; (**B**) Scanning electron microscopy images of diglycine, porous silica, and isolated solids after the desupersaturation experiments in the presence of porous silica.

The physicochemical and pharmacodynamic behaviours of pharmaceutical tablets are influenced by their mechanical strength [39,40]. A comparison of the tabletability, compressibility, and compactability profiles of the diglycine–silica–MCC composite tablets was conducted. This is the first study reporting the mechanical properties of the simplest peptide of glycine, diglycine. Tabletability is a function of tensile strength and compaction force. The tabletability profile of the diglycine–silica–MCC composite along with the pure MCC is presented in Figure 3 (top). As a widely used pharmaceutical excipient, MCC has been extensively studied and reported in the literature, and can be easily compressed with an excellent tensile strength. Hence, MCC was selected as the excipient to be used in this study [41]. Due to its easy compressibility, the tensile strength of MCC is >2 MPa at any compression force, while the tensile strength of the diglycine composite is >2 MPa after 2 kN, irrespective of the silica pore size. Pharmaceutical tablets exhibiting a tensile strength greater than 2 MPa are considered as passing the standard requirements for manufacturability, quality, and biopharmaceutical performance [42]. The tabletability parameter (C_p) of the tablets can be obtained using the Newton equation [32]. The C_p value of 1.61 exhibited by MCC is the highest (reported in Table 2) compared to that of the diglycine composite, indicating tablets for the diglycine composite are less stable compared to MCC. This is due to the presence of glycine crystals in the composite tablets.

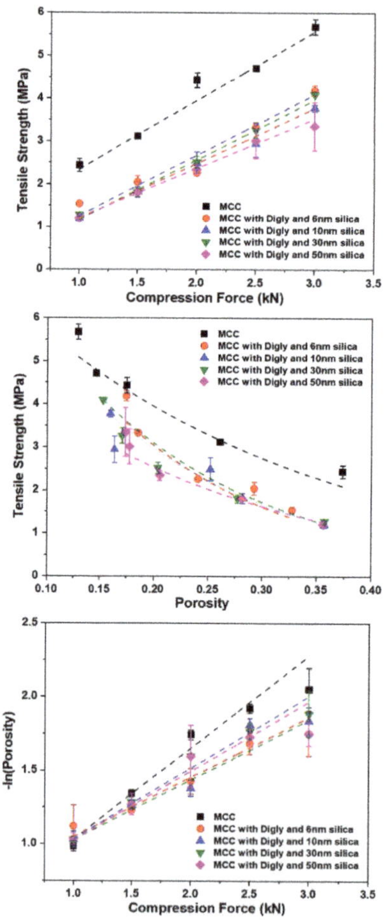

Figure 3. (**Top**) Tabletability, (**Middle**) compactability, and (**Bottom**) compressibility profiles of the 25% loading blend of diglycine–silica–MCC composite tablets prepared along with MCC alone (blue squares, PH101). Red circles, blue upwards triangles, green downwards triangles, and pink rhombuses represent diglycine crystallised using silica with pore sizes of 6 nm, 10 nm, 30 nm, and 50 nm as templates, respectively, with n $\geq$ 3 (n is the number of experiments).

Table 2. Summary of the mechanical parameters obtained from the curve fitting of tabletability, compressibility, and compactability curves for diglycine–silica–MCC composites and bare MCC tablets (C_p is the tabletability parameter, σ_{t0} is the tablet tensile strength at zero porosity, b is an empirical constant representing bonding capacity, k is the Heckle Coefficient, and P_y is the yield pressure).

Sample	Tabletability	Compactability		Compressibility	
	C_p	σ_{t0}	$-b$	k	P_y (kN)
MCC	1.61	7.98	3.30	0.54	1.84
Digly-6 nm pore size Silica-MCC	1.33	10.40	5.82	0.34	2.90
Digly-10 nm pore size Silica-MCC	1.26	7.95	5.16	0.43	2.32
Digly-30 nm pore size Silica-MCC	1.41	8.34	5.38	0.44	2.27
Digly-50 nm pore size Silica-MCC	1.10	7.60	5.19	0.38	2.64

Figure 3 (middle) presents a tensile strength vs. porosity graph, representing the compactability of the tablets. Compactability is the ability of a powder to reduce porosity under an applied pressure. The curve-fitting parameters obtained (reported in Table 2) using the Ryshkewitch–Duckworth equation [43] were used to analyse the compactability of the composite tablets. The tensile strength at zero porosity (σ_{t0}) is higher for all the composite samples except for the silica composite with a 50 nm pore size compared to that of MCC. This suggests that the porosity of the composite tablets is less than that of MCC. The composite with silica with a 50 nm pore size could potentially have more agglomerated samples, resulting in a high porosity in the tablets. Further, a higher value of the constant '*b*' represents a stronger bonding between the primary particles [34]. The value of '*b*' is higher for all the composite samples compared to that of MCC, suggesting that the interparticle bonding is higher in the composite tablets than the MCC tablets, which is due to the presence of silica in the composite tablets filling the interparticle distance, as also observed in the SEM micrographs (Figure 2B) with the silica particles associated with the diglycine crystals.

The compressibility of the powders was assessed using the Heckle equation. The values for the Heckle coefficient, k, which were obtained from the slope of the Heckle plot, and its inverse values, i.e., the yield pressure, (P_y), for the different composite systems along with that of pure MCC are presented in Table 2. The high value of k and the low value of P_y suggests the good plasticity of MCC. On the contrary, the diglycine composite exhibit an inverse behaviour to that of pure MCC, with low values for k and higher values for P_y, indicating poor plasticity and low compressibility properties. This is due to the presence of diglycine crystals, which exhibit poor compressibility, leading to high porosity in the tablets.

There were no differences observed in the flow properties of the formulations, and this potentially was due to the presence of 75% *w*/*w* microcrystalline cellulose (MCC) in the formulation mixture. This made the formulation mixture flow more like pure MCC. The silica tested in the study were mostly identical from outside in terms of their shape, size, and functionality, while the only difference was in their internal pore diameter. Due to the identical nature of the silica tested, there was no obvious difference in the flow properties. Moreover, variations in the particle size distribution of the crystallised diglycine will also influence the flow properties of the formulation mixture. Particles with a larger crystal size distribution are more likely to have poor flow properties due to the plate-shaped diglycine crystals resisting a smooth flow, whereas this may not be the problem with crystals of a smaller size distribution. However, this was not observed with the crystals in this work, as all the crystalline particles obtained from the templated crystallization had a similar size distribution, as can be seen in the SEM micrographs in Figure 2B. A similar phenomenon was observed by a group member in the past, confirming that elongated needles were more cohesive than hexagonal crystals, which is due to the combined effects of surface energy and surface area leading to poor flow behaviour due to the high cohesivity of the crystals [44].

The diglycine–silica–MCC tablets were subjected to the diffusion test using the dialysis tube to observe their diffusion rate from the dialysis membrane. Diglycine is readily dissolved in water based on its solubility data, which have already been published in the literature [31], and hence, performing the dissolution study is futile. In contrast, the low intrinsic permeability of peptides is a bottleneck in the oral delivery of peptides. The oral absorption of peptides in the body occurs through three possible pathways: [18]

i. paracellular, involving passing in between enterocyte cells, only allowing smaller peptides with a low molecular weight;
ii. transcellular, involving passage through the enterocyte cell with the help of other cells;
iii. ligand-assisted transport, allowing transportation through the intestinal mucosal membrane using a permeation enhancer.

The molecular weight, flexibility, and hydrophilic nature of smaller peptide-based drugs provide them the advantage of paracellular passage in between cells [18]. This phenomenon was later confirmed by Foger et al. [45] in a report that illustrated that the peptide permeability increases as the molecular weight decreases, but only for peptides with a molecular weight up to 1.4 kDa.

Figure 4 presents the diffusion of diglycine over a period through the permeation membrane with a molecular weight cut-off of 6–8 kDa. The diffusion rate experiments were performed using the diglycine composite tablets compressed at 2.5 kN, as the tablets had the desired tensile strength of more than 2 MPa at this compression force. All the tablets exhibited an extended-release profile, confirming their easy diffusion through the permeation membrane with approximately a 50% diffusion achieved in the first hour and a complete dissolution in 12–16 h. This steady diffusion suggests that permeation enhancers are not required for small peptides, as mentioned by Klepach et al. [18].

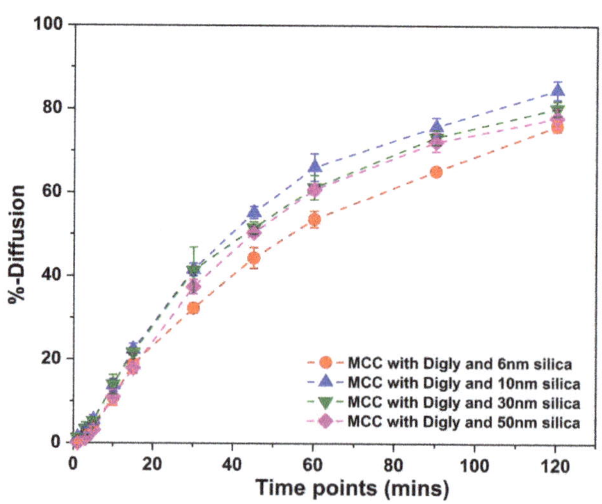

Figure 4. %diffusion of diglycine from the diglycine, silica, and MCC composite tablet compressed at 2.5 kN. The diffusion medium was water; volume = 100 mL; and stirring = 100 rpm.

Although diglycine was crystallised in the presence of different-pore-size silica, except for influencing the induction time of diglycine, the porous silica had a minimal impact on the mechanical properties of the tablets. The presence of silica improved the tablet properties by reducing the tablet porosity, thereby increasing the tensile strength of the tablets. Therefore, templated crystallization using pharmaceutical excipients as the templates could potentially not only improve the downstream processing time, but also improve the particle attributes, such as crystal size, shape, and form, also improving the mechanical properties of the tablets. However, more work is needed in this area to prove the ability of templates to crystallise larger peptides and formulate them in oral tablets to study their mechanical properties.

4. Conclusions

In this study, diglycine, the simplest glycine homopeptide, was crystallised in the presence of porous silica, presenting a template-assisted crystallization of peptides. The induction time of diglycine was significantly reduced in the presence of silica with 6 nm and 10 nm pore sizes due to the size complementarity in the cluster size of diglycine and the pores. The induction time of diglycine had a direct relationship with the silica pore size. The induction time increased with the increase in the silica pore size, but the presence of silica had a positive and discriminating effect. Further, our PXRD graphs suggested the crystallization of a stable α-form in the presence of porous silica, while all the crystals were seen to be associated with the silica particles in our SEM micrographs. The mechanical properties of the diglycine–silica–MCC tablets were accessed through tabletability, compactability, and compressibility curves. The mechanical properties of the tablets remained uninfluenced by the presence of diglycine–silica and presented similar tablet properties as those of MCC alone. Lastly, our diffusion analysis of the diglycine composite tablets presented an extended release with a constant flux of diglycine through the membrane. The oral formulation of peptides is not limited to diglycine, and hence more research is needed in the oral delivery of peptides.

Author Contributions: Conceptualization, V.V.; methodology, V.V., I.B. and V.K.; validation, V.V., I.B. and V.K.; formal analysis, V.V., I.B. and V.K.; investigation, V.V., I.B. and V.K.; resources, J.Y.Y.H.; data curation, V.V., I.B. and V.K.; writing—original draft preparation, V.V.; writing—review and editing, V.V., I.B. and J.Y.Y.H.; visualization, V.V.; supervision, J.Y.Y.H.; project administration, V.V.; funding acquisition, J.Y.Y.H. and V.V. All authors have read and agreed to the published version of the manuscript.

Funding: The results incorporated in this paper have received funding from the European Union's Horizon 2020 research and innovation program under the Marie Sklodowska-Curie grant, agreement no. 101026339. The results were collected at the Centre for Rapid Online Analysis of Reactions at Imperial College London (EPSRC grant nos. EP/R008825/1 and EP/V029037/1).

Institutional Review Board Statement: Not applicable.

Informed Consent Statement: Not applicable.

Data Availability Statement: Not applicable.

Acknowledgments: The authors would like to thank Bed Deadman and Christopher S. Roberts for their help with the crystallization experiments.

Conflicts of Interest: The authors declare no conflict of interest.

References

1. Wang, L.; Wang, N.; Zhang, W.; Cheng, X.; Yan, Z.; Shao, G.; Wang, X.; Wang, R.; Fu, C. Therapeutic peptides: Current applications and future directions. *Sig. Transduct. Target Ther.* **2022**, *7*, 48. [CrossRef]
2. Muttenthaler, M.; King, G.F.; Adams, D.J.; Alewood, P.F. Trends in peptide drug discovery. *Nat. Rev. Drug Discov.* **2021**, *20*, 309–325. [CrossRef] [PubMed]
3. Craik, D.J.; Fairlie, D.P.; Liras, S.; Price, D. The Future of Peptide-based Drugs. *Chem. Biol. Drug Des.* **2013**, *81*, 136–147. [CrossRef] [PubMed]
4. Ferrazzano, L.; Catani, M.; Cavazzini, A.; Martelli, G.; Corbisiero, D.; Cantelmi, P.; Fantoni, T.; Mattellone, A.; De Luca, C.; Felletti, S.; et al. Sustainability in peptide chemistry: Current synthesis and purification technologies and future challenges. *Green Chem.* **2022**, *24*, 975–1020. [CrossRef]
5. Isidro-Llobet, A.; Kenworthy, M.N.; Mukherjee, S.; Kopach, M.E.; Wegner, K.; Gallou, F.; Smith, A.G.; Roschangar, F. Sustainability Challenges in Peptide Synthesis and Purification: From R&D to Production. *J. Org. Chem.* **2019**, *84*, 4615–4628. [CrossRef]

6. Roque, A.C.A.; Pina, A.S.; Azevedo, A.M.; Aires-Barros, R.; Jungbauer, A.; Di Profio, G.; Heng, J.Y.Y.; Haigh, J.; Ottens, M. Anything but Conventional Chromatography Approaches in Bioseparation. *Biotechnol. J.* **2020**, *15*, 1900274. [CrossRef]
7. Guo, M.; Jones, M.J.; Goh, R.; Verma, V.; Guinn, E.; Heng, J.Y.Y. The Effect of Chain Length and Conformation on the Nucleation of Glycine Homopeptides during the Crystallization Process. *Cryst. Growth Des.* **2023**, *23*, 1668–1675. [CrossRef]
8. Guo, M.; Rosbottom, I.; Zhou, L.; Yong, C.W.; Zhou, L.; Yin, Q.; Todorov, I.T.; Errington, E.; Heng, J.Y.Y. Triglycine (GGG) Adopts a Polyproline II (pPII) Conformation in Its Hydrated Crystal Form: Revealing the Role of Water in Peptide Crystallization. *J. Phys. Chem. Lett.* **2021**, *12*, 8416–8422. [CrossRef]
9. Link, F.J.; Heng, J.Y.Y. Enhancing the crystallisation of insulin using amino acids as soft-templates to control nucleation. *CrystEngComm* **2021**, *23*, 3951–3960. [CrossRef]
10. Chen, W.; Park, S.J.; Kong, F.; Li, X.; Yang, H.; Heng, J.Y.Y. High Protein-Loading Silica Template for Heterogeneous Protein Crystallization. *Cryst. Growth Des.* **2020**, *20*, 866–873. [CrossRef]
11. Li, X.; Heng, J.Y.Y. Protein crystallisation facilitated by silica particles to compensate for the adverse impact from protein impurities. *CrystEngComm* **2021**, *23*, 8386–8391. [CrossRef]
12. Li, X.; Heng, J.Y.Y. The critical role of agitation in moving from preliminary screening results to reproducible batch protein crystallisation. *Chem. Eng. Res. Des.* **2021**, *173*, 81–88. [CrossRef]
13. Verma, V.; Mitchell, H.; Guo, M.; Hodnett, B.K.; Heng, J.Y.Y. Studying the impact of the pre-exponential factor on templated nucleation. *Faraday Discuss.* **2022**, *235*, 199–218. [CrossRef]
14. Verma, V.; Mitchell, H.; Errington, E.; Guo, M.; Heng, J.Y.Y. Templated Crystallization of Glycine Homopeptides: Experimental and Computational Developments. *Chem. Eng. Technol.* **2023**. [CrossRef]
15. Haddadzadegan, S.; Dorkoosh, F.; Bernkop-Schnürch, A. Oral delivery of therapeutic peptides and proteins: Technology landscape of lipid-based nanocarriers. *Adv. Drug Deliv. Rev.* **2022**, *182*, 114097. [CrossRef]
16. Jenkins, K., II. Needle phobia: A psychological perspective. *Br. J. Anaesth.* **2014**, *113*, 4–6. [CrossRef]
17. Karsdal, M.A.; Byrjalsen, I.; Riis, B.J.; Christiansen, C. Optimizing bioavailability of oral administration of small peptides through pharmacokinetic and pharmacodynamic parameters: The effect of water and timing of meal intake on oral delivery of Salmon Calcitonin. *BMC Clin. Pharmacol.* **2008**, *8*, 5. [CrossRef]
18. Klepach, A.; Tran, H.; Ahmad Mohammed, F.; ElSayed, M.E.H. Characterization and impact of peptide physicochemical properties on oral and subcutaneous delivery. *Adv. Drug. Deliv. Rev.* **2022**, *186*, 114322. [CrossRef]
19. Choonara, B.F.; Choonara, Y.E.; Kumar, P.; Bijukumar, D.; du Toit, L.C.; Pillay, V. A review of advanced oral drug delivery technologies facilitating the protection and absorption of protein and peptide molecules. *Biotechnol. Adv.* **2014**, *32*, 1269–1282. [CrossRef]
20. Maher, S.; Brayden, D.J. Formulation strategies to improve the efficacy of intestinal permeation enhancers. *Adv. Drug Deliv. Rev.* **2021**, *177*, 113925. [CrossRef]
21. Maher, S.; Ryan, B.; Duffy, A.; Brayden, D.J. Formulation strategies to improve oral peptide delivery. *Pharm. Pat. Anal.* **2014**, *3*, 313–336. [CrossRef]
22. Zhu, Q.; Chen, Z.; Paul, P.K.; Lu, Y.; Wu, W.; Qi, J. Oral delivery of proteins and peptides: Challenges, status quo and future perspectives. Acta Pharm. *Sin. B* **2021**, *11*, 2416–2448. [CrossRef]
23. Anselmo, A.C.; Gokarn, Y.; Mitragotri, S. Non-invasive delivery strategies for biologics. *Nat. Rev. Drug Discov.* **2019**, *18*, 19–40. [CrossRef]
24. Harrison, G.A. Insulin in Alcoholic Solution by the Mouth. *Br. Med. J.* **1923**, *2*, 1204. [CrossRef]
25. Basu, S.K.; Govardhan, C.P.; Jung, C.W.; Margolin, A.L. Protein crystals for the delivery of biopharmaceuticals. *Expert Opin. Biol. Ther.* **2004**, *4*, 301–317. [CrossRef]
26. Aguirre, T.A.S.; Teijeiro-Osorio, D.; Rosa, M.; Coulter, I.S.; Alonso, M.J.; Brayden, D.J. Current status of selected oral peptide technologies in advanced preclinical development and in clinical trials. *Adv. Drug Deliv. Rev.* **2016**, *106*, 223–241. [CrossRef]
27. Drucker, D.J. Advances in oral peptide therapeutics. *Nat. Rev. Drug Discov.* **2020**, *19*, 277–289. [CrossRef]
28. Lipinski, C.A.; Lombardo, F.; Dominy, B.W.; Feeney, P.J. Experimental and computational approaches to estimate solubility and permeability in drug discovery and development settings. *Adv. Drug Deliv. Rev.* **2001**, *46*, 3–26. [CrossRef]
29. Brayden, D.J.; Hill, T.A.; Fairlie, D.P.; Maher, S.; Mrsny, R.J. Systemic delivery of peptides by the oral route: Formulation and medicinal chemistry approaches. *Adv. Drug Deliv. Rev.* **2020**, *157*, 2–36. [CrossRef]
30. Verma, V.; Zeglinski, J.; Hudson, S.; Davern, P.; Hodnett, B.K. Dependence of Heterogeneous Nucleation on Hydrogen Bonding Lifetime and Complementarity. *Cryst. Growth Des.* **2018**, *18*, 7158–7172. [CrossRef]
31. Guo, M.; Chang, Z.H.; Liang, E.; Mitchell, H.; Zhou, L.; Yin, Q.; Guinn, E.J.; Heng, J.Y.Y. The effect of chain length and side chains on the solubility of peptides in water from 278.15 K to 313.15 K: A case study in glycine homopeptides and dipeptides. *J. Mol. Liq.* **2022**, *352*, 118681. [CrossRef]
32. Newton, J.M.; Rowley, G.; Fell, J.T.; Peacock, D.G.; Ridgway, K. Computer analysis of the relation between tablet strength and compaction pressure. *J. Pharm. Pharmacol.* **1971**, *23*, 195S–201S. [CrossRef] [PubMed]
33. Ryshkewitch, E. Compression Strength of Porous Sintered Alumina and Zirconia. *J. Am. Ceram. Soc.* **1953**, *36*, 65–68. [CrossRef]
34. Steendam, R.; Lerk, C.F. Poly(dl-lactic acid) as a direct compression excipient in controlled release tablets: Part I. Compaction behaviour and release characteristics of poly(dl-lactic acid) matrix tablets. *Int. J. Pharm.* **1998**, *175*, 33–46. [CrossRef]

35. de Andrade, D.F.; Zuglianello, C.; Pohlmann, A.R.; Guterres, S.S.; Beck, R.C.R. Assessing the In Vitro Drug Release from Lipid-Core Nanocapsules: A New Strategy Combining Dialysis Sac and a Continuous-Flow System. *AAPS PharmSciTech* **2015**, *16*, 1409–1417. [CrossRef]
36. Shah, U.V.; Williams, D.R.; Heng, J.Y.Y. Selective Crystallization of Proteins Using Engineered Nanonucleants. *Cryst. Growth Des.* **2012**, *12*, 1362–1369. [CrossRef]
37. Yang, H.; Belviso, B.D.; Li, X.; Chen, W.; Mastropietro, T.F.; Di Profio, G.; Caliandro, R.; Heng, J.Y.Y. Optimization of Vapor Diffusion Conditions for Anti-CD20 Crystallization and Scale-Up to Meso Batch. *Crystals* **2019**, *9*, 230. [CrossRef]
38. Gerard, C.J.J.; Briuglia, M.L.; Rajoub, N.; Mastropietro, T.F.; Chen, W.; Heng, J.Y.Y.; Di Profio, G.; ter Horst, J.H. Template-Assisted Crystallization Behavior in Stirred Solutions of the Monoclonal Antibody Anti-CD20: Probability Distributions of Induction Times. *Cryst. Growth Des.* **2022**, *22*, 3637–3645. [CrossRef]
39. Michrafy, A.; Michrafy, M.; Kadiri, M.S.; Dodds, J.A. Predictions of tensile strength of binary tablets using linear and power law mixing rules. *Int. J. Pharm.* **2007**, *333*, 118–126. [CrossRef]
40. Wu, C.-Y.; Best, S.M.; Bentham, A.C.; Hancock, B.C.; Bonfield, W. A simple predictive model for the tensile strength of binary tablets. *Eur. J. Pharm. Sci.* **2005**, *25*, 331–336. [CrossRef]
41. Pudasaini, N.; Upadhyay, P.P.; Parker, C.R.; Hagen, S.U.; Bond, A.D.; Rantanen, J. Downstream Processability of Crystal Habit-Modified Active Pharmaceutical Ingredient. *Org. Process Res. Dev.* **2017**, *21*, 571–577. [CrossRef]
42. Chen, H.; Aburub, A.; Sun, C.C. Direct Compression Tablet Containing 99% Active Ingredient&-A Tale of Spherical Crystallization. *J. Pharm. Sci.* **2019**, *108*, 1396–1400. [CrossRef] [PubMed]
43. Barralet, J.E.; Gaunt, T.; Wright, A.J.; Gibson, I.R.; Knowles, J.C. Effect of porosity reduction by compaction on compressive strength and microstructure of calcium phosphate cement. *J. Biomed. Mater. Res.* **2002**, *63*, 1–9. [CrossRef] [PubMed]
44. Shah, U.V.; Olusanmi, D.; Narang, A.S.; Hussain, M.A.; Gamble, J.F.; Tobyn, M.J.; Heng, J.Y.Y. Effect of crystal habits on the surface energy and cohesion of crystalline powders. *Int. J. Pharm.* **2014**, *472*, 140–147. [CrossRef] [PubMed]
45. Föger, F.; Kopf, A.; Loretz, B.; Albrecht, K.; Bernkop-Schnürch, A. Correlation of in vitro and in vivo models for the oral absorption of peptide drugs. *Amino Acids* **2008**, *35*, 233–241. [CrossRef] [PubMed]

pharmaceutics

Article

Numerical Simulation of Particle Dynamics in a Spiral Jet Mill via Coupled CFD-DEM

Satyajeet Bhonsale [1], Lewis Scott [2], Mojtaba Ghadiri [2] and Jan Van Impe [1,*]

[1] BioTeC+, Department of Chemical Engineering, Technology Campus Ghent, KU Leuven, Gebroeders de Smetstraat 1, 9000 Ghent, Belgium; satyajeetsheetal.bhonsale@kuleuven.be
[2] School of Chemical and Process Engineering, University of Leeds, Leeds LS2 9JT, UK; pm11lms@leeds.ac.uk (L.S.); M.Ghadiri@leeds.ac.uk (M.G.)
* Correspondence: jan.vanimpe@kuleuven.be

Abstract: Spiral jet mills are ubiquitous in the pharmaceutical industry. Breakage and classification in spiral jet mills occur due to complex interactions between the fluid and the solid phases. The study of these interactions requires the use of computational fluid dynamics (CFD) for the fluid phase coupled with discrete element models (DEM) for the particle phase. In this study, we investigate particle dynamics in a 50-mm spiral jet mill through coupled CFD-DEM simulations. The simulations showed that the fluid was significantly decelerated by the presence of the particles in the milling chamber. Furthermore, we study the particle dynamics and collision statistics at two different operating conditions and three different particle loadings. As expected, the particle velocity was affected by both the particle loading and operating pressure. The particles moved slower at low pressures and high loadings. We also found that particle–particle collisions outnumbered particle–wall collisions.

Keywords: spiral jet mills; discrete element models; computational fluid dynamics

Citation: Bhonsale, S.; Scott, L.; Ghadiri, M.; Van Impe, J. Numerical Simulation of Particle Dynamics in a Spiral Jet Mill via Coupled CFD-DEM. *Pharmaceutics* **2021**, *13*, 937. https://doi.org/10.3390/pharmaceutics13070937

Academic Editor: Anne Marie Healy

Received: 19 May 2021
Accepted: 19 June 2021
Published: 23 June 2021

1. Introduction

Particle size reduction is an important step in the design, development, and processing of active pharmaceutical ingredients (API). Spiral jet mills are the preferred comminution devices for ultra-fine grinding where particles less than 10 μm diameter are desired [1,2]. Since spiral jet mills were first patented in the 1930s [3], their design has remained relatively unchanged. Their design consists of a short cylindrical (or elliptical) milling chamber into which high velocity gas is pushed through several nozzles (called the grinding nozzles), which are at an angle to the mill perimeter. The gas jets entering through these nozzles create a vortex in the milling chamber.

Solid feed particles are fed to an injector, which delivers the feed to the vortex, wherein they are accelerated by the gas flow. The momentum gathered by the particles due to the high velocity gas jets leads to high energy particle–particle and particle–wall collisions, which cause breakage. The centrifugal forces in the vortex retain the coarse particles within the milling chamber. The centrifugal forces and the radial drag forces acting on the particles in the vortex depend on the particle size (x). As the size decreases due to breakage, centrifugal force ($\sim x^3$) reduces faster than the radial drag forces ($\sim x^2$).

When the radial drag force acting on a particle exceeds the centrifugal force, the particle is entrained out of the milling chamber via an outlet in the centre of the mill. Although the energy consumption of spiral jet mills is relatively high, they provide various advantages. Due to the absence of any moving parts and the self-classifying nature of the mill, contamination can be completely avoided [4]. Moreover, the expansion of fluid from the jet into the grinding chamber leads to a cooling effect, which makes spiral jet mills attractive for heat sensitive materials [5].

The popularity of the jet mills arises from the simplicity of its operation. In general, only three parameters are needed to control the operation of the jet mills: the injector nozzle pressure (IP), the grinding nozzle pressure (GP), and the solid feed rate (FR). Of these, the GP and FR have a significant impact on the milling performance, while the impact of IP is not large [6]. The kinetic energy within the jet mill is directly related to the GP. Higher kinetic energy leads to enhanced acceleration of the particles and hence higher impact collisions.

A large number of studies assessing the influence of the GP (or in some cases the gas flow rate) on milling performance arrived at the same conclusion: increasing the GP leads to a decrease in the output median particle size [7–11]. As the IP has a negligible impact on the mill performance, it is usually maintained at a pressure that is slightly higher than the GP. Increasing the FR makes the output particle size coarser. Although the frequency of collisions increases with increasing FR [12], the high particle concentration within the mill causes the fluid energy in the mill to dissipate faster [13]. This leads to low energy collisions.

A systematic mathematical description of the spiral jet mill is desirable for predictive purposes. A validated model can reduce the need for extensive experimentation with an expensive API. Moreover, a model-based process understanding is also a key element of the Quality by Design paradigm adopted by the Food and Drugs Authority in the United States. Despite the prevalence of spiral jet mills in the pharmaceutical industry, modelling studies are relatively sparse. Some of these studies rely on the force and energy balance approach [14–16], some on the population balance approach [17–19], and a few on computational fluid dynamics [15,20,21].

The energy and force balances equate the two opposing forces acting on the particles: the centrifugal force and the radial drag force. Based on the balance, a particle cut size that depends on the ratio of tangential and radial particle velocities (called the spin number) is derived. Under idealistic assumptions of Archimedes spiral flow, Tanaka [16] derived expressions that defined the tangential and radial velocity in a mill as a function of the mill parameters. Rodnianski et al. [15], on the other hand, used CFD simulations to obtain the spin number and radial velocity. They described the spin number as a general function of mill's geometric and operational parameters. An important conclusion from the CFD analysis of Rodnianski et al. [15] is that the gas flow rate does not affect the spin number. MacDonald et al. [14] built upon the previous results and derived a cut size equation by incorporating the energy balance. In all the above derivations, the particle tangential velocity was assumed to be the same as the gas velocity.

Population balance modelling (PBM) has also been used to describe breakage is described with empirical breakage distribution functions. As the fully described PBM is a complex integro-differential set of equations, numerical methods are commonly utilized to obtain a solution [22]. Gommeren et al. [17] presented a compartmentalized PBM describing three zones within the grinding chamber: the comminution zone, central (feed) zone, and the classifier zone. The model was then used to determine the residence time distribution, hold up, and closed-loop control. However, no discussion on the estimation of the parameters involved in the model was provided.

An highly empirical steady-state PBM was also described in Starkey et al. [18]. However, the algebraic equation set only considered six discrete size classes to describe the particle size distribution of the product. A major drawback in the PBM approach is the need for estimating the breakage parameters from experimental data. Although a variety of approaches have been proposed for this inverse problem, they cannot be applied directly to the spiral jet mill models. Bhonsale et al. [23] performed an identifiability analysis of a discretized spiral jet mill PBM, and showed that the convolution between classification and breakage in the jet mill led to non-identifiable parameters in the breakage kernels.

Computational fluid dynamics (CFD) relies on the numerical solution of the Navier–Stokes equations to resolve the fluid flow field within the mill. The CFD simulations of Kozawa et al. [20] showed that coarse particles near the upper wall could escape the mill easily. Similarly, Rodnianski et al. [15] reported the invariability of the spin number with gas mass flow rates. However, as the operation of the jet mill involves a complex interplay between the fluid and particle phases, simulations solely via CFD can be misleading. When the influence of particle phase cannot be ignored completely, a coupled CFD—Discrete Element Method (DEM) approach needs to be adopted. The DEM approach solves the Newtonian kinematic equations for individual particles to determine their trajectories. Given the computational limitations, the CFD-DEM approach cannot be used for very fine particles undergoing breakage. Thus, its application to modelling spiral jet mills requires simplifying assumptions.

Han et al. [24] reported the influence of the feed rate, feed nozzle angle, and the gas flow rate on the product particle size based on two dimensional CFD-DEM simulations. Levy and Kalman [21] presented three-dimensional simulations of particle motion in an industrial scale jet mill. Although particle breakage and particle–particle interactions are completely ignored, the simulations provide interesting insights into the flow field in the jet mill. Teng et al. [12] included particle–particle interactions but ignored particle breakage. With simulations involving only 1000 particles, they reported the influence of the GP on the particle velocity distribution. Along with an increase in the particle velocity, increasing the GP also led to an increased width of the particle velocity distribution.

Moreover, particle–particle collisions were shown to be the primary cause of breakage. By identifying the collision patterns, they concluded that the majority of collisions had a much larger tangential component, which would lead to abrasion rather than fragmentation. Brosh et al. [25] adopted the breakage model developed by Kalman et al. [26] to incorporate comminution in the simulations. To avoid an excessive number of particles in the simulation, particles that fell below 10 μm were removed from the simulation.

Bnà et al. [27] presented a thorough CFD-DEM simulation study. The cut size determined by their simulations is in good agreement with the previous studies by Dobson and Rothwell [4]. They also highlighted the influence of the product hold up inside the mill. They concluded that the fluid deceleration caused by the presence of the particle phase was responsible for the classification efficiency of the mill. However, only a one-way coupling between CFD-DEM was used. In such a coupling approach, the effect of the fluid phase on the particulate phase was considered; however, the effect of the particulate phase on the fluid was ignored.

Given the importance of hold up, Bnà et al. [27] recognized this limitation and underlined the need for a four-way coupling between CFD and DEM. Scott et al. [28] presented such a simulation study using a four-way CFD-DEM coupling. Their simulation reported a decrease in the tangential velocity component with increasing hold up. They also showed that most energy was dissipated along the bed surface and in front of each jet.

Although the use of CFD-DEM models for predictive purposes is limited by computational restrictions, they provide useful insight into the particle dynamics of a mill. In this paper, a coupled CFD-DEM simulation is used to analyse the particle dynamics in the spiral jet mill. Unlike the coupling used by Bnà et al. [27], the coupling used here considers both the influence of fluid on particles and particles on fluid. In the subsequent sections, the CFD-DEM approach is described, followed by the simulation results and conclusions.

2. Numerical Methods

The mill geometry used for the simulations is based on the Hosokawa AS50 spiral jet mill. However, the geometry is based on an in-house drawing made at the University of Leeds and shown in Figure 1 [28,29]. The milling chamber is 50 mm in diameter and has four jets angled at 50° from the radius. A special feature of the AS50 is its classifier design. The milling gas spirals up into the classifier section where a vortex finder reverses the flow direction. Figure 1a shows the CAD geometry of the mill used. Following Dogbe [29], an annular manifold for gas distribution was included around the milling chamber as this influences the fluid flow field within the chamber.

The numerical simulation of the process proceeds in three stages. The fluid field is resolved by CFD (using ANSYS Fluent v19), the particle phase is resolved by DEM (using EDEM 2019), and the coupling responsible for exchanging information on solid–fluid forces is achieved by EDEM's coupling tool. For the fluid field resolution, the gas is assumed to behave ideally. The *k*-ω shear stress transport (SST) [30] was used as the turbulence model. For its simplicity, the Morsi–Alexander correlation [31] was used to compute the fluid drag on the particles.

There are several other drag laws available in the literature, and the choice influences the results of the simulation. Although an evaluation of drag laws is out of the scope of this paper, a few other studies have made the comparisons in the context of coupled CFD-DEM [32–34]. The CFD simulations were carried out using the commercial software—ANSYS Fluent v19 with the first-order upwind approach used to solve the individual equations. The convergence tolerances were set at 10^{-4} for all the equations. All the boundary conditions were set to pressure type boundaries. For the 'feed hopper inlet' and the 'mill outlet', atmospheric conditions were assumed.

The 'injector inlet' and 'milling inlet' were set at the desired operational conditions. Two operating conditions were considered: 1 bar IP and GP, and 3 bar IP and GP (based on gauge pressure). A tetrahedral mesh was used with its size determined by the particle size was used in the DEM simulation. The mesh size was constrained to be 40% larger than the particle diameter [35]. A mesh convergence study was performed by recording the velocity gradient across the milling chamber. Following Norouzi et al. [35], the time step for the CFD simulations was set to 50 times the DEM time step (i.e., at 1×10^{-5} s).

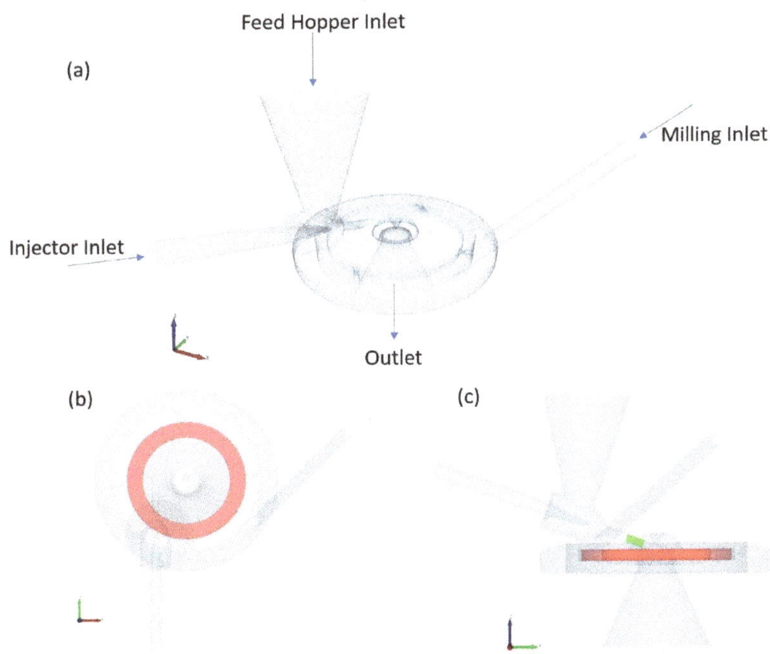

Figure 1. The isometric view (**a**) of the CAD geometry used for CFD-DEM simulations of the jet mill [28,29]. The red section in the top (**b**) and front (**c**) view of the geometry depicts the particle factory in which the initial particle bed is generated. The green section in the front view (**c**) depicts the particle factory through which particles are dynamically fed once the initial particle bed is dispersed.

For the DEM simulations, the particles were considered to be monosized perfect spheres of 200 μm diameter. The particle properties used in the simulations are listed in Table 1. The CFD-DEM coupling was handled by EDEM's coupling tool, which is based on the approach described by Tsuji et al. [36]. The mass of particles in the cell was decomposed into a weighted volume so that the pressure calculation could be performed. The respective velocity was then returned to EDEM to update the drag force acting on each individual particle. The Hertz–Midlin model was used to model the contact forces. The integration time step was fixed to 20% of the Rayleigh timestep. For the particle size and particle properties used in the simulation, the time step used was 2×10^{-7} s.

To evaluate the effect of the hold up, three different particle loadings were considered: 10,000 particles ($\approx$0.06 g), 40,000 particles ($\approx$0.25 g), and 100,000 particles ($\approx$0.63 g). In reality, the particles dropped into the feed hopper inlet are sucked into the milling chamber by the injector gas flow. To avoid extensive simulation times, a particle bed was pre-generated in a ring shape factory placed inside the milling chamber as depicted in red in Figure 1b. Once the particle bed is dispersed by the flow and has reached a steady state, 10,000 particles are fed to the mill via the injector factory (coloured green in Figure 1c) at a feed rate of 1.8 kg/h. It takes about 0.128 s to finish feeding the particles. The particle and collision data were collected for 0.15 s from the start of the feeding phase.

Table 1. The particle properties used in the DEM simulations [28,29,37].

Particle Property	Value
Diameter	200 μm
Density	1525 kg/m^3
Young's Modulus	2.7×10^8 Pa
Poisson's Ratio	0.35
Coefficient of Restitution	0.5
Coefficient of Static Friction	0.5
Coefficient of Rolling Friction	0.01

3. Results

3.1. Fluid and Particle Dynamics

Figure 2 illustrates the velocity magnitude within the milling chamber at the mid plane. These contours are plotted once the pre-generated particle bed is completely dispersed. Although the fluid velocity reaches a magnitude of around 300 m/s, the contour plot is clipped at 150 m/s to emphasize the lower velocity areas. The areas with low fluid velocities are coloured blue, and high fluid velocities are coloured red. Once the pre-generated particle bed is dispersed, particles form a bed on the mill periphery along which they circulate. For all three loadings, the periphery of the mill along which the particles circulate had the lowest velocity.

This particle bed is locally dispersed by the high velocity jet streams emitting from the nozzles. As higher particle loadings lead to a thicker circulating particle bed, the low velocity region around the wall increases in size. Moreover, the jets dispersing the particle bed loose energy much faster when the particle bed is thick. Thus, the length of the jet stream reduces with increasing particle load. This also leads to much lower fluid velocities in the entire mill. This decrease in jet penetration length was also reported by Scott et al. [28] for an even higher particle loading. It is evident that higher particle loadings lead to a larger dampening of the fluid velocity. The velocity magnitude increases radially toward the centre in all cases.

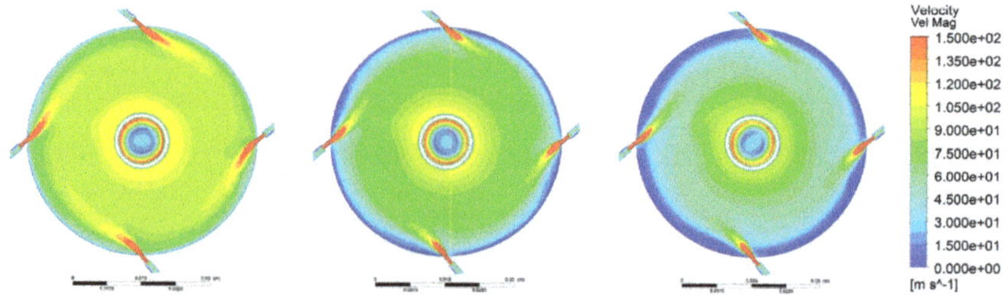

Figure 2. The velocity magnitude contours at the midplane for three particle loadings.

Snapshots of particle motion in the jet mill for the case with 100,000 particles and 3 bar pressure are presented in Figure 3. At the time of 0.0 s, the pre-generated particle bed is intact, and the CFD-DEM simulation is started. The particles start dispersing as the large tangential component of the velocity accelerates them towards the wall. A steady state is reached around 0.015 s, and a thick layer of particles is formed at the mill periphery, which moves along the wall at a very low velocity. Meanwhile, the jet streams disperse the particles when they approach the nozzles. Once the steady state has been achieved, 10,000 new particles are fed through a factory created in the injector nozzle at rate of 1.8 kg/h, taking roughly 0.128 s.

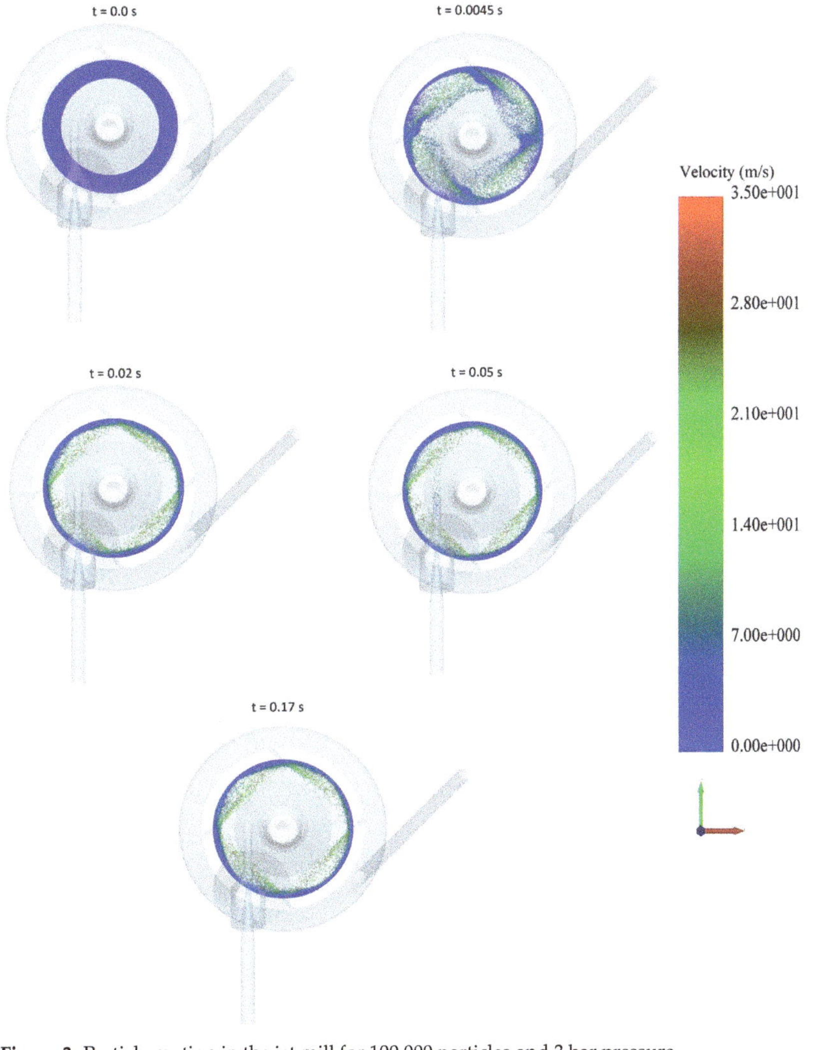

Figure 3. Particle motion in the jet mill for 100,000 particles and 3 bar pressure.

The evolution of the fluid field as the particles disperse is illustrated in Figure 4. As can be deduced from the blue ring in the contour plot at time 0.00035 s, the pre-generated particle bed slows the fluid notably. The particle bed, shown by the blue area, is pushed to the wall as it is dispersed. Once the steady state is reached, the velocity profile does not vary greatly. Even when new particles are fed, the velocity profile stays consistent. Thus, it can be said that the particles being entrained into the milling chamber from the feeder do not affect the fluid field in the mill. However, over time, as the particles build up in the mill, the fluid behaviour will be affected. This is evident from Figure 2. The presence of a particle bed circulating along the wall periphery has been reported in previous experimental and numerical studies [6,12,28].

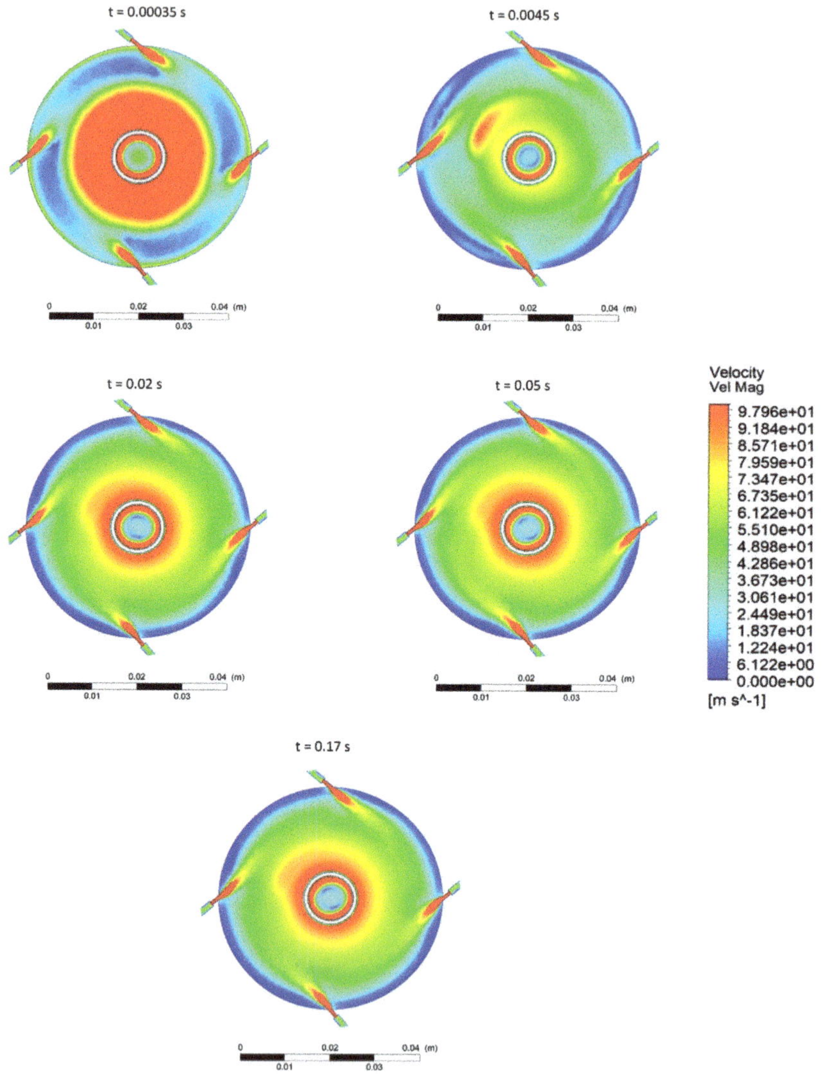

Figure 4. Evolution of fluid field in the jet mill along the mid plane for 100,000 particles with an operating pressure of 3 bar.

Figure 5 displays heat maps of the particle velocities. The figure maps every particle for 0.15 s from the time particle feeding starts. All the particles are coloured according to their velocity at the given time. The prominence of the slow moving particle bed on the mill periphery is evident for all three particle loadings. Similar to the observations from the fluid fields, the increase in the thickness of this bed with particle loading is clear. In all cases, particles with the highest velocity lie along the jet trajectory.

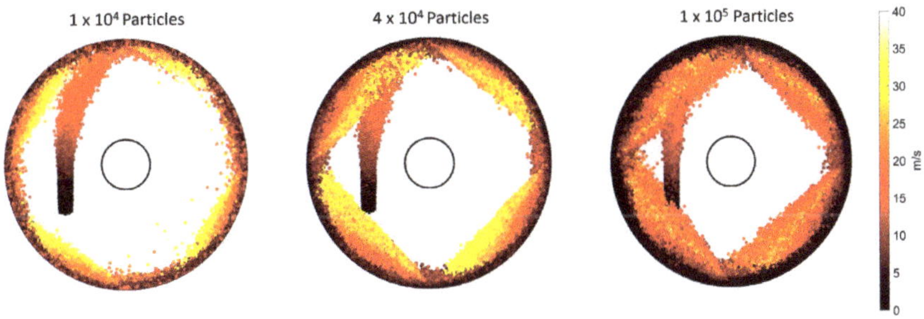

Figure 5. Particle velocity heat map for 0.15 s from the start of the feeding.

Figure 6a,b illustrate the effect of particle loading on the particle velocity distribution. For both operating pressures, the particle velocity distribution shifts to the left with increasing particle loading. The average and maximum particle velocities also decrease with increasing particle loading. At low particle loadings, the mean free path (distance travelled without colliding) of a particle is much larger. Thus, the particle can be accelerated over a larger distance. Moreover, more fluid kinetic energy per unit particle mass is available for the acceleration.

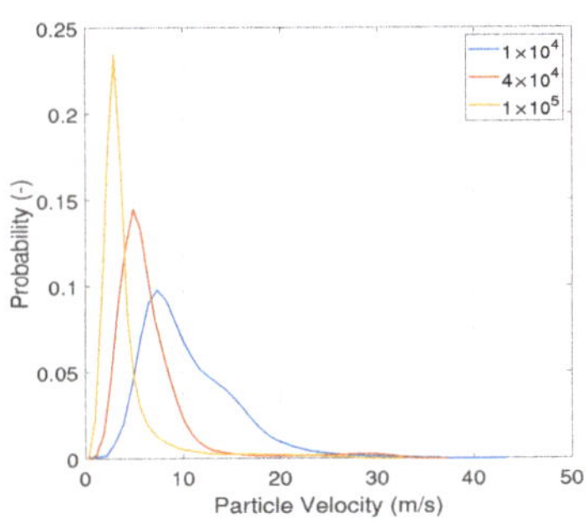

(**a**) Particle velocity distribution for 3 bar and three particle loadings

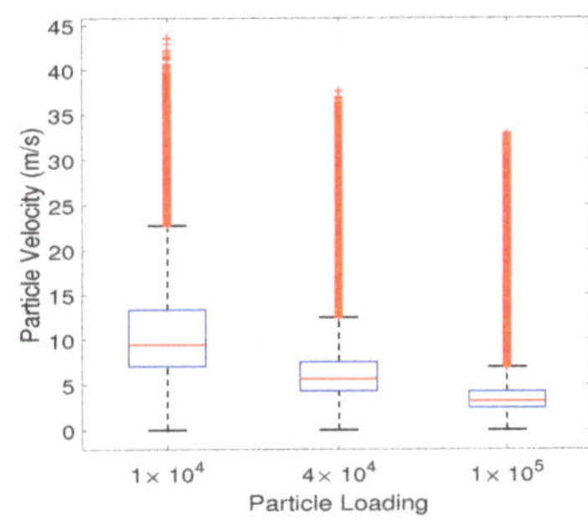

(**b**) Box plot for 3 bar and three particle loadings

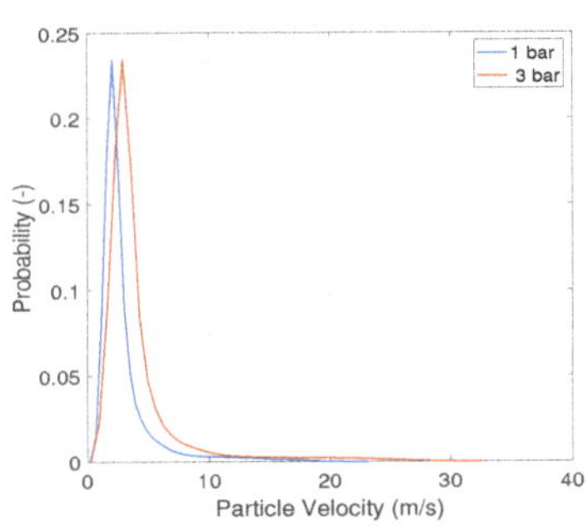

(**c**) Particle velocity distribution for 100,000 particles and two operating pressures

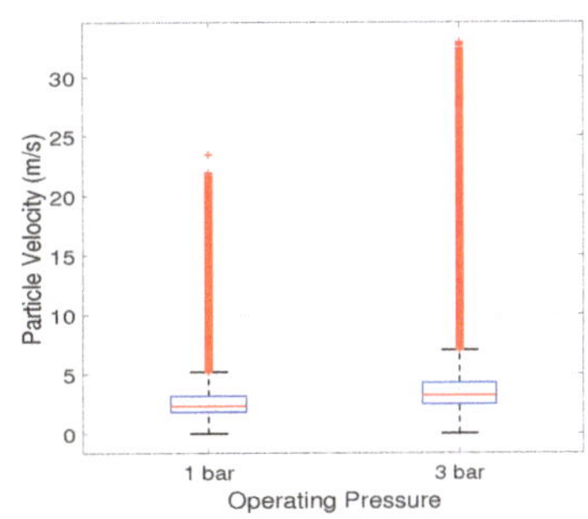

(**d**) Box plot for 100,000 particles and two operating pressures

Figure 6. Probability distribution and box plot for particle velocity with increasing particle loading and two operating pressures

The effect of operating pressure can be discerned from Figure 6c,d wherein the particle velocity distribution and box plot are depicted for 100,000 particles at the two operating pressures. Increasing the operating pressure leads to a slight rightwards shift in the particle velocity distribution while a much larger increase is noticed in the average and maximum particle velocity. The increase in the velocity distribution is due to the much higher kinetic energy provided by the fluid at higher operating pressures.

The motion of the particles fed into the jet mill via the injector is illustrated by the streamlines plotted in Figure 7. The image on the left depicts the three particle loadings at 3 bar operating pressure, and the image on the right depicts the streamlines for the two operating pressures with 100,000 particles. With higher particle loading, the particles are ejected from the bed by the jet stream move more towards the centre of the mill. Similarly, the particles travel closer to the classifier at higher operating pressure. In all cases, no particles escape the mill via the classifier.

As Bnà et al. [27] indicated, particle classification is a function of the fluid deceleration caused by the particle phase. At the current particle concentrations, very small particles could be entrained. The cut size can be derived by a force balance between the centrifugal and radial forces [4,14–16]. However, the particle size used in the current study is much larger. Thus, no classification was observed from the simulations. Even with around 250,000 particles (some as small as 160 μm), Scott et al. [28] could not observe classification in their four-way coupled simulations. This highlights the importance of the solid hold up within the mill on the final particle size of the milled and classified product.

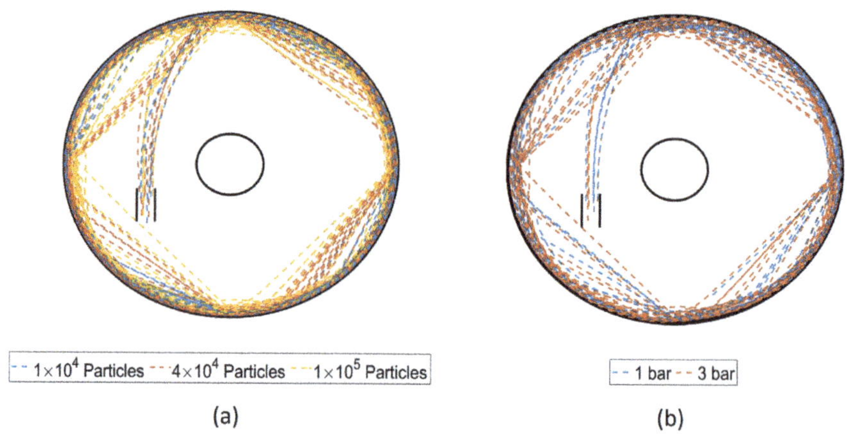

Figure 7. Streamlines of first five particles fed to the mill while describing their motion in the mill for 0.15 s after the feeding. Case (**a**) depicts the streamlines for three particle loadings at 3 bar operating pressure, and case (**b**) depicts the streamlines at two operating conditions and particle loading of 100,000 particles.

3.2. *Particle Collision Statistics*

Breakage in jet mills occurs when the particles accelerated by the fluid collide with other particles or the wall. If the dissipated energy due to the collisions is higher than a threshold, then the particle breaks. Particles that do not break may experience significant weakening. Previous studies proposed the use of a "fatigue function" to account for this weakening [26,38]. Although this study does not explicitly consider particle breakage, this section presents some results on the collision statistics of particles in the jet mill for all the case studies. All the collision statistics were collected for a time of 0.15 s after the feeding of 10,000 particles begins.

In Figure 8, the particle–particle and particle–wall collisions are presented as a percentage of the total number of collisions for the given case. The black dots (connected by the solid black lines) lying on the bars corresponding to the simulation cases represent the total number of collisions for those cases as a percentage of the maximum number of collisions observed across all the case studies. In general, increased particle loading led to an increase in the number of collisions.

In all the cases considered, particle–particle collisions were prevalent. At low particle loading, a significant fraction of the collisions occurred between the particle and the wall. The effect of the operating pressure on the number of collisions was prevalent at high loadings. For 100,000 particles, 1 bar operating pressure led to only around 60% of the collisions as 3 bar pressure. For 10,000 particles, increased pressure also led to a slight increase in the fraction of particle–wall collisions.

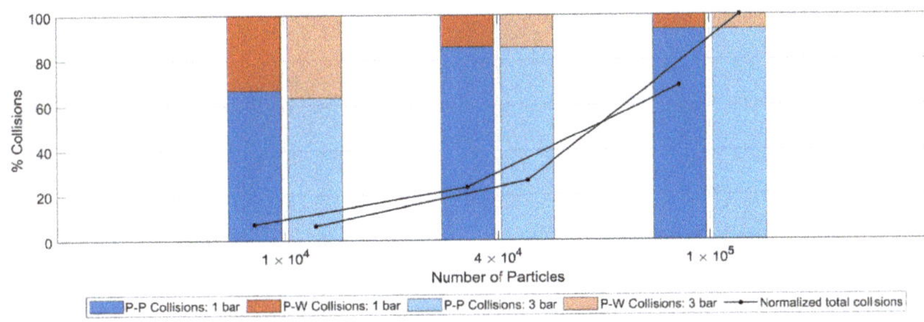

Figure 8. The number of particle–particle (P–P) and particle–wall (P–W) collisions as a percentage of the total collisions for three loadings and two operating pressures. The total number of collisions as a normalized percent of the maximum collisions across all conditions. Collision data collected over 0.15 s after the start of particle feeding.

Figure 9 reports the number of collisions per particle for increasing particle loading and the two operating pressures. Again, increased pressure and increased loading both led to an increase in the number of collisions experienced by a particle.

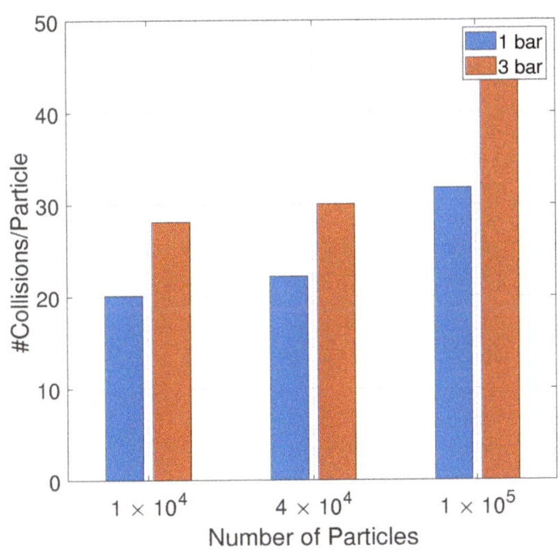

Figure 9. The total number of collisions per particle for the three loadings and two operating pressures over 0.15 s after the start of the feeding phase.

The distribution of collision velocities is illustrated in Figure 10. For both particle–particle and particle–wall collisions, increasing particle load leads to a reduction in the impact velocities. This can be explained by the fact that, at lower particle concentrations, each particle travels on a longer mean free path, thus, accelerating to higher velocities before the collision. The high impact collisions at low loading and high operating pressure will lead to a higher degree of breakage. This is corroborated by experimental studies that showed that finer product was obtained at low feed rates and high operating pressures [7,8,11,12].

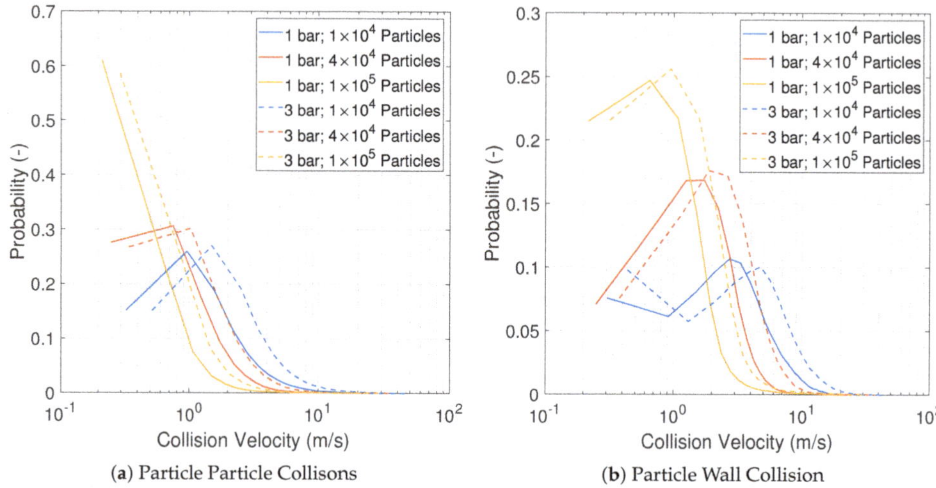

Figure 10. Collision velocity distribution for particle–particle (P–P) and particle–wall (P–W) collision for all cases. Collision data collected over 0.15 s after the start of particle feeding.

As the study by Bnà et al. [27] involved only a one-way coupling in which particles mostly followed the fluid streamlines, they reported particle–wall collision velocities of around 40–100 m/s. These were much higher than the collision velocities observed in this study. Moreover, in contrast to the findings in this study, they reported no significant effect of particle loading on the particle–wall collision velocity distribution. This shows the importance of the fluid deceleration by particle phase for breakage as well as classification.

Figure 11 shows the scatter plots of the tangential component against the normal component of the particle–particle and particle wall collision velocities. As already mentioned, in all cases, the particle–particle collisions are prevalent. Increasing the particle loading led to a decrease in the velocity magnitudes of both the components while increasing the operating pressure led to an increase. At higher particle loadings, the particle–wall collisions were characterized by a high tangential component, while the particle–particle collisions had a much higher normal component.

This contradicts the findings of both Bnà et al. [27] and Teng et al. [12]. Both studies reported a higher tangential component for all collisions. As mentioned, the study by Bnà et al. [27] used only a one-way coupling scheme and, thus, ignored the fluid deceleration by the particle phase. The study by Teng et al. [12], although a two-way coupling, only considered 1000 particles. Such a low particle concentration cannot decelerate fluid to the extent observed in the current study.

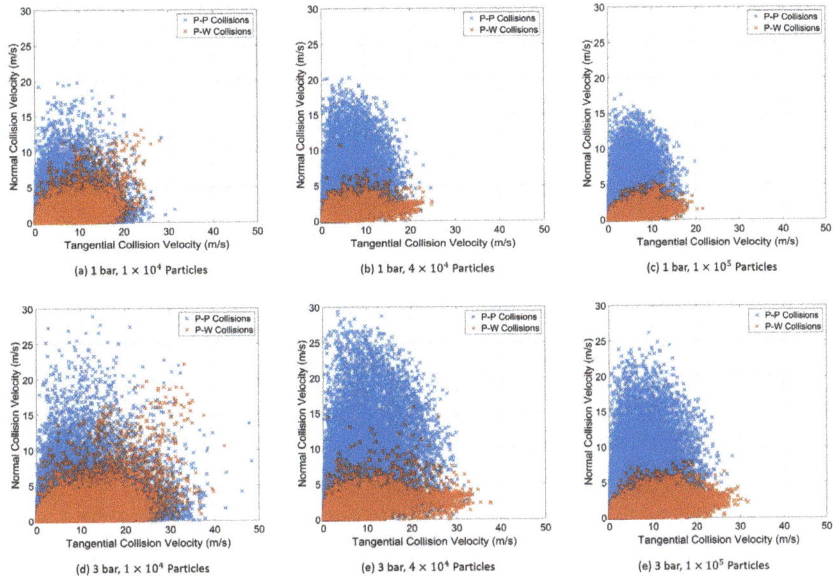

Figure 11. Scatter of the tangential component and normal component of the collision velocity for all cases studied. Collision data collected over 0.15 s after the start of particle feeding.

4. Conclusions

We analysed the effect of particle loading and the operating pressure on the fluid and particle dynamics in a spiral jet mill through coupled CFD-DEM simulations. Three particle loadings and two operating pressures were considered in the study.

We found that the particles significantly decelerated the fluid. The particles dispersed from the pre-generated bed formed a new bed along the mill periphery along which they moved at very slow velocities. The reduction in the fluid velocities was at the maximum in this area. The velocity profile increased monotonically along the radius from the wall toward the mill centre. The particle loading and operating pressure had a much higher impact on the tangential velocity than on the radial velocity. At increasing particle loading and decreasing pressure, the particles experienced more radial forces, which led to the entrainment of coarser particles.

The particle velocity also followed the same trend as the fluid velocity. The particles in the bed moving along the mill periphery were ejected with force by the jet streams. The highest particle velocities were observed directly in front of the jets. Higher operating pressure and low loading led to higher particle velocities and, subsequently, higher impact velocities. In all cases, substantially more particle–particle collisions were reported compared with particle–wall collisions. Most high impact collisions occurred at the surface of the circulating particle bed.

The results show that the particle loading had a profound effect on the fluid field, which, in turn, influenced both the breakage and classification. The computational method and the results presented provide a valuable tool-process optimisation for industrial applications of spiral jet mills.

Author Contributions: Conceptualization, S.B., L.S., M.G. and J.V.I.; Data curation, S.B.; Formal analysis, S.B. and L.S.; Investigation, S.B. and L.S.; Methodology, S.B., L.S., M.G. and J.V.I.; Project administration, M.G. and J.V.I.; Software, S.B., L.S. and M.G.; Supervision, M.G. and J.V.I.; Validation, S.B., L.S., M.G. and J.V.I.; Visualization, S.B. and L.S.; Writing—original draft, S.B.; Writing—review & editing, S.B., L.S., M.G. and J.V.I. All authors have read and agreed to the published version of the manuscript.

Funding: This work was supported by KU Leuven Center-of-Excellence Optimization in Engineering (OPTEC) and projects G086318N and G0B4121N of the Fund for Scientific Research Flanders (FWO). The APC was funded by FWO G0B4121N.

Acknowledgments: The authors would like to thank Mehrdad Pasha and Wei Pin Goh for their suggestions and input. We thank DEM Solutions (now Altair), Edinburgh, UK for providing a special license for the EDEM software used in this work.

Conflicts of Interest: The authors declare no conflict of interest.

References

1. Nakach, M.; Authelin, J.R.; Chamayou, A.; Dodds, J. Comparison of various milling technologies for grinding pharmaceutical powders. *Int. J. Miner. Process.* **2004**, *74*, S173–S181. [CrossRef]
2. Parrott, E.L. Milling of Pharmaceutical Solids. *J. Pharm. Sci.* **1974**, *63*, 813–829. [CrossRef] [PubMed]
3. Andrews, N. Method and Apparatus for Providing Material in Finely Divided Form. U.S. Patent US002032827, 3 March 1936.
4. Dobson, B.; Rothwell, E. Particle size reduction in a fluid energy mill. *Powder Technol.* **1969**, *3*, 213–217. [CrossRef]
5. Liu, G. Hammer Milling and Jet Milling Fundamentals. *Chem. Eng. Prog.* **2017**, *113*, 48–54.
6. Teng, S.; Wang, P.; Zhu, L.; Young, M.W.; Gogos, C.G. Experimental and numerical analysis of a lab-scale fluid energy mill. *Powder Technol.* **2009**, *195*, 31–39. [CrossRef]
7. Katz, A.; Kalman, H. Preliminary Experimental Analysis of a Spiral Jet Mill Performance. *Part. Part. Syst. Charact.* **2007**, *24*, 332–338. [CrossRef]
8. Midoux, N.; Hošek, P.; Pailleres, L.; Authelin, J. Micronization of pharmaceutical substances in a spiral jet mill. *Powder Technol.* **1999**, *104*, 113–120. [CrossRef]
9. Müller, F.; Polke, R.; Schädel, G. Spiral jet mills: Hold up and scale up. *Int. J. Miner. Process.* **1996**, *44–45*, 315–326. [CrossRef]
10. Ramanujam, M.; Venkateswarlu, D. Studies in fluid energy grinding. *Powder Technol.* **1969**, *3*, 92–101. [CrossRef]
11. Tuunila, R.; Nyström, L. Effects of grinding parameters on product fineness in jet mill grinding. *Miner. Eng.* **1998**, *11*, 1089–1094. [CrossRef]
12. Teng, S.; Wang, P.; Zhang, Q.; Gogos, C. Analysis of Fluid Energy Mill by gas-solid two-phase flow simulation. *Powder Technol.* **2011**, *208*, 684–693. [CrossRef]
13. Luczak, B.; Müller, R.; Kessel, C.; Ulbricht, M.; Schultz, H.J. Visualization of flow conditions inside spiral jet mills with different nozzle numbers—Analysis of unloaded and loaded mills and correlation with grinding performance. *Powder Technol.* **2019**, *342*, 108–117. [CrossRef]
14. MacDonald, R.; Rowe, D.; Martin, E.; Gorringe, L. The spiral jet mill cut size equation. *Powder Technol.* **2016**, *299*, 26–40. [CrossRef]
15. Rodnianski, V.; Krakauer, N.; Darwesh, K.; Levy, A.; Kalman, H.; Peyron, I.; Ricard, F. Aerodynamic classification in a spiral jet mill. *Powder Technol.* **2013**, *243*, 110–119. [CrossRef]
16. Tanaka, T. Scale-Up Theory of Jet Mills on Basis of Comminution Kinetics. *Ind. Eng. Chem. Process Des. Dev.* **1972**, *11*, 238–241. [CrossRef]
17. Gommeren, H.; Heitzmann, D.; Kramer, H.; Heiskanen, K.; Scarlett, B. Dynamic modeling of a closed loop jet mill. *Int. J. Miner. Process.* **1996**, *44-45*, 497–506. [CrossRef]
18. Starkey, D.; Taylor, C.; Morgan, N.; Winston, K.; Svoronos, S.; Mecholsky, J.; Powers, K.; Iacocca, R. Modeling of continuous self-classifying spiral jet mills part 1: Model structure and validation using mill experiments. *AIChE J.* **2014**, *60*, 4086–4095. [CrossRef]
19. Starkey, D.; Taylor, C.; Siddabathuni, S.; Parikh, J.; Svoronos, S.; Mecholsky, J.; Powers, K.; Iacocca, R. Modeling of continuous self-classifying spiral jet mills part 2: Powder-dependent parameters from characterization experiments. *AIChE J.* **2014**, *60*, 4096–4103. [CrossRef]
20. Kozawa, K.; Seto, T.; Otani, Y. Development of a spiral-flow jet mill with improved classification performance. *Adv. Powder Technol.* **2012**, *23*, 601–606. [CrossRef]
21. Levy, A.; Kalman, H. Numerical Study of Particle Motion in Jet Milling. *Part. Sci. Technol.* **2007**, *25*, 197–204. [CrossRef]
22. Bhonsale, S.S.; Telen, D.; Stokbroekx, B.; Impe, J.V. Comparison of numerical solution strategies for population balance model of continuous cone mill. *Powder Technol.* **2019**, *345*, 739–749. [CrossRef]
23. Bhonsale, S.S.; Stokbroekx, B.; Van Impe, J. Assessment of the parameter identifiability of population balance models for air jet mills. *Comput. Chem. Eng.* **2020**, *143*, 107056. [CrossRef]
24. Han, T.; Kalman, H.; Levy, A. DEM Simulation of Particle Comminutionin Jet Milling. *Part. Sci. Technol.* **2002**, *20*, 325–340. [CrossRef]
25. Brosh, T.; Kalman, H.; Levy, A.; Peyron, I.; Ricard, F. DEM–CFD simulation of particle comminution in jet-mill. *Powder Technol.* **2014**, *257*, 104–112. [CrossRef]
26. Kalman, H.; Rodnianski, V.; Haim, M. A new method to implement comminution functions into DEM simulation of a size reduction system due to particle–wall collisions. *Granul. Matter* **2009**, *11*, 253–266. [CrossRef]

27. Bnà, S.; Ponzini, R.; Cestari, M.; Cavazzoni, C.; Cottini, C.; Benassi, A. Investigation of particle dynamics and classification mechanism in a spiral jet mill through computational fluid dynamics and discrete element methods. *Powder Technol.* **2020**, *364*, 746–773. [CrossRef]
28. Scott, L.; Borissova, A.; Burns, A.; Ghadiri, M. Influence of holdup on gas and particle flow patterns in a spiral jet mill. *Powder Technol.* **2021**, *377*, 233–243. [CrossRef]
29. Dogbe, S. Predictive Milling of Active Pharmaceutical Ingredients and Excipients. Ph.D. Thesis, School of Chemical and Process Engineering, The University of Leeds, Leeds, UK, 2016.
30. Menter, F.R. Two-equation eddy-viscosity turbulence models for engineering applications. *AIAA J.* **1994**, *32*, 1598–1605. [CrossRef]
31. Morsi, S.A.; Alexander, A.J. An investigation of particle trajectories in two-phase flow systems. *J. Fluid Mech.* **1972**, *55*, 193–208. [CrossRef]
32. Agrawal, V.; Shinde, Y.; Shah, M.T.; Utikar, R.P.; Pareek, V.K.; Joshi, J.B. Effect of drag models on CFD–DEM predictions of bubbling fluidized beds with Geldart D particles. *Adv. Powder Technol.* **2018**, *29*, 2658–2669. [CrossRef]
33. Marchelli, F.; Hou, Q.; Bosio, B.; Arato, E.; Yu, A. Comparison of different drag models in CFD-DEM simulations of spouted beds. *Powder Technol.* **2020**, *360*, 1253–1270. [CrossRef]
34. Zhou, L.; Zhang, L.; Bai, L.; Shi, W.; Li, W.; Wang, C.; Agarwal, R. Experimental study and transient CFD/DEM simulation in a fluidized bed based on different drag models. *RSC Adv.* **2017**, *7*, 12764–12774. [CrossRef]
35. Norouzi, H.R.; Zarghami, R.; Sotudeh-Gharebagh, R.; Mostoufi, N. *Coupled CFD-DEM Modeling: Formulation, Implementation and Application to Multiphase Flows*; Wiley: Newark, NJ, USA, 2016.
36. Tsuji, Y.; Tanaka, T.; Ishida, T. Lagrangian numerical simulation of plug flow of cohesionless particles in a horizontal pipe. *Powder Technol.* **1992**, *71*, 239–250. [CrossRef]
37. Dogbe, S.; Ghadiri, M.; Hassanpour, A.; Hare, C.; Wilson, D.; Storey, R.; Crosley, I. Fluid-particle energy transfer in spiral jet milling. *EPJ Web Conf.* **2017**, *140*, 09040. [CrossRef]
38. Rodnianski, V.; Levy, A.; Kalman, H. A new method for simulation of comminution process in jet mills. *Powder Technol.* **2019**, *343*, 867–879. [CrossRef]

pharmaceutics

Article

Numerical Investigation of the Particle Dynamics in a Rotorgranulator Depending on the Properties of the Coating Liquid

Philipp Grohn [1,*], Stefan Heinrich [2] and Sergiy Antonyuk [1]

1 Institute of Particle Process Engineering, University of Kaiserslautern-Landau, Gottlieb-Daimler-Straße 44, 67663 Kaiserslautern, Germany
2 Institute of Solids Process Engineering and Particle Technology, Hamburg University of Technology, Denickestraße 15, 21073 Hamburg, Germany
* Correspondence: philipp.grohn@mv.uni-kl.de

Abstract: In the pharmaceutical industry, the coating of particles is a widely used technique to obtain desired surface modifications of the final product, e.g., controlled release of the active agents. The production of round, coated particles is particularly important, which is why fluidized bed rotor granulators (FBRG) are often used for this process. In this work, Computational Fluid Dynamics (CFD) coupled with the Discrete Element Method (DEM) is used to investigate the wet particle dynamics, depending on the properties of the coating liquid in a FBRG. The DEM contact model was extended by liquid bridge model to account for capillary and viscous forces during wet contact of particles. The influence of the relative contact velocity on the maximum length of the liquid bridge is also considered in the model. Five different cases were compared, in which the particles were initially wetted, and the liquid loading as well as the surface tension and viscosity of the liquid were changed. The results show that increasing viscosity leads to a denser particle bed and a significant decrease in particle rotational velocities and particle motion in the poloidal plane of the FBRG. Reducing the liquid loading and surface tension results in increased particle movement.

Keywords: CFD-DEM simulation; wet particles; capillary force; viscous force; fluidized bed rotor granulator

Citation: Grohn, P.; Heinrich, S.; Antonyuk, S. Numerical Investigation of the Particle Dynamics in a Rotorgranulator Depending on the Properties of the Coating Liquid. *Pharmaceutics* **2023**, *15*, 469. https://doi.org/10.3390/pharmaceutics15020469

Academic Editor: Colin Hare

Received: 14 December 2022
Revised: 20 January 2023
Accepted: 27 January 2023
Published: 31 January 2023

1. Introduction

For various products in the pharmaceutical, chemical and food industries, the coating of particles is an important processing step in order to obtain desired surface modification of the final product [1,2]. Numerous coating equipment exists for this purpose. The coating devices can be distinguished according to their method of introducing kinetic energy into a particle bed, between a purely mechanical input (e.g., mixer, disc and drum granulators) and a fluidization induced by the energy of the process gas flow (e.g., fluidized bed or spouted bed systems). Particularly in the pharmaceutical industry, fluidized bed rotor granulators (FBRG) are widely used to produce round coated pellets for oral drug delivery with a narrow size distribution, high strength, smooth surface and high sphericity [3–7]. This is achieved by the special design of a FBRG. It consists of a rotating circular rotating base plate and a stationary cylindrical wall. The fluidization gas flow passes through an annular gap between the rotating plate and the cylindrical wall. This combination enables the individual process steps of spheronization, coating and drying to be carried out in the same unit [8].

Although the technology of a FBRG is widely used, the particle dynamics are still not fully understood due to the complex micro mechanisms in the process. Several experimental studies can be found in the literature that describe some mechanisms during the granulation process [4,6,9–11]. However, the knowledge in this field is mainly empirical and all the particle interactions are not yet fully understood. A detailed knowledge of the particle motion on the micro level is required to better understand the coating process in the rotor granulator. Numerical simulations are particularly suitable for this purpose [5,8]. The widely used Euler–Lagrange approach can be applied to simulate the multiphase flow, where Computational Fluid Dynamics (CFD) is coupled with the Discrete Element Method (DEM) [12,13]. In CFD, the flow field of the gas in the process is calculated treating the fluid phase as a continuum. For the DEM, the interactions of each particle are determined based on contact models describing the physical properties of the particles, such as adhesion, and their mechanical behavior under slow, fast and repeated loading. In two-way CFD-DEM coupling, both the influence of the gas phase on the particle phase and the influence of the particle phase on the gas phase are considered [13–17], while in one-way coupling, only the influence of the gas phase on the particles is taken into account [5].

The particle dynamics in a rotor granulator were first investigated by Muguruma et al. [18] numerically with DEM and experimentally with Particle Tracking Velocimetry (PTV). They studied the influence of liquid on particle motion, but considered only capillary and not viscous contact forces, and did not vary the properties of the liquid. Weis et al. [19,20] used DEM simulations to obtain the particle dynamics and mixing behavior, as well as the contact frequency of the particles in a spheronizer that, in contrast to a rotor granulator, works without fluidization air and usually at higher rotation velocities of the structured friction plate. In addition, they extended the DEM approach to consider particle rounding during this process. Recently, Grohn et al. [21] investigated numerically the multiphase flow of cylindrical particles in a FBRG with CFD-DEM simulations. A significant influence of the particle shape on the particle dynamics was found. Neuwirth et al. [4,22] performed an experimental study of the particle dynamics in a FBRG under dry and wet conditions using magnetic particle tracking (MPT). The comparison with the CFD-DEM simulations showed good agreement with the experiments for the dry case. However, the experiments were performed with particles of 6-mm diameter, which are not representative for real applications in FBRG. This particle size was required by the MPT measurement equipment available at that time. In our last study [8], the dynamics of initially wetted particles in the FBRG was investigated numerically by CFD-DEM simulations and experimentally by an improved MPT measurement system. In the numerical simulations, the capillary forces due to the presence of liquid on particles were considered based on the model of Israelachvili [23] and the viscous forces were calculated according to the models of Lian et al. [24] and Popov [25]. In addition, a new model was implemented to describe the velocity-dependent rupture length of liquid bridges [8,26]. With the improved MPT equipment, the particle dynamics of spherical particles with a minimum diameter of 2.8 mm could be measured. It was possible to validate contact models used in simulations of dry particles and particles wetted with water, and a good agreement was found.

Since in real applications, both the liquid spray rate and thus, the liquid loading of the particles and the properties of the coating solution vary, the influence of these parameters on particle dynamics and contact behavior in the FBRG are investigated in this work using the model previously validated in [8]. On the one hand, the influence of the liquid loading of 1 vol.-% and 5 vol.-% with water is investigated. On the other hand, the liquid properties are varied three times at a constant liquid loading of 5 vol.-%. The basis of this liquid is a coating solution frequently used in the pharmaceutical industry, consisting of distilled water with 6 mass-% PHARMACOAT® 606 (hydroxypropyl methylcellulose, Shin-Etsu Chemical Co., Ltd., Chiyoda-ku, Tokyo, Japan) [5]. This coating solution is characterized by a reduced surface tension of 42.5 $mN{\cdot}m^{-1}$ compared to water and a strongly increased viscosity of 61.9 mPa·s. The three other variants thus result from: a reduction of the surface tension to the value of the coating solution while the viscosity of water remains unchanged, the surface tension of water remains unchanged but the viscosity is increased to the value of the coating solution, and both the surface tension and the viscosity are changed to the values of the coating solution. To analyze the influence of the studied liquid parameters on the particle dynamics, the distributions of solid volume fraction, tangential, poloidal and rotating particle velocities are compared. For a deeper understanding of the process, the particle contact phenomena, such as the resulting aggregate size, are investigated with DEM.

2. Model Description

2.1. CFD Modeling

In the CFD, the gas flow field is calculated by solving the volume-averaged Navier–Stokes equations [27]. For this purpose, the flow domain for the CFD simulation must first be discretized by mesh cells. In order to take the influence of the particulate phase on the gas flow into account, the volume fraction of the gas phase ε_g in each mesh cell is included in the volume-averaged Navier–Stokes equation. Therefore, the governing equations of the mass and momentum conversation can be given as follows:

$$\frac{\partial(\varepsilon_g \rho_g)}{\partial t} + \nabla \cdot \left(\varepsilon_g \rho_g \vec{u}_g\right) = 0\,, \tag{1}$$

$$\frac{\partial\left(\varepsilon_g \rho_g \vec{u}_g\right)}{\partial t} + \nabla \cdot \left(\varepsilon_g \rho_g \vec{u}_g \vec{u}_g\right) = -\varepsilon_g \nabla p + \nabla \cdot \left(\varepsilon_g \vec{\tau}_g\right) - \vec{S}_p + \varepsilon_g \rho_g \vec{g}\,, \tag{2}$$

where, p represents the pressure, ρ_g describes the density, $\vec{u}_g$ and $\vec{\tau}_g$ are the velocity and the stress tensor of the gas phase, respectively. For the calculation of the volume fraction of the gas phase ε_g in each CFD mesh cell, the volume fraction x_i of each particle volume $V_{p,i}$ within the cell were determined using the so-called sample points approach. The principle of this method, where the volume of all particles z in a grid cell with the volume V_{cell} is approximated by cubic sample volumes, was first presented by Hoomans et al. [28]. In the used framework of CFDEM® coupling [29], the particle is divided into 29 non-overlapping regions of equal volume, each with one sample point [30]. At each time step, the algorithm checks which of the sample cubes are located in which mesh cell:

$$\varepsilon_g = 1 - \left(\sum_{i=0}^{z} x_i V_{p,i}\right) \frac{1}{V_{cell}}\,. \tag{3}$$

To consider the interactions between the particulate phase and the gas phase, the momentum balance is extended by the momentum sink term $\vec{S}_p$. The momentum sink term can be determined from the drag forces $\vec{F}_{d,i}$ of all particles n_p in the mesh cell with the volume V_{cell}:

$$\vec{S}_p = \frac{1}{V_{cell}} \cdot \sum_{i=0}^{n_p} \vec{F}_{d,i}\,. \tag{4}$$

Various gas–solid models can be found in the literature that describe the drag forces acting on the particles in a fluidized bed [12,31]. As in our previous work [8], the drag forces are calculated according to the widely used model of Di Felice [32], which describes the entire porosity range and for particle Reynold numbers $Re_{p,i}$ up to 10^6 with a continuous function. Here, the drag force counteracts the relative velocity of a particle in a fluid $(\vec{u}_g - \vec{u}_p)$:

$$\vec{F}_{d,i} = \frac{1}{8} C_{D,i}(Re_p) \rho_g \pi d_{p,i}^2 \left(\vec{u}_g - \vec{u}_p\right) \left|\vec{u}_g - \vec{u}_p\right| \varepsilon_g^{2-\beta}\,. \tag{5}$$

The drag force is considerably influenced by the drag coefficient $C_{D,i}$. This coefficient is related to the Reynolds number of the particles, which takes into account the superficial velocity differences between particles and the surrounding fluid [33,34]:

$$C_{D,i} = \left(0.63 + \frac{4.8}{\sqrt{Re_{p,i}}}\right)^2, \tag{6}$$

$$Re_{\mathrm{p,i}} = \frac{\varepsilon_{\mathrm{g}} \rho_{\mathrm{g}} d_{\mathrm{p,i}} \left| \vec{u}_{\mathrm{g}} - \vec{u}_{\mathrm{p}} \right|}{\eta_{\mathrm{f}}}, \tag{7}$$

where, η_{f} is the dynamic viscosity of the fluid. The influence of the particle concentration in a mesh cell on the drag force in Equation (5) is modeled with a function [32]:

$$\beta = 3.7 - 0.65 \exp\left[-\frac{\left(1.5 - \log_{10}(\mathrm{Re}_{\mathrm{p,i}})\right)^2}{2} \right]. \tag{8}$$

2.2. DEM Modeling

In order to simulate the motion of particles, the equations of motion for translation and rotation according to Newton and Euler are solved. For this purpose, the Discrete Element Method (DEM), firstly described by Cundall and Strack [35], is applied, which also allows for an investigation of the mechanical interactions between particles and between particles and walls. For the consideration of the influence of the gas phase on the particles in the simulations, the multiphase flows have to be calculated with the two-way CFD-DEM coupling. Therefore, in this coupling method, the model of Cundall and Strack [35] is extended to incorporate the drag force, pressure gradient force $\vec{F}_{\nabla \mathrm{p,i}}$ and viscous force in gas $\vec{F}_{\vec{\tau},\mathrm{i}}$ [12,33,36]:

$$m_{\mathrm{p,i}} \frac{\mathrm{d}\vec{v}_{\mathrm{p,i}}}{dt} = \vec{F}_{\mathrm{d,i}} + \vec{F}_{\nabla \mathrm{p,i}} + \vec{F}_{\vec{\tau},\mathrm{i}} + \vec{F}_{\mathrm{g,i}} + \sum_{\mathrm{j}=0}^{\mathrm{k}} \vec{F}_{\mathrm{c,ij}} + \vec{F}_{\mathrm{vis,ij}} + \vec{F}_{\mathrm{cap,ij}}, \tag{9}$$

$$J_{\mathrm{p,i}} \frac{\mathrm{d}\vec{\omega}_{\mathrm{p,i}}}{\mathrm{d}t} = \sum_{\mathrm{j}=0}^{\mathrm{k}} \left(\vec{M}_{\mathrm{t,ij}} + \vec{M}_{\mathrm{r,ij}} \right). \tag{10}$$

The gravitational force $\vec{F}_{\mathrm{g,i}}$ and the sum of the contact forces $\vec{F}_{\mathrm{c,ij}}$, which act on the particle due to interactions with other particles j or the walls, are modeled to determine the translational velocity of each particle $\vec{v}_{\mathrm{p,i}}$ with the mass $m_{\mathrm{p,i}}$. Similar to our latest work [8], the particles are initially wetted. Therefore, viscous forces $F_{\mathrm{vis,ij}}$ and capillary forces $F_{\mathrm{cap,ij}}$ act during a particle contact. Both forces are described in detail in the following Section 2.2.2. The sum of the torques $\vec{M}_{\mathrm{t,ij}}$ caused by the tangential forces acting on the particle and the torques $\vec{M}_{\mathrm{r,ij}}$ due to rolling friction if the particle rotates are calculated to determine the angular velocity $\vec{\omega}_{\mathrm{p,i}}$ of each particle with the moment of inertia $J_{\mathrm{p,i}}$.

2.2.1. Contact Forces

The contact forces are calculated according to the well-known Hertz–Mindlin model, which is described in detail in the literature [37–39]. The contact force is decomposed into a normal and tangential component, index n and t, respectively, and are expressed as:

$$F_{\mathrm{c,n,ij}} = -k_{\mathrm{n}} \delta_{\mathrm{n}}^{\frac{3}{2}} n_{\mathrm{ij}} - \eta_{\mathrm{n}} u_{\mathrm{p,n,ij}}, \tag{11}$$

$$F_{\mathrm{c,ij,t}} = \begin{cases} -k_{\mathrm{t}} \delta_{\mathrm{t}} t_{\mathrm{ij}} - \eta_{\mathrm{t}} u_{\mathrm{p,t,ij}} & \text{if } \left| -k_{\mathrm{t}} \delta_{\mathrm{t}} t_{\mathrm{ij}} - \eta_{\mathrm{t}} u_{\mathrm{p,t,ij}} \right| \leq \mu \left| F_{\mathrm{c,n,ij}} \right| \\ -\mu \left| F_{\mathrm{c,n,ij}} \right| t_{\mathrm{ij}} & \text{if } \left| -k_{\mathrm{t}} \delta_{\mathrm{t}} t_{\mathrm{ij}} - \eta_{\mathrm{t}} u_{\mathrm{p,t,ij}} \right| > \mu \left| F_{\mathrm{c,n,ij}} \right| \end{cases}, \tag{12}$$

where k describes the spring stiffness coefficient, δ represents the displacement, n_{ij} as well as t_{ij} are normal and tangential unit vectors, $u_{\mathrm{p,n,ij}}$ and $u_{\mathrm{p,t,ij}}$ are the normal and tangential component of the contact velocity vector of the particle with another particle or a wall. The sliding friction coefficient is represented by μ. The energy dissipation due to viscoelastic deformation behavior of the material is taken into account by the damping factor $\eta_{n,t}$ for the normal and tangential direction:

$$\eta_{\mathrm{n}} = 2\alpha \sqrt{m^* k_{\mathrm{n}}} \delta_{\mathrm{n}}^{1/4}, \tag{13}$$

$$\eta_t = 2\alpha\sqrt{m^* k_t \delta_t^{1/4}}, \tag{14}$$

where, α represents a function of the restitution coefficient, and m^* is the reduced mass of the contact partners. A more detailed description can be found in Heinrich et al. [17] and Salikov et al. [37].

2.2.2. Capillary and Viscous Forces

During a wet particle contact, additional adhesive forces need to be considered. Capillary forces act due to the surface tension of the liquid, and viscous forces act due to the relative motion of particles and liquid in the liquid bridge. A wet contact begins as soon as the liquid layers on the contact partners touch and ends as soon as the maximum liquid bridge length is reached. During this contact time, the capillary and viscous forces must be taken into account [8,26]. In the literature, several models describe the capillary forces for symmetric pendular bridges [18,23,38–40] on the basis of the total liquid bridge energy and the pressure difference across the liquid bridge. The numerical simulation method validated in our last study [8] is also used in this work. Therefore, the capillary forces are calculated according to Israelachvili [23]. Thus, the capillary force acting between two particles $F_{cap,pp}$ and the capillary force acting between a particle and a wall $F_{cap,pw}$ are given by the following expressions:

$$F_{cap,pp} = \frac{-4\pi R^* \gamma cos(\theta)}{1 + \left(\sqrt{1 + \frac{V_b}{\pi R^* h^2}} - 1\right)^{-1}}, \tag{15}$$

$$F_{cap,pw} = \frac{-8\pi R^* \gamma cos(\theta)}{1 + \left(\sqrt{1 + \frac{V_b}{\pi R^* h^2}} - 1\right)^{-1}}, \tag{16}$$

where, γ represents the surface tension, θ is the wetting angle, V_b describes the volume of the liquid bridge and h is the shortest distance between the particles or the particle and the wall. R^* represents the effective contact radius, which is expressed as:

$$R^* = \frac{r_i r_j}{r_i + r_j}, \tag{17}$$

where, r_i and r_j are the radii of the two contact partners. The assumption is made that the liquid on the particles forms a uniform thin film over the particle surface. During a wet particle contact, a liquid bridge is formed between the contact partners in the rebound phase. Shi and McCarthy's [41] distribution model is used to determine the liquid volume of these liquid bridges for particle–particle contacts. The distribution model ensures that the liquid on the particle surface contributes to only one liquid bridge (Figure 1a). However, this approach is only valid for monodisperse systems. The liquid volume $V_{b,i}$ that particle i contributes to the liquid bridge is then calculated as follows:

$$V_{b,i} = \frac{L_i}{2} \cdot \left(1 - \sqrt{1 - \frac{r_j^2}{(r_i + r_j)^2}}\right), \tag{18}$$

where, L_i is the total liquid volume present on particle i. The contributed liquid volume from particle j is determined in a similar manner:

$$V_{b,j} = \frac{L_j}{2} \cdot \left(1 - \sqrt{1 - \frac{r_i^2}{(r_i + r_j)^2}}\right). \tag{19}$$

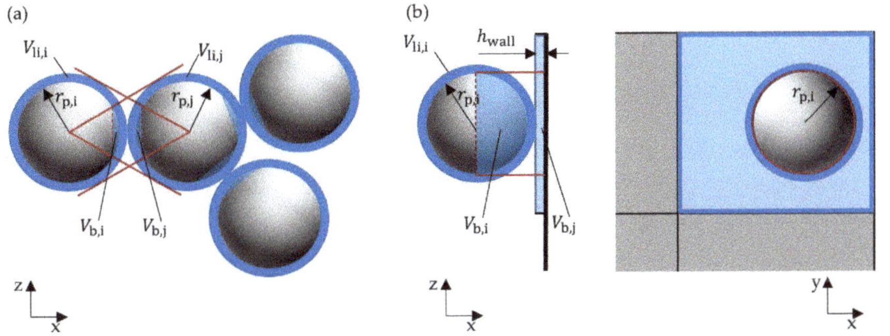

Figure 1. Determination of (**a**) the volume of liquid used to form the liquid bridge according to the model of Shi and McCarthy [41] and (**b**) the volume of liquid passing from a wetted grid cell of the wall to the liquid bridge.

Figure 1b shows the contact case between a particle and a wetted wall. The volume of liquid contributed by the wall depends on the virtual liquid layer thickness h_{wall} on the surface grid cell of the wall geometry in contact. The layer thickness is calculated by the liquid volume associated with the wetted wall grid cell divided by its surface area. Often, the 2D surface grid cells of the geometry are of different sizes and often much larger than the particle surfaces. Therefore, it is assumed that only the liquid in the region corresponding to the projection area of the contacting particle needs to be considered. In Figure 1b, this area is marked with a red circle. Thus, the amount of liquid in the wall grid cell that contributes to the formation of the liquid bridge is expressed as:

$$V_{\text{b,j}} = h_{\text{wall}} \pi r_{\text{i}}^2 \,. \tag{20}$$

The final volume of the liquid bridge is then the sum of both contributed liquid volumes:

$$V_{\text{b}} = V_{\text{b,i}} + V_{\text{b,j}} \,. \tag{21}$$

As the particles rebound, the liquid bridge is stretched until it ruptures at a critical distance between the contact partners. This critical distance, also called the maximum liquid bridge length, is described by various models [38,39,42–45]. All models have in common that they do not consider the significant influence of the impact velocity on the bridge length, which was found in our recent experimental study [26]. In this work, three different experimental setups were developed to investigate the maximum liquid bridge length in a velocity range from 0.0001 $\text{s}\cdot\text{m}^{-1}$ to 4 $\text{s}\cdot\text{m}^{-1}$ for particle–particle as well as particle–wall contact. Based on our experimental results, we extended the model of Mikami et al. [39] to account for the strong influence of the impact velocity $u_{\text{im,ij}}$ on the maximum liquid bridge length. For particle–wall contact, the maximum bridge length was expressed as:

$$l_{\text{max,pw}} = (0.95 + 0.22\theta) V_{\text{b}}^{0.32} (1 + C_{\text{pw}} u_{\text{im,ij}})^{\frac{2}{3}} \,, \tag{22}$$

where, C_{pw} represents a constant parameter with was found in [26] to have a value of 4.424 $\text{s}\cdot\text{m}^{-1}$. For particle–particle contact, the maximum liquid bridge length was calculated as:

$$l_{\text{max,pp}} = (0.99 + 0.62\theta) V_{\text{b}}^{0.34} (1 + C_{\text{pp}} u_{\text{im,ij}})^{\frac{2}{3}} \,, \tag{23}$$

where, C_{pw} is a constant parameter with the value of 6.266 $\text{s}\cdot\text{m}^{-1}$. The end of the contact is indicated by the rupture of the liquid bridge. The volume of the liquid bridge is then distributed evenly among the contact partners.

In addition to the capillary forces, viscous forces are also considered. They slow down the contact velocity during the approach phase as well as the velocity of the rebound phase after contact. Based on the Reynolds lubrication theory [46], Adams and Perchard [47] developed a model to describe the viscous force in the normal direction, which is often used in DEM simulations [24,41,48,49]. In the model, two particles are assumed to be in a liquid layer and move with a relative velocity in the normal direction. The viscous forces in normal direction $F_{\mathrm{vis,n}}$ between the particles can then be calculated by the Reynolds lubrication equation:

$$F_{\mathrm{vis,n}} = \frac{6\pi\eta_{\mathrm{l}}R^{*2}u_{\mathrm{p,ij,n}}}{h}, \quad (24)$$

where, η_{l} is the dynamic viscosity of the liquid, $u_{\mathrm{p,ij,n}}$ represents the relative velocity of the colliding particles or particle with a wall in normal direction and h is the shortest distance between the surfaces of the contact partners. Similar to DEM studies from other authors [48,50,51], a minimum distance between the contact partners is set for the calculation of the viscous forces as it is physically limited by the roughness of the respective surfaces. The viscous force in tangential direction is calculated according to the model of Popov [25]. It describes the tangential force acting on a spherical particle moving along a plate wetted with a liquid film:

$$F_{\mathrm{vis,t}} = 2\pi\eta_{\mathrm{f}}R^{*}u_{\mathrm{p,ij,t}}\ln\left(1 + \frac{R^{*}}{2h}\right), \quad (25)$$

where, $u_{\mathrm{p,ij,t}}$ describes the relative velocity of the colliding particles or particle with a wall in tangential direction.

3. Simulation Setup

3.1. Geometry of the Fluidized Bed Rotor Granulator

In this study, a FBRG is investigated, whose dimensions are inspired by the commercially used rotor granulator Rotor 300 (Glatt GmbH, Binzen, Germany). In Figure 2, the geometry of the apparatus is shown. The diameter of the cylindrical process chamber is 295 mm; thus, the radius R_{FBRG} is 147.5 mm (Figure 2b). FBRG has an unstructured rotating plate with a diameter of 268 mm located in the middle of the apparatus. In addition, the gas flows vertically into the particle bed via a two-millimeter-wide annular gap between the rotor plate and the apparatus wall. Due to the small inflow surface of the annular gap, the gas enters the process chamber with a much higher velocity than in conventional fluidized beds. The direction of the air flow in the apparatus is shown by blue arrows and the rotation of the plate is represented by green arrows. In real applications, an additional nozzle is often placed above the plate to coat the particles, but in the simulated cases in this study, only initially wetted particles are examined.

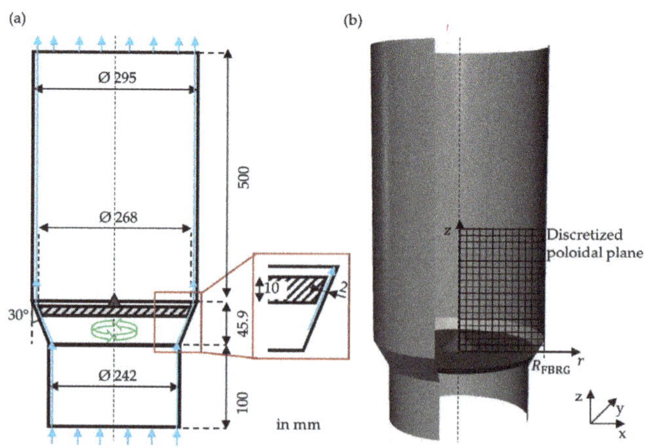

Figure 2. (**a**) Dimensions of the fluidized bed rotor granulator (blue arrows indicate the air flow and the green arrows the direction of rotation of the plate) and (**b**) its STL 3D geometry for the simulations.

3.2. Simulation Setup

For the CFD, the geometry of the FBRG was discretized into around 100,000 hexagonal mesh cells. The open-source software OpenFOAM® [52] was used to solve the Navier–Stokes equations with a pressure implicit with splitting of operator algorithm (PISO method) [53]. Turbulence was included using a k-ε turbulence model [54] and the CFD time-step was 5×10^{-6} s. The fluidization gas was air at 20 °C. An operation point was simulated, where the inlet flow was 200 $m^3 \cdot h^{-1}$, which corresponded to an inlet velocity of 1.21 $m \cdot s^{-1}$. The air velocity in the annular gap was about 32 $m \cdot s^{-1}$, which is 20 times higher than the minimal fluidization velocity of the particles. The rotor plate rotated with 100 rpm. The CFD simulation parameters are listed in Table 1.

Table 1. Gas properties for the CFD simulation.

Parameters of the Gas	Unit	Value
Fluid	-	air
Fluid temperature	°C	20
Fluid kinematic viscosity	$kg \cdot m^{-1} \cdot s^{-1}$	1.58×10^{-5}
Fluid density	$kg \cdot m^{-3}$	1.2
Inlet gap velocity	$m \cdot s^{-1}$	32

The particulate phase was calculated with DEM using the open-source software LIGGGHTS® [55] and coupled with the CFD by the open-source software CFDEM® coupling [29]. Similar to previous studies, round particles with a diameter of 2.8 mm consisting of a ceramic core and a shell of polyvinyl butyral (PVB) were investigated [5,8,26,56]. Initially wetted particles with a total mass of 1 kg were generated above the rotor plate. It was assumed that the liquid on the particles was evenly distributed on the particle surface with a layer of equal thickness. Similar to our previous work [5,8], different setups were used to obtain the particle properties needed for the contact model in DEM. With a free-fall device [26,56], the restitution coefficient was determined. A Nanoindenter (Hysitron TI Premier, Bruker Corporation, Billerica, Massachusetts, USA) was used to measure the Young's modulus, and with a Texture Analyser® (TA.XTplus, Stable Micro Systems, Godalming, United Kingdom), the static as well as the rolling friction coefficients were obtained [34]. The contact angle of water was determined with a camera setup and evaluated by a MATLAB script [26]. The DEM time-step was 1×10^{-7} s. The initial liquid loading of the particles in the bed is varied from 1 vol.-% to 5 vol.-% distilled water. In addition, the surface tension and viscosity of the liquid are varied three times at 5 vol.-%. The basis is a coating solution frequently used in the pharmaceutical industry, consisting of distilled water with 6 mass-% PHARMACOAT® 606 (hydroxypropyl methylcellulose, Shin-Etsu Chemical Co., Ltd., Chiyoda-ku, Tokyo, Japan) [8]. This coating solution is characterized by a reduced surface tension of 42.5 $mN \cdot m^{-1}$ compared to water and a strongly increased viscosity of 61.9 mPa·s. The three other cases thus result from a reduction in surface tension with no change in the viscosity of the coating liquid, no change in the surface tension of the coating liquid but an increase in viscosity, and both the change in surface tension and viscosity to the values of the coating solution. The DEM parameters can be seen in Table 2 and the performed simulation cases are listed in Table 3.

Table 2. Material properties for the DEM simulation.

Parameter	Unit	Value/Variation Range
Particle	-	ceramic cores coated with PVB
Particle bed mass	kg	1.0
Particle density	kg·m^{-3}	3485
Particle diameter	mm	2.8
Young's modulus particle	GPa	1.69
Young's modulus walls	GPa	3.0
Poisson ratio	-	0.3
Restitution coefficient	-	0.89
Static friction particle–particle	-	0.23
Static friction particle–wall	-	0.46
Rolling friction particle–particle	-	0.075
Rolling friction particle–wall	-	0.065
Liquid volume on particle	vol-%	1–5
Surface tension	mN·m^{-1}	42.5–72.8
Liquid dynamic viscosity	Pa·s	0.001–0.619
Liquid density	kg·m^{-3}	1000
Contact angle particle-particle	°	25
Contact angle particle–wall/rotor	°	45

Table 3. The five simulation cases with different liquid loading and properties at a fluidization flow of 200 m^3·h^{-1}, a rotation velocity of the rotor plate of 100 rpm and a bed mass of 1 kg.

Case	Liquid Properties
(a) 1.0 vol.-%	1 vol.-%, γ = 72.8 mN·m^{-1}, η = 1 mPas
(b) 5.0 vol.-%	5 vol.-%, γ = 72.8 mN·m^{-1}, η = 1 mPas
(c) γ = 42.5 mN·m^{-1}	5 vol.-%, γ = 42.5 mN·m^{-1}, η = 1 mPas
(d) η = 61.9 mPa·s	5 vol.-%, γ = 72.8 mN·m^{-1}, η = 61.9 mPas
(e) γ = 42.5 mN·m^{-1}, η = 61.9 mPa·s	5 vol.-%, γ = 42.5 mN·m^{-1}, η = 61.9 mPas

4. Results

The particle dynamics and contact behavior in the fluidized bed rotor granulator obtained with CFD-DEM simulations for the five cases with different liquid loading, as well as different liquid viscosity and surface tension, are compared in the following sections.

4.1. Particle Solid Volume Fraction

First, the poloidal distribution of the solid volume fraction is analyzed (Figure 3). Figure 2b shows how the internal volume of the apparatus was discretized into 2.8 mm × 2.8 mm squares in the axial and radial directions to calculate the solid volume fraction. Every 10 ms during the steady-state periods of the simulations from 1.5 s to 2 s, the volume of particles located within this regular poloidal discretization grid was determined and averaged. In the last step, it was divided by the volume of the associated ring cell. Weis et al. [19] also evaluated the particle dynamics in a spheronizer in the same way.

The zones with high concentrations are located in the center of the particle bed, with a maximum in the area of 10 mm above the rotor plate. The increase in liquid loading from 1 vol.-% (Figure 3a) to the second case with 5 vol.-% water (Figure 3b) leads to an increase of the region with a high solid volume fraction. In the third case (Figure 3c), the decrease in surface tension also results in a reduction of the zone with high particle concentration. The poloidal distributions of the solid volume fraction in the first and third cases are very similar (Figure 3a,c). It can be seen that the two cases with a high viscosity (Figure 3d,e) differ the most from the other three cases. Here, the region with high solid volume fraction is the largest. Due to high viscosity, this region has expanded towards the apparatus wall and thus, there is also a high particle concentration in the area between 20 mm and 30 mm above the annular gap near the wall. Thereby, the particle bed has expanded the lowest in the case of the increased viscosity at constant surface tension of water (Figure 3d) and is therefore the densest. However, the particle concentration is highest at high axial positions. The reason is that the particles adhere to the wall over time and, unlike in the other cases, very rarely come off. Therefore, the concentration here increases over time. It can be clearly seen that with high liquid loading, surface tension and viscosity, and thus, high liquid bridge forces, the particle bed becomes denser.

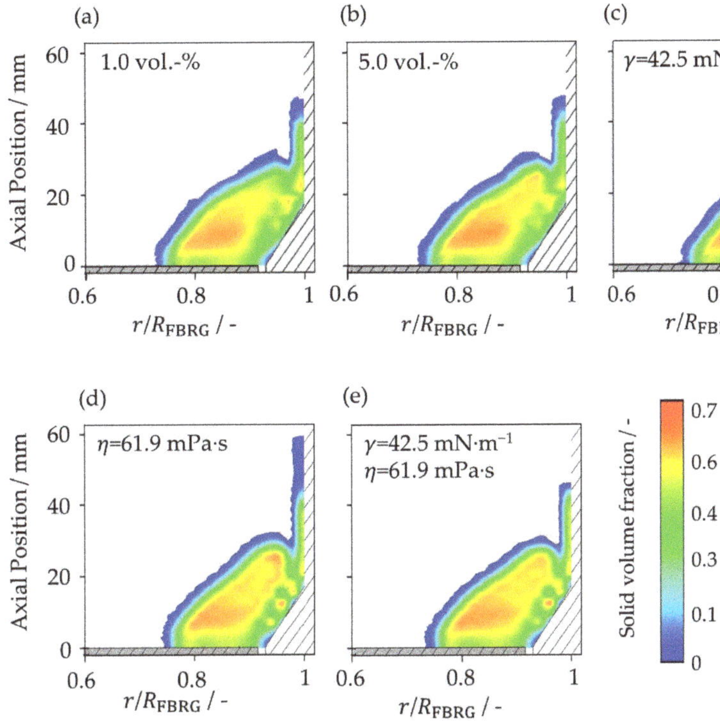

Figure 3. Poloidal distribution of solid volume fraction for studied cases (Table 3): (**a**) liquid loading of 1 vol.-%, (**b**) liquid loading of 5 vol.-%, (**c**) surface tension of 42.5 mN·m^{-1}, (**d**) viscosity of 61.9 mPa·s, and (**e**) surface tension of 42.5 mN·m^{-1} and viscosity of 61.9 mPa·s.

4.2. Particle Velocity

Figure 4 shows the differential distributions of the absolute particle velocities in the fluidized bed at different liquid loading conditions and different liquid properties. All distributions are multimodal. The first peak is at very low velocities of less than 0.01 m·s^{-1}. These are particles that are in contact with the stationary wall of the process chamber. A second peak is between 0.5 m·s^{-1} and 0.6 m·s^{-1}. These are particles located in the upper half of the particle bed. The third peak at 1.2 m·s^{-1} represents the particles interacting with the rotor plate. Again, a clear difference can be seen between the cases with high viscosity (curves (d) and (e)) and the other three variants with lower viscosity (curves (a)–(c)). The distributions for the first three cases are very similar (curves (a)–(c)). The liquid bridge forces decrease with lower liquid bridge volume as well as lower surface tension; therefore, less kinetic energy is dissipated during the particle contacts. Nevertheless, the average velocity of the particles with 1 vol.-% loading is 4.0% lower than that of the particles with 5 vol.-%. Also compared to the case with 5 vol.-% of water, the average particle velocity decreases by about 4.1% in the third case with a surface tension of 42.9 mN·m^{-1}. The reason is the increased slip of the particles on the rotating plate. As a result, the energy input from the rotor to the bed is lower and the mean velocity of the particles in the bed decreases slightly. This leads to a small difference between the three cases.

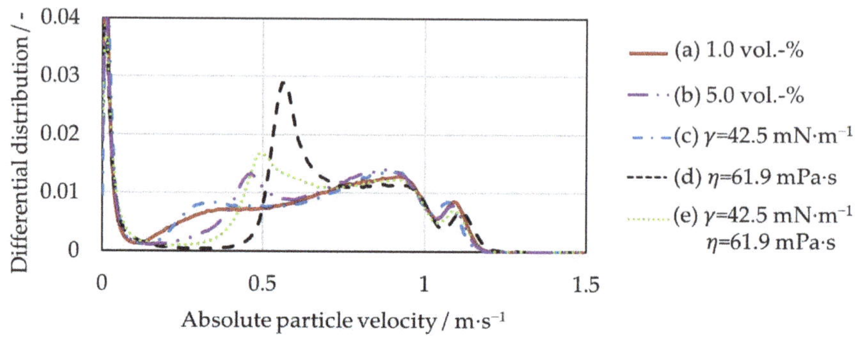

Figure 4. Differential distributions of the absolute particle velocity for a liquid loading of 1 vol.-%, a liquid loading of 5 vol.-%, a surface tension of 42.5 mN·m^{-1}, a viscosity of 61.9 mPa·s, and a surface tension of 42.5 mN·m^{-1} and a viscosity of 61.9 mPa·s.

The increased viscosity of 61.9 mPa·s in the fourth case (curve (d)) again leads to a lower slip, slightly increasing the mean particle velocity by 2.3%, compared to 5 vol.-% water. It can be seen that the proportion of velocities between 0.5 m·s^{-1} and 0.6 m·s^{-1} increases significantly. In the last case (curve (e)), the positive effect that the increased viscosity has on the slip of the particles on the rotation plate is counteracted by the negative effect of the lower surface tension. As a result, the average particle velocity only changes by less than 1%.

The tangential velocity distribution in the poloidal plane as a function of radial and axial position at a different liquid loading and at different liquid properties is shown in Figure 5. Although in all five cases, the acting liquid bridge forces differ, their velocity profiles are quite similar. The highest tangential velocities can be seen in the region directly above the rotating plate caused by transfer of momentum into the particle bed. Due to liquid bridges and, therefore, acting adhesive forces, there is a significant reduction in particle velocity near the wall in all cases investigated, as the particles repeatedly adhere to the wall. Thus, the particles are strongly decelerated and have a tangential velocity of 0 m·s^{-1} directly at the wall. It can be seen that with an increase in the liquid bridge forces in cases (b), (d) and (e), the zone with low tangential velocities near to the stationary apparatus wall decreases. The particles are more strongly connected to each other, which improves the energy input through the rotor plate to the entire particle bed.

More pronounced differences between the five cases can be seen in the poloidal velocity distribution (Figure 6). The poloidal velocity is the velocity component composed of the z-velocity and the radial velocity [8]. It can be clearly seen that the poloidal velocities of the particles are significantly lower than their tangential velocities. The direction of particle motion in the poloidal plane is evident from the velocity vectors. Due to the rotation of the plate and the axial acceleration above the annular gap caused by the fluidization air, the particles obtain a circular movement in the poloidal plane of the particle bed. The highest poloidal velocities can be observed near the wall directly above the annular gap due to the high inflow velocity of the fluidization air. In addition, the particles have a high poloidal velocity at the surface of the particle bed, where the particles fall down by gravity. In contrast to the center of the particle bed, as well as near the apparatus wall, the particles move very slowly. The particles in the two cases with increased viscosity (Figure 6d,e) have the lowest poloidal velocities. The increased viscous forces, due to the higher viscosity of the liquid, lead to higher energy dissipation, and as a result, the poloidal velocity of the particles decreases. For the first and third case (Figure 7a,c), it can be seen that the poloidal particle velocities are slightly higher compared to the case with 5 vol.-% water (Figure 7b).

Here, the lower capillary forces due to the smaller amount of liquid or the lower surface tension are responsible for the reduced energy dissipation.

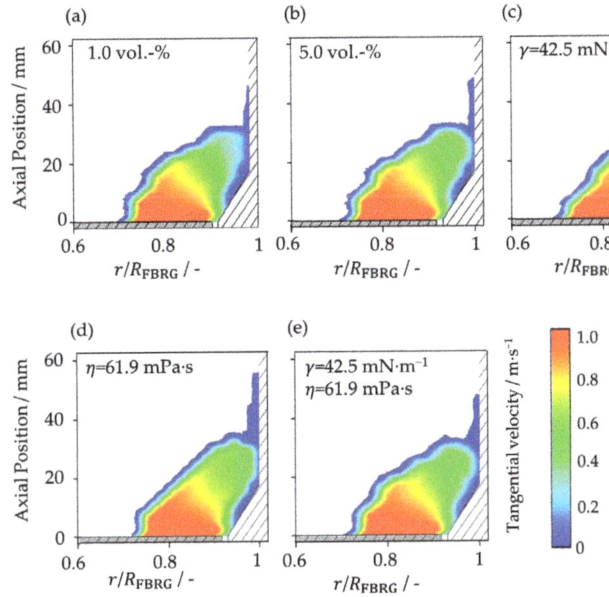

Figure 5. Poloidal distribution of tangential particle velocity at (**a**) liquid loading of 1 vol.-%, (**b**) liquid loading of 5 vol.-%, (**c**) surface tension of 42.5 mN·m^{-1}, (**d**) viscosity of 61.9 mPa·s, and (**e**) surface tension of 42.5 mN·m^{-1} and viscosity of 61.9 mPa·s.

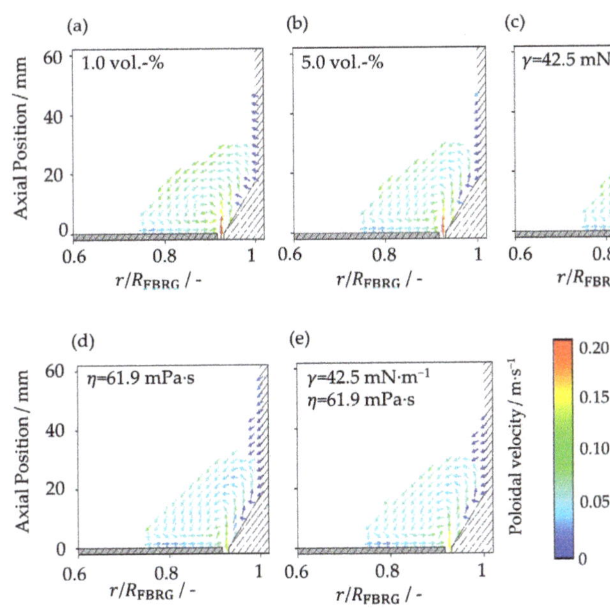

Figure 6. Poloidal distribution of poloidal particle velocity at (**a**) liquid loading of 1 vol.-%, (**b**) liquid loading of 5 vol.-%, (**c**) surface tension of 42.5 mN·m^{-1}, (**d**) viscosity of 61.9 mPa·s, and (**e**) surface tension of 42.5 mN·m^{-1} and viscosity of 61.9 mPa·s.

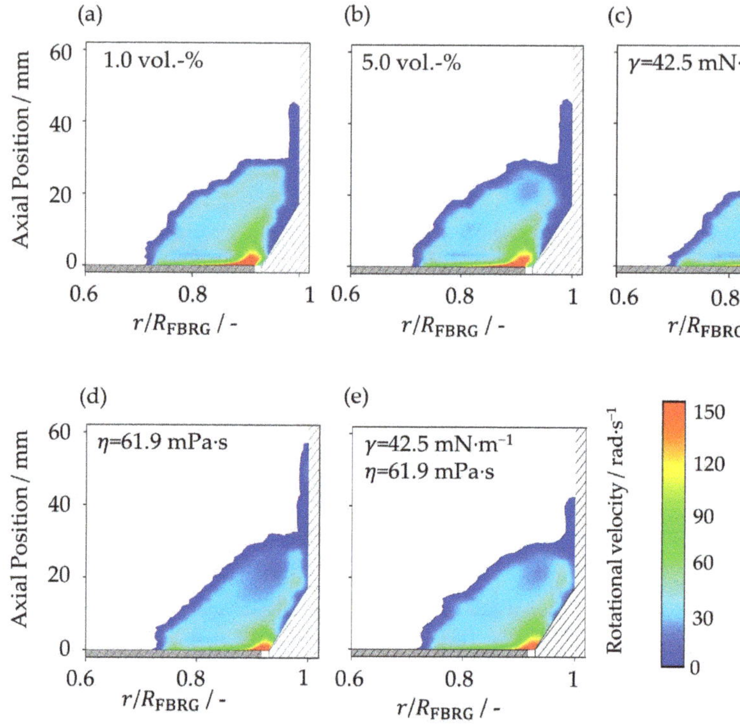

Figure 7. Poloidal distribution of rotational particle velocity at (**a**) liquid loading of 1 vol.-%, (**b**) liquid loading of 5 vol.-%, (**c**) surface tension of 42.5 mN·m^{-1}, (**d**) viscosity of 61.9 mPa·s, and (**e**) surface tension of 42.5 mN·m^{-1} and viscosity of 61.9 mPa·s.

For a useful description of the particle dynamics, the particle rotation number (PRN) and the radial movement proportion (RMP) can be calculated. Both parameters were developed in our previous study [8]. The particle rotation number is defined as the number of 360-degree rotations of the all particles around the central vertical axis of the process chamber per second. If, in a later case, a nozzle is installed in the wall of the process chamber to coat the particles, this key number describes how many times per second the particles pass the wet zone of the nozzle. The second number RMP describes the proportion of the radial velocity to the total velocity of the particles in the xy-plane. The higher this value, the more the kinetic energy of the particles goes into their radial motion in the bed. The values of the PRN and RMP for all simulated cases are given in Table 4. Due to the increased slip at lower liquid loading and lower surface tension (cases (a) and (c), the PRN decreases by 3.9% compared to the case (b) with 5 vol.-% of water. In contrast, it increases by 14.5% with increased viscosity (cases (d) and (e). All five cases have a significantly lower rotational speed than the rotor plate, which rotates at 1.67 s^{-1}. This is mainly due to the fact that it is an unstructured plate, where the energy transfer is not as good as with structured plates [57]. Looking at the RMP, it is clear that a reduction in liquid loading leads to an increase in radial motion. This is even more pronounced for the case with reduced surface tension. In both cases, the lower capillary forces compared to 5 vol.-% water lead to a greater freedom of movement of the particles and thus to an increased poloidal velocity. As already seen in Figure 6, the velocities are lower in the poloidal plane when the viscosity is increased to 61.9 mPa·s. The reason is due to the significant increase in viscous forces, the particles are slowed down more during contacts. As a consequence, the RMP also decreases more significantly. A lower surface tension in case e) with 6 mass-% PHARMACOAT®

606 solution, and thus lower capillary forces, lead to reduced energy dissipation, which again slightly increases the RMP compared to the fourth case.

Table 4. The particle rotation number and the radial movement proportion for the five simulation cases.

Case	PRN in 1/s	RMP in %
(a) 1.0 vol.-%	0.73	6.86
(b) 5.0 vol.-%	0.76	6.01
(c) $\gamma = 42.5\ \text{mN}\cdot\text{m}^{-1}$	0.73	7.21
(d) $\eta = 61.9\ \text{mPa}\cdot\text{s}$	0.87	4.58
(e) $\gamma = 42.5\ \text{mN}\cdot\text{m}^{-1}$, $\eta = 61.9\ \text{mPa}\cdot\text{s}$	0.87	5.14

4.3. Rotational Particle Velocity

Another important kinematic parameter to analyze the particle dynamics is the rotational particle velocity, which describes the rotation of the particles around their center of mass. Figure 7 shows the poloidal distribution for the rotational velocity of the particles as a function of radial and axial position at different liquid loading and at different liquid properties. Directly above the rotor plate near the annular gap, the highest rotational velocities can be seen. This is due to the fact that, on the one hand, the circumferential velocity of the rotor is highest in this zone and, on the other hand, the solid volume fraction is low due to the inflowing fluidization air (Figure 3). The highest rotation velocities of the particles can be obtained in the case with 1 vol.-% water (Figure 7a). Increasing the amount of liquid as well as the viscosity increases the liquid bridge forces, which reduces the rotation velocity of the particles, whereas reduced surface tension counteracts this.

In the fourth case (Figure 7d), where the liquid bridge forces are highest, the region with high rotational velocities is smallest. The reason for this is the comparatively densest particle bed, since the mean free path length of the particles is smallest there. This in turn leads to higher adhesion rates and thus to a reduction in the rotational velocity of the particles.

4.4. Particle Contacts

In Table 5, the contact rates and average numbers of contact partners can be compared for the cases with different liquid loading of water as well as different liquid properties. All cases lead to a decrease in the average contact rate compared to case (b) with 5 vol.-% water. However, no significant differences can be identified between the cases. In general, a lower liquid loading as well as surface tension results in smaller liquid bridge forces. As a consequence, the particle bed is slightly less densely packed and the mean free path length, and thus the time until contact occurs again, is increased. At the same time, the average number of contact partners decreases, since the adhesive forces that lead to the formation of aggregates decrease. In the fourth case (d), with a high viscosity and unchanged surface tension, the viscous forces slow down the particle contact velocity, while capillary forces remain strongly attractive. This leads to an increase in the average number of contact partners compared to the case (b) with 5 vol.-% water, since the aggregates remain longer stable. However, this reduces the average contact rate of the individual particles. At high viscosity and lower surface tension in case (e) with 6 mass-% PHARMACOAT® 606 solution, both effects counteract each other. The result is a minimally higher contact rate of the particles compared to the third case (c) and a slightly lower average number of contact partners compared to the fourth case.

Table 5. Contact rate and average number of contact partners for the five simulation cases.

Case	Contact Rates/-	Average Numbers of Contact Partners/-
(a) 1.0 vol.-%	2.14	3.23
(b) 5.0 vol.-%	2.96	3.45
(c) γ = 42.5 mN·m^{-1}	2.21	3.34
(d) η = 61.9 mPa·s	2.56	3.76
(e) γ = 42.5 mN·m^{-1}, η = 61.9 mPa·s	2.24	3.43

For a more detailed analysis of the aggregates formed during the process, the differential distributions of the number of simultaneous contact partners for all five cases are shown in Figure 8. While the distributions at 1 vol.-% water (Figure 8a) and a reduced surface tension (Figure 8c) differ only slightly from the case with 5 vol.-% water, significant differences can be seen between the two cases with increased viscosity. In the fourth case (Figure 8d), the proportion of aggregates consisting of two or three particles decreases by 3.7%, while the proportion of aggregates consisting of more than seven particles increases by 4.8%. In the last case studied (Figure 8e), the proportion of aggregates consisting of two or three particles increases by 4.1%, and the proportion of aggregates of more than nine particles increases slightly as well. In all cases, the large aggregates form in the upper region of the particle bed, where the poloidal and tangential particle velocities are lowest. Since the poloidal velocities are lowest in the cases with increased viscosity, the large aggregates can exist here for the longest time before the particles separate from each other again due to shear forces.

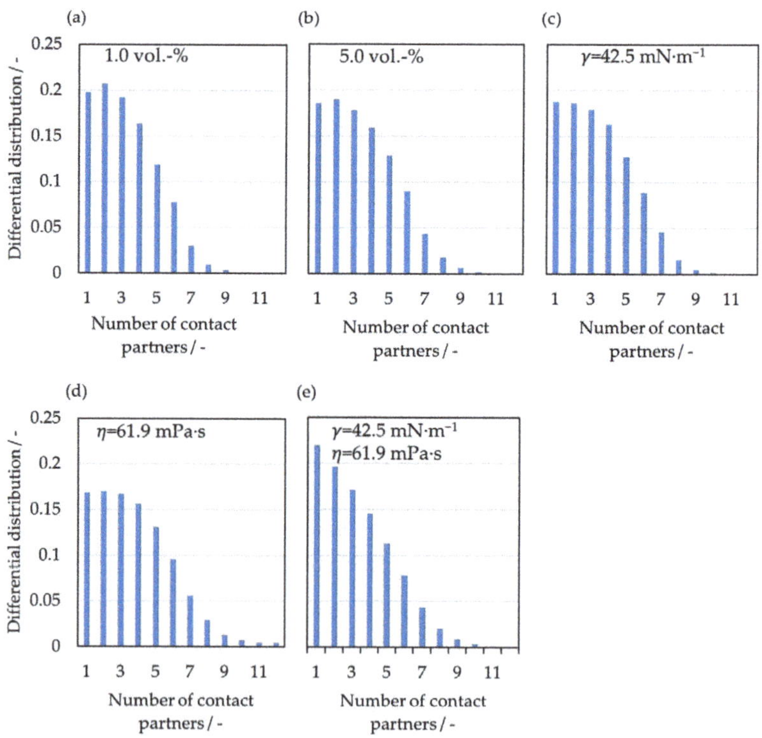

Figure 8. Differential distribution of the number of simultaneous contact partners for a (**a**) liquid loading of 1 vol.-%, (**b**) liquid loading of 5 vol.-%, (**c**) surface tension of 42.5 mN·m^{-1}, (**d**) viscosity of 61.9 mPa·s, and (**e**) surface tension of 42.5 mN·m^{-1} and viscosity of 61.9 mPa·s.

In Figure 9, the differential distribution of the time-averaged contact velocities can be seen. For this purpose, the contact velocities were counted in intervals of 0.01 m·s^{-1} and divided by the total number of contacts. Since the distribution of the contact velocities is wide, the velocity range is shown up to 0.2 m·s^{-1} for clearer visualization. In addition, the shown range represents 98% of the occurring impact velocities. As expected, the significantly higher viscosity in the last two cases (d) and (e) leads to an increase in the proportion of low contact velocities of less than 0.05 m·s^{-1}, while the proportion of higher contact velocities decreases accordingly. The reason is the stronger viscous forces, which always oppose the motion of the particles and thus slow them down in the approach phase during contact. This can also be confirmed by the capillary number, which, in these cases, is in the range of 10^{-2}, whereas in the other cases (a)–(c), it is in the range of 10^{-4}. The reduction of the liquid loading (curve a)) or the surface tension (curve (c)) leads to a decrease in the attractive capillary forces. Therefore, the particles' movement, for example PRN and RMP, in the bed is higher (Table 4). As a result, the fraction of contact velocities of more than 0.05 m·s^{-1} is slightly greater than for case b) with 5 vol.-% water, but less than for the two cases with high viscosity.

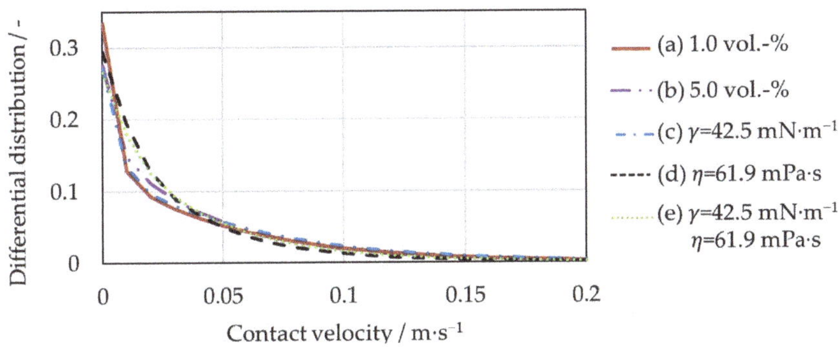

Figure 9. Differential distribution of the contact velocities for the five simulation cases.

Table 6 summarizes all average contact velocities with variation of liquid loading and liquid properties. As could already be seen from the differential distribution of the contact velocities (Figure 9), with increasing coating liquid viscosity, the fraction of low contact velocities grows. This can also be seen when looking at the average contact velocities. The highest contact velocities occur during interactions of the particles with the rotating rotor plate. In wet particle contacts, both contact forces due to viscoelastic deformation and contact forces due to liquid bridges are existent. In the FBRG, however, the contacts in which only a liquid bridge force acts predominate. Table 6 shows that the mean particle particle contact forces, mainly contributed by liquid bridge forces, increase in both normal and tangential directions for the highly viscous cases (d) and (e). In the normal direction, the contact force increases by 14.3% for the fourth case (d) and by 4.1% for the fifth case (e) compared to the case (b) with 5 vol.-% water. When comparing the tangential force for particle–particle contacts, an even more significant increase of 76.9% in the fourth case (d) and 65.4% in the fifth case (e) is noticeable. The reason is that the viscous force increases proportionally with the viscosity of the liquid (Equations (24) and (25)). Since the capillary force is also proportionally dependent on the surface tension (Equations (15) and (16)), the mean contact force decreases when the surface tension is reduced. The average particle–particle contact forces in the normal and tangential directions in the case (a) with reduced liquid loading and in the case (c) with lower surface tension are smaller than for 5 vol.-% water (case (b)). The ratio of the normal to the tangential contact force changes significantly for the last two cases (d) and (e) and decreases. This influence is most clearly seen for the particle–rotor contacts. Thus, the high viscosity with simultaneously reduced surface tension of the 6 mass-% PHARMACOAT® 606 solution (case (e)) leads to a ratio of normal to tangential contact force of less than one. The explanation for this is the lower capillary forces acting only in the normal direction, with a simultaneous sharp increase in the viscous forces in the tangential direction. The results for case (a) with a reduced liquid loading and case (c) with a reduced surface tension differ only slightly from the second case with 5 vol.-% water.

Table 6. Average contact velocities and forces for the five simulation cases in contacts between particles (P–P), particles with the cylindrical wall (P–W) and particles with rotor plate (P–R).

Contact Partners	Contact Velocity/m·s^{-1}	Normal Contact Force/mN	Tangential Contact Force/mN
(a) 1.0 vol.-%			
P–P	0.040	1.40	0.25
P–W	0.046	1.52	0.62
P–R	0.208	4.01	1.36
(b) 5.0 vol.-%			
P–P	0.041	1.47	0.26
P–W	0.066	1.58	0.63
P–R	0.217	3.89	1.34
(c) γ = 42.5 mN·m^{-1}			
P–P	0.045	1.24	0.22
P–W	0.079	1.18	0.46
P–R	0.230	3.29	1.16
(d) η_l = 61.9 mPa·s			
P–P	0.033	1.68	0.46
P–W	0.039	1.32	0.77
P–R	0.205	3.59	2.88
(e) γ = 42.5 mN·m^{-1}, η_l = 61.9 mPa·s			
P–P	0.037	1.53	0.43
P–W	0.069	1.12	0.82
P–R	0.222	3.27	3.73

5. Conclusions

In this work, the dynamics of wet particles in a fluidized bed rotor granulator was investigated using CFD-DEM simulation. A liquid bridge model was implemented in DEM to account for the acting physical adhesion mechanisms due to the capillary and viscous forces. In general, the dynamics of the wet particles are affected by the rotation of the plate; therefore, the particles are located near the apparatus wall. The particle concentration is highest in the center of the particle bed and lowest directly above the annular gap. In addition, in the region above the annular gap, due to the inflowing gas, the poloidal velocity of the particles is highest. For the tangential and rotational velocities, the region with high velocities is mainly directly above the rotation plate.

The following findings regarding the influence of liquid loading and liquid properties on the particle dynamics and interactions were obtained:

- Increasing the viscosity to the value of a 6 mass-% PHARMACOAT® 606 coating solution results in a denser particle bed. In addition, the particle rotation velocities and the particle movement in the poloidal plane are reduced.
- A reduced liquid loading in the bed as well as a reduced surface tension of the coating liquid lead to lower capillary forces, and thus, to increased particle movement.
- The fraction of high contact velocities increases at low liquid loading or low surface tension, while it decreases at high viscosity. On the other hand, the average contact force increases significantly with high viscosity.
- Based on the proportional dependence of capillary force on surface tension or viscosity force on viscosity, it was found that an increase in viscosity leads to an increase in aggregate size, whereas a reduction in surface tension results in a decrease.

Author Contributions: Conceptualization, methodology, investigation, data curation and analysis, project administration, writing—review and editing: P.G.; resources, investigation, supervision, funding acquisition: S.H.; methodology, investigation, supervision, funding acquisition, writing—review and editing: S.A. All authors have read and agreed to the published version of the manuscript.

Funding: This research was funded by Forschungsgemeinschaft (DFG, German Research Foundation), grant number AN 782/12-1 and HE 4526/25-1, what the authors gratefully acknowledge.

Institutional Review Board Statement: Not applicable.

Informed Consent Statement: Not applicable.

Data Availability Statement: The data presented in this study are available on request from the corresponding author.

Conflicts of Interest: The authors declare no conflict of interest.

References

1. Mörl, L.; Heinrich, S.; Peglow, M. Chapter 2 Fluidized bed spray granulation. In *Granulation*; Elsevier: Amsterdam, The Netherlands, 2007; pp. 21–188, ISBN 9780444518712.
2. Tsotsas, E.; Mujumdar, A.S. *Modern Drying Technology*; Wiley: Hoboken, NJ, USA, 2011; Volume 3, ISBN 9783527315581.
3. Korakianiti, E.S.; Rekkas, D.M.; Dallas, P.P.; Choulis, N.H. Sequential optimization of a pelletization process in a fluid bed rotor granulator. *J. Drug Deliv. Sci. Technol.* **2004**, *14*, 207–214. [CrossRef]
4. Neuwirth, J.; Antonyuk, S.; Heinrich, S.; Jacob, M. CFD–DEM study and direct measurement of the granular flow in a rotor granulator. *Chem. Eng. Sci.* **2013**, *86*, 151–163. [CrossRef]
5. Grohn, P.; Lawall, M.; Oesau, T.; Heinrich, S.; Antonyuk, S. CFD-DEM Simulation of a Coating Process in a Fluidized Bed Rotor Granulator. *Processes* **2020**, *8*, 1090. [CrossRef]
6. Langner, M.; Kitzmann, I.; Ruppert, A.-L.; Wittich, I.; Wolf, B. In-line particle size measurement and process influences on rotary fluidized bed agglomeration. *Powder Technol.* **2020**, *364*, 673–679. [CrossRef]
7. Jacob, M. Chapter 9 Granulation equipment. In *Granulation*; Elsevier: Amsterdam, The Netherlands, 2007; pp. 417–476, ISBN 9780444518712.
8. Grohn, P.; Oesau, T.; Heinrich, S.; Antonyuk, S. Investigation of the influence of wetting on the particle dynamics in a fluidized bed rotor granulator by MPT measurements and CFD-DEM simulations. *Powder Technol.* **2022**, *408*, 117736. [CrossRef]
9. Vuppala, M.K.; Parikh, D.M.; Bhagat, H.R. Application of Powder-Layering Technology and Film Coating for Manufacture of Sustained-Release Pellets Using a Rotary Fluid Bed Processor. *Drug Dev. Ind. Pharm.* **1997**, *23*, 687–694. [CrossRef]
10. Kristensen, J.; Schaefer, T.; Kleinebudde, P. Direct pelletization in a rotary processor controlled by torque measurements. II: Effects of changes in the content of microcrystalline cellulose. *AAPS PharmSci* **2000**, *2*, 45. [CrossRef]
11. Gu, L.; Liew, C.V.; Heng, P.W.S. Wet spheronization by rotary processing—A multistage single-pot process for producing spheroids. *Drug Dev. Ind. Pharm.* **2004**, *30*, 111–123. [CrossRef]
12. Deen, N.G.; van Sint Annaland, M.; van der Hoef, M.A.; Kuipers, J.A.M. Review of discrete particle modeling of fluidized beds. *Chem. Eng. Sci.* **2007**, *62*, 28–44. [CrossRef]
13. Fries, L.; Dosta, M.; Antonyuk, S.; Heinrich, S.; Palzer, S. Moisture Distribution in Fluidized Beds with Liquid Injection. *Chem. Eng. Technol.* **2011**, *34*, 1076–1084. [CrossRef]
14. Salikov, V.; Heinrich, S.; Antonyuk, S.; Sutkar, V.S.; Deen, N.G.; Kuipers, J.A.M. Investigations on the spouting stability in a prismatic spouted bed and apparatus optimization. *Adv. Powder Technol.* **2015**, *26*, 718–733. [CrossRef]
15. Breuninger, P.; Weis, D.; Behrendt, I.; Grohn, P.; Krull, F.; Antonyuk, S. CFD–DEM simulation of fine particles in a spouted bed apparatus with a Wurster tube. *Particuology* **2019**, *42*, 114–125. [CrossRef]
16. Sutkar, V.S.; Deen, N.G.; Salikov, V.; Antonyuk, S.; Heinrich, S.; Kuipers, J.A.M. Experimental and numerical investigations of a pseudo-2D spout fluidized bed with draft plates. *Powder Technol.* **2015**, *270*, 537–547. [CrossRef]
17. Heinrich, S.; Dosta, M.; Antonyuk, S. Multiscale Analysis of a Coating Process in a Wurster Fluidized Bed Apparatus. In *Mesoscale Modeling in Chemical Engineering Part I*; Elsevier: Amsterdam, The Netherlands, 2015; pp. 83–135, ISBN 9780128012475.
18. Muguruma, Y.; Tanaka, T.; Tsuji, Y. Numerical simulation of particulate flow with liquid bridge between particles (simulation of centrifugal tumbling granulator). *Powder Technol.* **2000**, *109*, 49–57. [CrossRef]
19. Weis, D.; Evers, M.; Thommes, M.; Antonyuk, S. DEM simulation of the mixing behavior in a spheronization process. *Chem. Eng. Sci.* **2018**, *192*, 803–815. [CrossRef]
20. Weis, D.; Grohn, P.; Evers, M.; Thommes, M.; García, E.; Antonyuk, S. Implementation of formation mechanisms in DEM simulation of the spheronization process of pharmaceutical pellets. *Powder Technol.* **2021**, *378*, 667–679. [CrossRef]
21. Grohn, P.; Schaedler, L.; Atxutegi, A.; Heinrich, S.; Antonyuk, S. CFD-DEM Simulation of Superquadric Cylindrical Particles in a Spouted Bed and a Rotor Granulator. *Chem. Ing. Tech.* **2023**, *95*, 244–255. [CrossRef]

22. Neuwirth, J. *Charakterisierung und Diskrete-Partikel-Modellierung des Strömungs- und Dispersionsverhaltens im Rotorgranulator*, 1st ed.; Cuvillier Verlag: Göttingen, Germany, 2017; ISBN 9783736994768.
23. Israelachvili, J.N. *Intermolecular and Surface Forces*; Elsevier: Amsterdam, The Netherlands, 2011; ISBN 9780123919274.
24. Lian, G.; Thornton, C.; Adams, M.J. Discrete particle simulation of agglomerate impact coalescence. *Chem. Eng. Sci.* **1998**, *53*, 3381–3391. [CrossRef]
25. Popov, V.L. (Ed.) *Contact Mechanics and Friction: Physical Principles and Applications*, 1st ed.; Springer: Berlin, Germany, 2010; ISBN 978-3-642-10802-0.
26. Grohn, P.; Oesau, T.; Heinrich, S.; Antonyuk, S. Investigation of the influence of impact velocity and liquid bridge volume on the maximum liquid bridge length. *Adv. Powder Technol.* **2022**, *33*, 103630. [CrossRef]
27. Moukalled, F. *The Finite Volume Method in Computational Fluid Dynamics: An Advanced Introduction with OpenFOAM® and Matlab*, 1st ed.; Springer: Cham, Switzerland, 2016; ISBN 9783319168746.
28. Hoomans, B.P.B.; Kuipers, J.A.M.; Briels, W.J.; van Swaaij, W.P.M. Discrete particle simulation of bubble and slug formation in a two-dimensional gas-fluidised bed: A hard-sphere approach. *Chem. Eng. Sci.* **1996**, *51*, 99–118. [CrossRef]
29. Goniva, C.; Kloss, C.; Deen, N.G.; Kuipers, J.A.M.; Pirker, S. Influence of rolling friction on single spout fluidized bed simulation. *Particuology* **2012**, *10*, 582–591. [CrossRef]
30. Kanitz, M.; Grabe, J. The influence of the void fraction on the particle migration: A coupled computational fluid dynamics–discrete element method study about drag force correlations. *Int. J. Numer. Anal. Methods Geomech.* **2021**, *45*, 45–63. [CrossRef]
31. Zhao, J.; Shan, T. Coupled CFD–DEM simulation of fluid–particle interaction in geomechanics. *Powder Technol.* **2013**, *239*, 248–258. [CrossRef]
32. Di Felice, R. The voidage function for fluid-particle interaction systems. *Int. J. Multiph. Flow* **1994**, *20*, 153–159. [CrossRef]
33. Zhou, Z.Y.; Kuang, S.B.; Chu, K.W.; Yu, A.B. Discrete particle simulation of particle–fluid flow: Model formulations and their applicability. *J. Fluid Mech.* **2010**, *661*, 482–510. [CrossRef]
34. Hesse, R.; Krull, F.; Antonyuk, S. Experimentally calibrated CFD-DEM study of air impairment during powder discharge for varying hopper configurations. *Powder Technol.* **2020**, *372*, 404–419. [CrossRef]
35. Cundall, P.A.; Strack, O.D.L. A discrete numerical model for granular assemblies. *Géotechnique* **1979**, *29*, 47–65. [CrossRef]
36. Crowe, C.T.; Schwarzkopf, J.D.; Sommerfeld, M.; Tsuji, Y. *Multiphase Flows with Droplets and Particles*; CRC Press: Boca Raton, FL, USA, 2011; ISBN 9780429106392.
37. Salikov, V.; Antonyuk, S.; Heinrich, S.; Sutkar, V.S.; Deen, N.G.; Kuipers, J.A.M. Characterization and CFD-DEM modelling of a prismatic spouted bed. *Powder Technol.* **2015**, *270*, 622–636. [CrossRef]
38. Lian, G.; Thornton, C.; Adams, M.J. A Theoretical Study of the Liquid Bridge Forces between Two Rigid Spherical Bodies. *J. Colloid Interface Sci.* **1993**, *161*, 138–147. [CrossRef]
39. Mikami, T.; Kamiya, H.; Horio, M. Numerical simulation of cohesive powder behavior in a fluidized bed. *Chem. Eng. Sci.* **1998**, *53*, 1927–1940. [CrossRef]
40. Rabinovich, Y.I.; Esayanur, M.S.; Moudgil, B.M. Capillary forces between two spheres with a fixed volume liquid bridge: Theory and experiment. *Langmuir* **2005**, *21*, 10992–10997. [CrossRef] [PubMed]
41. Shi, D.; McCarthy, J.J. Numerical simulation of liquid transfer between particles. *Powder Technol.* **2008**, *184*, 64–75. [CrossRef]
42. Pitois, O.; Moucheront, P.; Chateau, X. Rupture energy of a pendular liquid bridge. *Eur. Phys. J. B* **2001**, *23*, 79–86. [CrossRef]
43. Antonyuk, S.; Heinrich, S.; Deen, N.; Kuipers, H. Influence of liquid layers on energy absorption during particle impact. *Particuology* **2009**, *7*, 245–259. [CrossRef]
44. Gollwitzer, F.; Rehberg, I.; Kruelle, C.A.; Huang, K. Coefficient of restitution for wet particles. *Phys. Rev. E Stat. Nonlin. Soft Matter Phys.* **2012**, *86*, 11303. [CrossRef]
45. Schmelzle, S.; Asylbekov, E.; Radel, B.; Nirschl, H. Modelling of partially wet particles in DEM simulations of a solid mixing process. *Powder Technol.* **2018**, *338*, 354–364. [CrossRef]
46. Reynolds, O. IV. On the theory of lubrication and its application to Mr. Beauchamp tower's experiments, including an experimental determination of the viscosity of olive oil. *Phil. Trans. R. Soc.* **1886**, *177*, 157–234. [CrossRef]
47. Adams, M.J.; Perchard, V. The cohesive forces between particles with interstitial liquid. In *Institution of Chemical Engineering Symposium*; Institution of Chemical Engineers: Rugby, UK, 1985; pp. 147–160.
48. Nase, S.T.; Vargas, W.L.; Abatan, A.A.; McCarthy, J.J. Discrete characterization tools for cohesive granular material. *Powder Technol.* **2001**, *116*, 214–223. [CrossRef]
49. Tang, T.; He, Y.; Tai, T.; Wen, D. DEM numerical investigation of wet particle flow behaviors in multiple-spout fluidized beds. *Chem. Eng. Sci.* **2017**, *172*, 79–99. [CrossRef]
50. Anand, A.; Curtis, J.S.; Wassgren, C.R.; Hancock, B.C.; Ketterhagen, W.R. Predicting discharge dynamics of wet cohesive particles from a rectangular hopper using the discrete element method (DEM). *Chem. Eng. Sci.* **2009**, *64*, 5268–5275. [CrossRef]
51. Washino, K.; Miyazaki, K.; Tsuji, T.; Tanaka, T. A new contact liquid dispersion model for discrete particle simulation. *Chem. Eng. Res. Des.* **2016**, *110*, 123–130. [CrossRef]
52. Weller, H.G.; Tabor, G.; Jasak, H.; Fureby, C. A tensorial approach to computational continuum mechanics using object-oriented techniques. *Comput. Phys.* **1998**, *12*, 620. [CrossRef]
53. Issa, R.I. Solution of the implicitly discretised fluid flow equations by operator-splitting. *J. Comput. Phys.* **1986**, *62*, 40–65. [CrossRef]

54. Launder, B.E.; Spalding, D.B. The numerical computation of turbulent flows. *Comput. Methods Appl. Mech. Eng.* **1974**, *3*, 269–289. [CrossRef]
55. Kloss, C.; Goniva, C.; Hager, A.; Amberger, S.; Pirker, S. Models, algorithms and validation for opensource DEM and CFD-DEM. *Prog. Comput. Fluid Dyn.* **2012**, *12*, 140–152. [CrossRef]
56. Oesau, T.; Grohn, P.; Pietsch-Braune, S.; Antonyuk, S.; Heinrich, S. Novel approach for measurement of restitution coefficient by magnetic particle tracking. *Adv. Powder Technol.* **2022**, *33*, 103362. [CrossRef]
57. Weis, D. Beschreibung des Sphäronisationsprozesses von Pharmazeutischen Pellets Mittels Numerischer Methoden. Ph.D. Thesis, TU Kaiserslautern, Kaiserslautern, Germany, 2021.

pharmaceutics

MDPI

Article

A Hybrid Model to Predict Formulation Dependent Granule Growth in a Bi-Component Wet Granulation Process

Indu Muthancheri and Rohit Ramachandran *

Department of Chemical and Biochemical Engineering, Rutgers, The State University of New Jersey, Piscataway, NJ 08854, USA; im225@scarletmail.rutgers.edu
* Correspondence: rohit.r@rutgers.edu

Abstract: In this study, a hybrid modeling framework was developed for predicting size distribution and content uniformity of granules in a bi-component wet granulation system with components of differing hydrophobicities. Two bi-component formulations, (1) ibuprofen-USP and micro-crystalline cellulose and (2) micronized acetaminophen and micro-crystalline cellulose, were used in this study. First, a random forest method was used for predicting the probability of nucleation mechanism (immersion and solid spread), depending upon the formulation hydrophobicity. The predicted nucleation mechanism probability is used to determine the aggregation rate as well as the initial particle distribution in the population balance model. The aggregation process was modeled as Type-I: Sticking aggregation and Type-II: Deformation driven aggregation. In Type-I, the capillary force dominant aggregation mechanism is represented by the particles sticking together without deformation. In the case of Type-II, the particle deformation causes an increase in the contact area, representing a viscous force dominant aggregation mechanism. The choice between Type-I and II aggregation is determined based on the difference in nucleation mechanism that is predicted using the random forest method. The model was optimized and validated using the granule content uniformity data and size distribution data obtained from the experimental studies. The proposed framework predicted content non-uniform behavior for formulations that favored immersion nucleation and uniform behavior for formulations that favored solid-spreading nucleation.

Keywords: wet granulation; multicomponent; population balance model; content uniformity

Citation: Muthancheri, I.; Ramachandran, R. A Hybrid Model to Predict Formulation Dependent Granule Growth in a Bi-Component Wet Granulation Process. *Pharmaceutics* **2021**, *13*, 2063. https://doi.org/10.3390/pharmaceutics13122063

Academic Editor: Colin Hare

Received: 17 October 2021
Accepted: 25 November 2021
Published: 2 December 2021

1. Introduction

The modeling of wet granulation process provides better insight into its process dynamics, which then can be used for efficient process control and scale-up if required [1,2]. The simultaneous, interacting mechanisms in the wet granulation process make it a difficult unit operation to model. Various techniques are used for the modeling of granulation (based on time scale and length scale). The population balance model (PBM) or PBM coupled with particle-level models such as discrete element method (DEM) and continuous fluid dynamics (CFD) are extensively used for this purpose [3–5]. Although PBM captures multi-dimensional properties of granules such as size, composition, and porosity, the model predictions are restricted to the experimental design space due to the large number of fitting parameters involved. DEM is a physics-based model that tracks the particle positions based on collisions between particles and therefore is not impacted by design space changes. However, DEM is unable to independently simulate particle property changes resulting from various mechanisms in the wet granulation process. A coupled or hybrid DEM-PBM is found to be useful for comprehensively representing wet granulation at all scales (micro-meso scale using DEM and meso-macro scale using PBM). The macro-scale models the overall particle property change, the meso-scale represents property change within a granule ensemble, and the micro-scale represents individual particle or constituent powder dynamics.

A significant number of modeling studies have been conducted for the prediction of a single-component wet granulation process, and this includes both nucleation and growth kinetics [6–9]. Although most of the wet granulation process involves more than one solid component, there are not many models reported that consider a multi-component process. Building a formulation-dependent constitutive equation requires the understanding of the effects of primary particle properties such as maximum pore saturation, material yield strength, cohesiveness, and primary particle size, on the final product granule attributes.

Matsoukas et al. [10,11] and Marshall Jr et al. [12] applied a multi-component PBM for modeling a bi-component aggregation using a composition-dependent aggregation kernel. In their work, two components, solute and solvent, are considered. The internal coordinates of the granule or particle were determined by the total mass of the granule and the mass of solute in it. In the Matsoukas et al. [10] model, the solute is soluble in the solvent, and the bivariate distribution of number density is shown as the product of the size distribution, with a Gaussian compositional distribution. In an insoluble system, the Gaussian compositional distribution function that is used to represent the relative concentration of one component to another may not be applicable. Matsoukas et al. [11] represented a size-dependent aggregation kernel with a composition-dependent multiplicative factor. They defined an adjustable interaction parameter that describes the attraction or repulsion between the two components. The model was able to demonstrate the extent of blending with positive and negative interaction parameters.

The granule growth (aggregation/coalescence), however, is expressed nearly always as a semi-empirically based [4,13] kernel or as a function of the particle collision information obtained from DEM modeling [14,15]. Some of the limitations of the existing coalescence models for representing granule growth are that they neglect capillary force interaction and use the static yield strength analyses of the powder bed to calculate the granule strength during the collision. A multi-dimensional population balance [16] that accounts for size, solid content, surface liquid, and deformability needs to be used to couple "aggregation" and "layering" granule growth mechanisms.

Another mechanism to consider is the nucleation process at the start of the liquid addition phase during granulation, that impacts the granule growth mechanism and the final granule quality attributes. There are two important aspects of nucleation modeling: the kinetics of nuclei formation and the physical attributes such as size, porosity, and content uniformity of the nuclei. In the case of immersion nucleation, kinetic models were developed by Hounslow et al. [17], Hapgood et al. [18]. These kinetic models provide the nuclei size distribution as model output and which can be used for predicting the final granule size distribution [19]. The time scale of the nucleation process is relatively faster than the rest of the granulation rate mechanisms [20], and thus, it is not necessary to incorporate the dynamics of the nucleation process into the granulation modeling. However, the internal properties of nuclei such as size, deformability, surface liquid content, and content uniformity affect the granulation growth kinetics, and the final granule quality attributes [21,22]. Thus, it is important to have an experimental or modeling framework to predict the properties of the initial nuclei to simulate granule properties other than size distribution using PBM.

Physics-based models for simulating nuclei require complicated multi-phase simulations. Washino et al. [23] presented a coupled DEM and constrained interpolation profile (CIP) for simulating the nuclei during the wet granulation process. The effect of surface tension on the liquid binder flow was modeled depending on the relative position of the fluid interfaces to the solid particles, i.e., the model on the outside, inside, or on the surface of the powder bed corresponds to a free surface, capillary action, and bed surface wetting. Washino et al. [24] showed the CFD-DEM simulation of nuclei generation in a dynamic powder bed. These studies consider a particle system with good wettability or spreading coefficient [25]. In the case of powder mixture with hydrophobic powders, the nuclei formed are solid-spread nuclei, and the physics-based approach developed for immersion nuclei is not suitable for such systems. Due to the complexities of nucleation mechanisms that can be either an immersion or solid-spread nuclei based on the wettability of the constituent material on the droplet vicinity during binder addition, the currently available models fall short of representing the nucleation mechanism and nuclei property in such systems.

Hybrid modeling has been demonstrated to have various advantages of improving process understanding with the incorporation of empirically based statistical models with mechanistic models [26–28]. A hybrid model consisting of both statistical correlations and physics-based models is often used to simplify the computational efforts and incorporate complex mechanisms into the model. Such models can overcome the disadvantages of both purely data-driven and physics-based models [27]. A hybrid model such as PBM with artificial neural network (ANN) was developed to substitute for the high-fidelity PBM-DEM model by Barrasso et al. [29].

The objectives of this study are as follows:

1. Incorporate the nuclei particle characteristics in the population balance model based on the classification model result from Muthancheri et al. [30].
2. Develop a composition-dependent PBM framework for bi-component wet granulation process with a large binder droplet for predicting the granule quality attributes with change in percentage formulation.

2. Model Development

2.1. Population Balance Model

The population balance equation as shown in Equations (1) and (3) are used in this work to predict the particle size distribution, liquid distribution, and component distribution or the content uniformity [31]. In this work, the liquid volume of the granules is considered to be a lumped parameter under the assumption that all granules of the same size with the same composition of solids and pore volume have the same average liquid content. Such a reduced-order model was compared with higher-order models and was reported to have a significant time saving without compromising much on accuracy in previous studies [32].

$$\frac{\partial F}{\partial t} + \frac{\partial}{\partial s_1}\left(F\frac{ds_1}{dt}\right) + \frac{\partial}{\partial s_2}\left(F\frac{ds_2}{dt}\right) + \frac{\partial}{\partial p}\left(F\frac{dp}{dt}\right) = \Re_{nuc} + \Re_{agg} + +\Re_{brk} \tag{1}$$

$$\frac{\partial L_i}{\partial t} = \Re_{nuc,l_i} + \Re_{agg,l_i} + \Re_{brk,l_i} + F\frac{dl_i}{dt} \tag{2}$$

$$\frac{\partial L_e}{\partial t} = \Re_{nuc,l_e} + \Re_{agg,l_e} + \Re_{brk,l_e} + F\frac{dl_e}{dt} \tag{3}$$

where $F = F(s_1, s_2, p, t)$ is the number of particles with API volume s_1 and excipient volume s_2, pore volume p at time t. $L_e = L_e(s_1, s_2, p, t)$ and $L_i = L_i(s_1, s_2, p, t)$ is the average external liquid volume and average internal liquid volume of particles with API volume s_1 and excipient volume s_2, pore volume p at time t, respectively. Table 1 represents the dependent and independent variables. The rate mechanisms $\frac{ds_1}{dt}$, $\frac{ds_2}{dt}$, $\frac{dl_i}{dt}$, $\frac{dl_e}{dt}$, $\Re_{nuc}$, $\Re_{agg}$, and $\Re_{brk}$ are detailed in the next section.

The relationship between the variables in Table 1 are summarized in the following equations. The total granule volume v is obtained by Equation (4).

$$v = s_1 + s_2 + l_e + p \tag{4}$$

The surface area of the particles can be derived from the granule total volume as shown in Equation (5)

$$a = \pi^{\frac{1}{3}}(6v)^{\frac{2}{3}} \tag{5}$$

The porosity and content uniformity can be calculated using Equations (6) and (7), respectively.

$$\epsilon = \frac{p}{v} \tag{6}$$

$$q = s_1/v \tag{7}$$

Table 1. Variables in the PBM to describe granule.

Description	Notation
Independent variables	
API solid volume	s_1
Excipient solid volume	s_2
External liquid volume	l_e
Internal liquid volume	l_i
Pore volume	p
Dependent variables	
Total granule volume	v
Surface area	a
Porosity	ϵ
Content uniformity	q

2.2. Mechanisms Involved in the Model

The rate mechanisms during wet granulation determine the final characteristics of a granule. The following rate mechanisms are considered in the current model:

1. Immersion nucleation
2. Solid-spread nucleation of hydrophobic API
3. Granule surface wetting during liquid addition
4. Granule surface growth due to solid-spread nuclei
5. Hydrophilic excipient layering
6. Particle aggregation
7. Particle breakage
8. Compaction

2.2.1. Immersion Nucleation

Immersion nucleation kinetics are assumed to be instantaneous, with the nuclei being formed as soon as the drop hits the powder bed. The number of immersion nuclei generated N_{im} is calculated by Equation (8).

$$N_{im} = P_{im}\frac{\dot{Q}_{spray} \times \Delta t}{v_d} \tag{8}$$

where Q_{spray} is the volumetric spray rate of binder liquid and v_d is the volume of a single drop calculated from the nozzle opening. P_{im} is the probability of immersion nucleation to happen for the given percentage composition of API powder bed (calculated from the classification model from Muthancheri et al. [30]). All the immersion nuclei is assigned to the first bin of same percentage composition as that of the powder bed and pore volume close to the volume of single droplet.

The mass of API (M_{s_1}) and excipient (M_{s_2}) available is calculated at every time step based on the mass balance of nuclei generated and excipient particle layering. The volume of immersion nuclei (v_{im}) at the end of nucleation can be estimated from the equation derived by Hounslow et al. [17] (Equation (9)).

$$v_{im} = v_d\left(1 + \frac{1-\phi_{cp}}{\phi_{cp}}\right) \tag{9}$$

$$v_{im,s} = v_d\left(\frac{1-\phi_{cp}}{\phi_{cp}}\right) \tag{10}$$

where ϕ_{cp} is the critical-packing liquid volume fraction, which is kept constant at 0.2 in this study. $v_{im,s}$ is the solid volume in the nuclei. Total mass of solid component $M_{im,si}$, where $i = 1, 2$ for API and excipient, respectively, utilized to form immersion nuclei at Δt can be calculated as follows:

$$M_{im,si} = f_i N_{im} m_{im,s} \tag{11}$$
$$= f_i N_{im} \rho_s v_{im,s} \tag{12}$$

where $m_{im,s}$ is the solid mass in a single immersion nuclei and ρ_s is the weighted true density of the solid components. f_i is the fraction of solid component present in the powder bed ($i = 1, 2$ for API and excipient, respectively).

2.2.2. Solid-Spread Nucleation

In the current model, the hydrophobic API is considered to form solid-spread nuclei, and the number of solid-spread nuclei (N_{ss}) is calculated as shown in Equation (13).

$$N_{ss} = (1 - P_{im}) \frac{\dot{Q}_{spray} \times \Delta t}{v_d} \tag{13}$$

Assuming that in the solid-spread nucleation the liquid drop (diameter d_d) is surrounded by hydrophobic API particles of diameter (d_p) (Figure 1), the approximate solid-spread nuclei volume (v_{ss}) and the volume of API particles in a solid-spread nuclei ($v_{ss,s1}$) can be calculated as follows:

$$v_{ss} = \frac{\pi}{6}\left(d_d + 2d_p\right)^3 \tag{14}$$
$$v_{ss,s1} = \frac{\pi}{6}\left(d_d + 2d_p\right)^3 - v_d \tag{15}$$

The total mass of API particles that form solid-spread nuclei at time Δt can be calculated as shown in Equation (17).

$$M_{ss,s1} = f_1 N_{ss} m_{ss,s1} \tag{16}$$
$$= f_1 N_{ss} \rho_{s1} v_{ss,s1} \tag{17}$$

where $m_{ss,s}$ is the solid mass in a single solid-spread nuclei and ρ_{s1} is the weighted true density of the API.

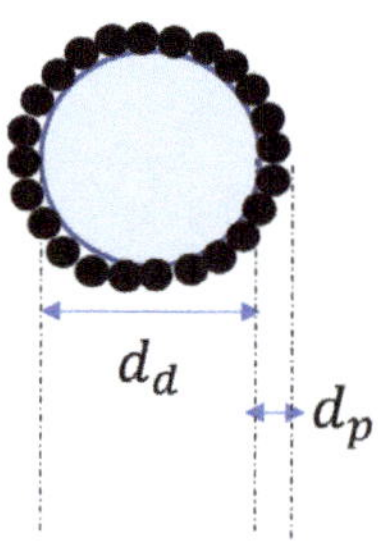

Figure 1. Solid-spread nuclei schematic.

2.2.3. Granule Surface Wetting during Liquid Addition

Rewetting of the granule surface during granulation is incorporated in the model in the $\frac{dl_e}{dt}$ term. The rewetting depends on the volume of granule in comparison with the total volume of granule present in the system.

$$\left.\frac{dl_e}{dt}\right|_{rewetting} = \frac{\dot{Q}_{spray} \times v}{\sum_{s_1}\sum_{s_2}\sum_{p} Fv} \quad (18)$$

2.2.4. Granule Surface Growth Due to Solid-Spread Nuclei

As the pore volume of solid-spread nuclei are larger than that of the immersion nuclei, they are not directly added to the PBM equation. Instead the rate of change of number density of solid-spread nuclei (F_{ss}) is tracked using a differential equation. When two solid-spread nuclei aggregate, the resulting volume is divided proportionally between the F_{ss} and F distribution using two class PBM approaches discussed by Jeong and Choi [33]. The depletion of solid-spread nuclei (F_{ss}) due to nuclei–nuclei aggregation and nuclei–granule surface growth is formulated as shown in Jeong and Choi [33] (Equation (19)).

$$\left.\frac{ds_1}{dt}\right|_{surfacegrowth} = k_{sg}v^{2/3} \quad (19)$$

Planchette et al. [34] studied the transition of liquid marble onto solid surfaces. They studied three mechanisms involved when solid-spread nuclei collide on a surface. The drop extension of the solid-spread nuclei is related to the impact velocity as $(D_{max} - D)/D \approxeq 0.12\sqrt{(We)}$, where D_{max} is the diameter of the disk shape the solid-spread nuclei takes before rupture, D is the diameter of the solid-spread nuclei ($=d_d + 2d_p$), and We is the Weber number during collision. This equation gives the minimum size ($=f(D_{max})$) of the granule upon which, when the solid-spread nuclei collide, the impact results in the surface growth of s_1 particles, as shown in Figure 2.

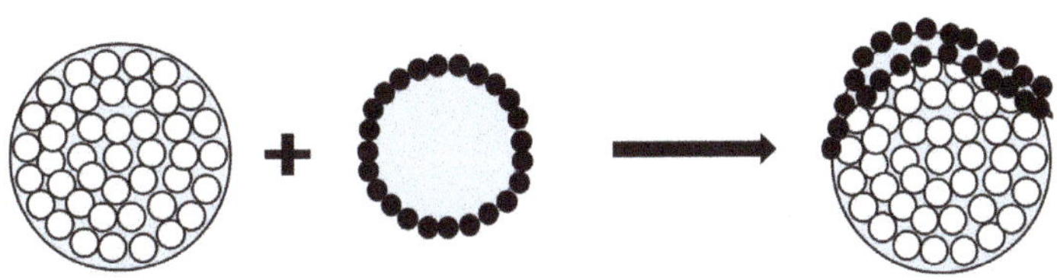

Figure 2. Solid-spread nuclei interaction with granule resulting in surface growth.

2.2.5. Hydrophilic Excipient Layering

Hydrophilic excipient layering is modeled as an increase in granule size as fine excipient powder particles adhere to the wet surfaces. The rate of increase in excipient volume of a granule ($\frac{ds_2}{dt}$) is assumed to be proportional to the granule surface area (Equation (5)) and the total mass of excipient powder left (m_{s2}) in the granulator as shown in the following equations.

$$\frac{ds_2}{dt} = \frac{k_{layer} \times a \times m_{s2}}{\rho_{s2}} \quad (20)$$

It is modeled such that only the granules with surface wetness ($l_e \neq 0$) experience layering. As a result, the increase in consolidation has a secondary effect on layering. The depletion of excipient fines is given by Equation (21).

$$\frac{dm_{s2}}{dt} = \rho_{s2}\int_0^{s1}\int_0^{s2}\int_{p_0}^{p} F\frac{ds_2}{dt}ds_1 ds_2 dp \quad (21)$$

2.2.6. Particle Aggregation

Goodson et al. [16] developed a PBM framework for characterizing the granule based on three properties: (1) size (big or small), (2) liquid content (wet or dry), and (3) strength (hard or soft). Figure 3 shows the two extremes of granule interactions. Two granules of high strength, less deformability, and lower capillary number are assumed to conserve surface area upon collision (Type I) and two granules of low strength, high deformability, and higher capillary number are assumed to conserve pore volumes upon collision.

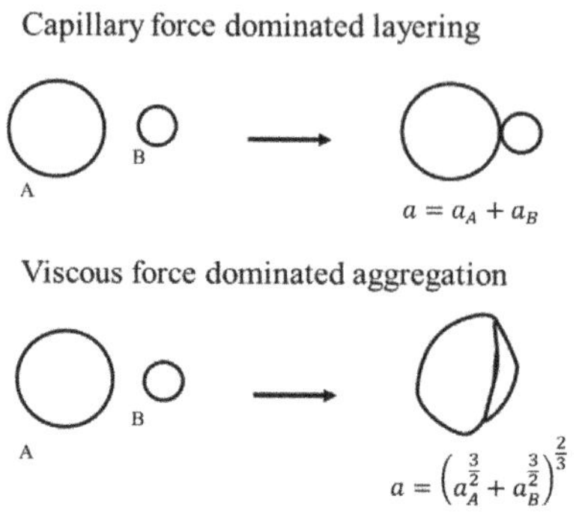

Figure 3. Representation of soft and hard granule growth.

The solid volumes are conserved during the aggregation. The resulting aggregate solid volume is a sum of the two colliding particle solid volume. The pore volume and liquid volume undergo a different transformation rule. Depending on the deformability of the colliding particles, the final granule pore volume is an interpolation of the two extremes in Figure 3. The resulting pore volume can be written as,

$$a = \zeta(a_A^{\frac{3}{2}} + a_B^{\frac{3}{2}}) + (1-\zeta)(a_A + a_B),\ 0 < \zeta < 1 \tag{22}$$

$$p = \frac{a^{\frac{3}{2}}}{6\sqrt{\pi}} - s_1 - s_2 - l_e \tag{23}$$

where ζ represents the relative softness or deformability of the granule (a function of coefficient of restitution, $\zeta = 1 - e_{coag}$ from Equation (36)). This internal coordinate representation was shown to provide differences in critical properties such as granule porosity, despite the similar granule size predictions [16].

The rate of particle aggregation is calculated as shown below.

$$R_{agg}(s_1, s_2, p, t) = R_{agg}^{form}(s_1, s_2, p, t) - R_{agg}^{dep}(s_1, s_2, p, t) \tag{24}$$

where R_{agg}^{form} and R_{agg}^{dep} are the rates of formation of larger particles (Equation (26)) and rate of depletion of smaller particles (Equation (27)), respectively.

$$R_{agg}^{form}(s_1, s_2, p, t) = \frac{1}{2}\int_0^{s_1}\int_0^{s_2}\int_{p_0}^{p} \beta(s_1 - s_1', s_2 - s_2', p - p', s_1', s_2', p') \times F(s_1 - s_1', s_2 - s_2', p - p', t)F(s_1', s_2', p', t)ds_1'ds_2'dp' \tag{25}$$

$$R_{agg}^{dep}(s_1, s_2, p, t) = F(s_1, s_2, p, t)\int_0^{s_{1,max}-s_1}\int_0^{s_{2,max}-s_2}\int_{p_0}^{p_{max}-p} \beta(s_1, s_1', s_2, s_2', p, p') \times F(s_1', s_2', p', t)ds_1 ds_2 dp \tag{26}$$

The rate of aggregation R_{agg,l_i} and R_{agg,l_e} in Equations (2) and (3) are the rates at which the internal and external liquid volumes are transferred between particles due to aggregation. Similar to the formation and depletion of what was discussed above, these rates can be calculated as shown in Equations (28) and (29).

$$R_{agg,l_i}(s_1,s_2,p,t) = \frac{1}{2}\int_0^{s_1}\int_0^{s_2}\int_{p_0}^{p}\beta(s_1-s_1^{'},s_2-s_2^{'},p-p^{'},s_1^{'},s_2^{'},p^{'})\times$$
$$F(s_1-s_1^{'},s_2-s_2^{'},p-p^{'},t)F(s_1^{'},s_2^{'},p^{'},t)$$
$$(l_i(s_1-s_1^{'},s_2-s_2^{'},p-p^{'},t)+l_i(s_1^{'},s_2^{'},p^{'},t)+l_{e\to i})ds_1^{'}ds_2^{'}dp^{'}$$
$$-L_i(s_1,s_2,p,t)\int_0^{s_{1,max}-s_1}\int_0^{s_{2,max}-s_2}\int_{p_0}^{p_{max}-p}$$
$$\beta(s_1,s_1^{'},s_2,s_2^{'},p,p^{'})\times F(s_1^{'},s_2^{'},p^{'},t)ds_1ds_2dp \tag{27}$$

$$R_{agg,l_e}(s_1,s_2,p,t) = \frac{1}{2}\int_0^{s_1}\int_0^{s_2}\int_{p_0}^{p}\beta(s_1-s_1^{'},s_2-s_2^{'},p-p^{'},s_1^{'},s_2^{'},p^{'})\times$$
$$F(s_1-s_1^{'},s_2-s_2^{'},p-p^{'},t)F(s_1^{'},s_2^{'},p^{'},t)$$
$$(l_e(s_1-s_1^{'},s_2-s_2^{'},p-p^{'},t)+l_e(s_1^{'},s_2^{'},p^{'},t)-l_{e\to i})ds_1^{'}ds_2^{'}dp^{'}$$
$$-L_e(s_1,s_2,p,t)\int_0^{s_{1,max}-s_1}\int_0^{s_{2,max}-s_2}\int_{p_0}^{p_{max}-p}$$
$$\beta(s_1,s_1^{'},s_2,s_2^{'},p,p^{'})\times F(s_1^{'},s_2^{'},p^{'},t)ds_1ds_2dp \tag{28}$$

The transfer of external liquid volume to internal liquid volume due to aggregation (represented as $l_{e\to i}$ in Equations (28) and (29)) can be computed as discussed by Braumann et al. [31].

$$l_{e\to i} = \left[l_e(s_1-s_1^{'},s_2-s_2^{'},p-p^{'},t)l_e(s_1^{'},s_2^{'},p^{'},t)\right]^{\frac{1}{2}} \tag{29}$$

$$\times\left[1-\sqrt{1-\left(\frac{\sqrt[3]{v(s_1-s_1^{'},s_2-s_2^{'},p-p^{'},t)}-l_e(s_1-s_1^{'},s_2-s_2^{'},p-p^{'},t)}{\sqrt[3]{v(s_1-s_1^{'},s_2-s_2^{'},p-p^{'},t)}+\sqrt[3]{v(s_1^{'},s_2^{'},p^{'},t)}}\right)^2}\right]^{\frac{1}{2}} \tag{30}$$
$$\times\left[1-\sqrt{1-\left(\frac{\sqrt[3]{v(s_1^{'},s_2^{'},p^{'},t)}-l_e(s_1^{'},s_2^{'},p^{'},t)}{\sqrt[3]{v(s_1-s_1^{'},s_2-s_2^{'},p-p^{'},t)}+\sqrt[3]{v(s_1^{'},s_2^{'},p^{'},t)}}\right)^2}\right]^{\frac{1}{2}}$$

The aggregation kernel β depends on the properties of the colliding particle $\mathbf{A}(s_1,s_2,p)$ and $\mathbf{B}(s_1^{'},s_2^{'},p^{'})$ (as shown in Figure 3). $\beta(\mathbf{A},\mathbf{B}) = \beta_0\beta^*\mathbf{A},\mathbf{B}$. β_0 is independent of the colliding particle properties and is an optimized parameter in this work. β^* is the efficiency of particle collision which can be determined based on the following model proposed by Balakin et al. [35]. The model accounts for both capillary and viscous forces during particle collision. The efficiency is determined as a ratio of the total work of forces within the liquid bridge to the kinetic energy of the particle.

$$\beta^* = \frac{W_c+W_d}{E_k}\Psi \tag{31}$$

where W_c and W_d are the work of the capillary and dissipative forces, respectively. E_k is the kinetic energy calculated from the mean relative velocity (v_r), mass (m), and coefficient of restitution (e, Equation (37)) of particle before collision, as shown in Equation (32). The relative velocity is calculated from the granular temperature (Θ), as given by Equation (33).

$$E_k = \frac{1}{2} m e^2 v_r^2 \tag{32}$$

$$v_r = \frac{3}{2}\sqrt{\pi\Theta} \tag{33}$$

$$\Theta = \frac{(5\pi/96)\gamma^2 d^2}{12\phi_p^2\left(1 - \left(\frac{\phi_p}{1-\phi_p}\right)^2\right)} \tag{34}$$

where ϕ_p is the volume fraction of particle in the granulator, γ is the shear rate, and d is the granule diameter. The work of dissipative force is calculated using the following equation.

$$W_d = \frac{3\pi\mu\tilde{d}^2 e_{coag} v_r}{4} ln\left(\frac{h}{h_a}\right) \tag{35}$$

where $\tilde{d}$ is the harmonic mean diameter of the colliding two particles and μ is the viscosity of the binder. h is the binder layer thickness calculated from the external liquid content (l_e) and h_a represents the granule surface asperities. The resulting aggregated particle coefficient of restitution (e_{coag}) is represented as a function of coefficient of restitution of the constituent material properties by Braumann et al. [31] by the following equations:

$$e_{coag} = \sqrt{e_{\mathbf{A}} e_{\mathbf{B}}} \tag{36}$$

$$e_i = \frac{\sum_\alpha e_\alpha m_\alpha}{\sum_\alpha m_\alpha}, i \in \{\mathbf{A},\mathbf{B}\} \tag{37}$$

where $\alpha \in \{s_1, s_2, p\}$, m is the mass of colliding granule and e is the ratio of rebound energy to impact energy. It takes a value between 0 (totally plastic impact) and 1 (totally elastic impact). e for pore is assumed to be 0.

The work of capillary force (W_c) is experimentally determined from the regime map analysis carried out Muthancheri and Ramachandran [22]. Figure 4 plots the capillary number which is the ratio between capillary force and viscous force as a function of API fraction. The equation determined from the experiment analysis is used to provide a composition dependent work of capillary force in the model.

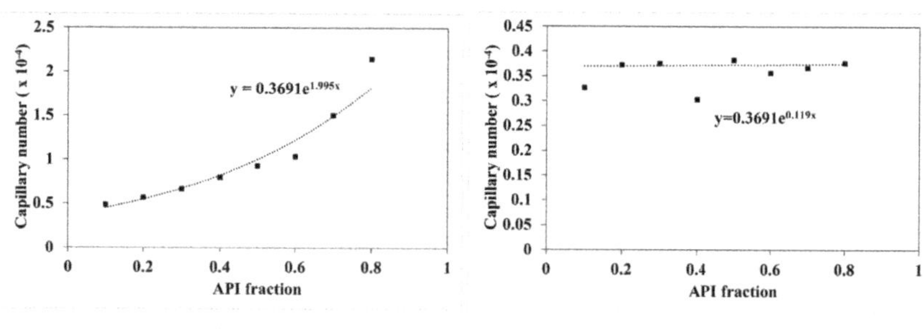

Figure 4. Change in capillary numbers with increase in API fraction. (**a**) Ibuprofen and MCC101 formulation, (**b**) Acetaminophen (APAP) and MCC101 formulation.

2.2.7. Compaction

Compaction of granules occur during collision and result in porosity reduction of the granules.

$$\Delta\epsilon = \begin{cases} k_{con} U(\epsilon - \epsilon_{min}) & \text{if } \epsilon - \Delta\epsilon \geq \epsilon_{min} \\ 0 & \text{otherwise} \end{cases} \tag{38}$$

where k_{con} is the consolidation rate constant, U is the particle collision velocity, and ϵ_{min} is the minimum porosity. Two conditions are modeled in this study. A non-squeeze case: if no internal liquid is transferred to the external surface. In this case, there is only pore volume reduction due to consolidation (Equation (39)). The next scenario is a squeeze case. Some liquid is transferred from internal to external liquid volume (Equation (40)). This occurs if the porosity after consolidation is smaller than a critical porosity. In this scenario, the pore volume is completely occupied by internal liquid volume ($l_i = p$).

$$\frac{dp}{dt} = -\left(\frac{1}{1-(\epsilon - \Delta\epsilon)}(s_1 + s_2 + l_e) - v\right) \tag{39}$$

$$\frac{dp}{dt} = -((1 + (\epsilon - \Delta\epsilon))(s_1 + s_2 + l_e) + (\epsilon - \Delta\epsilon)l_i - v) \tag{40}$$

2.2.8. Particle Breakage

The particle breakage occurs when a large particle disintegrates into two or more daughter (or smaller) particles. The net particle breakage rate is modeled using the method described in Barrasso and Ramachandran [32]. The details of the model are provided in the Appendix A.

2.3. Hybrid Modeling

Figure 5 depicts the hybrid modeling framework. The data-driven classification model presented in the previous literature [30] is used to predict the probability of nucleation (P). Based on that probability, the number of immersion nuclei and solid-spread nuclei is determined. The porosity and size of the initial nuclei effects the granule growth through the aggregation kernel and API surface growth discussed in Section 2.2.

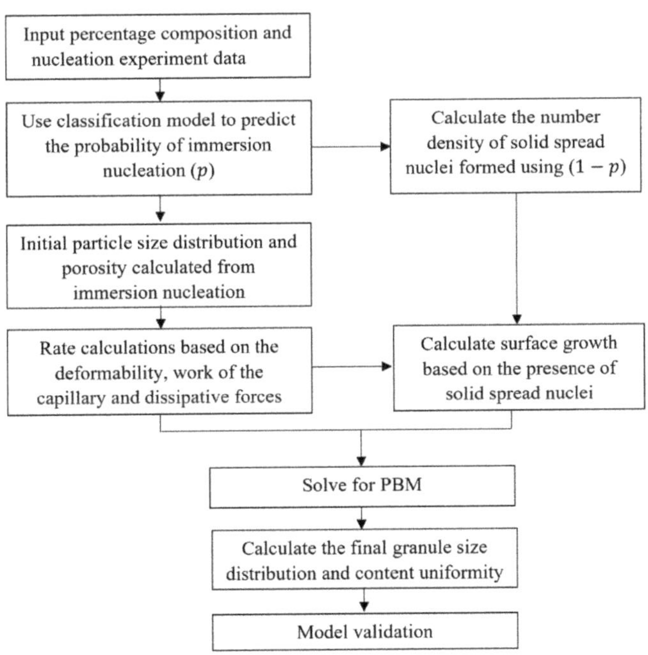

Figure 5. Hybrid modeling framework.

2.4. Numerical Solution

The ordinary differential equations (ODE) obtained after discretization for different particle size combinations are integrated simultaneously using the first-order explicit Euler integration technique, which is popularly used to solve PBMs [5,9,36–38]. The numerical stability of a PBM is complex due to the presence of multiple dimensions and the inherent possibility of instability involved with the time-step of the integration. The integration time-step was thus chosen, such that the rate of particles leaving a particular size class (bin) is not higher than the number of particles in that size class at any time-step based on the Courant–Friedrichs–Lewy (CFL) condition [8,39]. The partial derivatives with respect to internal coordinate volume ($u \in (s_1, s_2, p)$) and time (t) were discretized using a non-linear grid ($u_i = u \times (4)^{i-1}$). Here, i represents the bin number in one dimension, and u indicates the volume of particle in the smallest bin. The smallest particle size is 31.5 µm, and the largest particle size is 6000 µm. There is a total of 20 bins or grid points. A cell average technique discussed by Chaudhury et al. [7] is utilized in this study to distribute the particles that are formed in the intermediate range of two bins, into the adjacent bins. The computations were performed in MATLAB 2020a on an Intel(R) Core(TM) i7-8700 CPU (3.20 GHz) with 16 GB RAM.

2.5. Sensitivity Analysis

A sensitivity study was performed to investigate the effect of adjustable parameters on the simulation results. The optimized values of parameters were perturbed from −20% to 20% with a step size of 10%, and the results were compared to the base value. The sensitivity is measured using Equation (41).

$$Sensitivity = \left| \frac{Y_0^j(t) - Y_i^j(t)}{Y_0^j(t)} \right| \tag{41}$$

where $Y_i^j(t)$ is the value of granule property of interest in the ith perturbation of the jth parameter and $Y_0^j(t)$ is the base value for the jth parameter.

Figure 6 illustrates that the sensitivity of parameters on d_{10}, d_{50}, and d_{90} simulation. It shows that growth parameters (k_{layer} and k_{sg}) are much less sensitive than the aggregation and consolidation parameters when the variables are perturbed ±20%. The average diameter is found to be highly sensitivity toward the coefficient of restitution of API (e_{s1}). The study shows a decrease in average diameter with an increase in e_{s1}. A decrease in e_{s1} indicates that the API is very deformable, resulting in smaller average granule size. The aggregation rate constant, β_0, has a positive impact on the granule size, showing an increase in the rate constant increasing the average granule size. The consolidation rate equation has a negative term (Equations (39) and (40)), which means the increase in k_{con} results in a decrease in consolidation rate. In Figure 6c, it can be seen that a decrease in consolidation rate to 20% results in larger granules.

Similarly, Figure 7 shows the effect of the adjustable parameters on the average porosity and API content of the granules. The sensitivity of parameters to granule API content is similar to that of the granule size. Aggregation rate constant, coefficient of restitution, and consolidate rate were found to be most significant in impacting the granule API content. A decrease in e_{s1} results in a decrease in aggregation rate and thus results in granules with less s_1 or API content. Average porosity of granule is most impacted by the consolidation rate. It can be seen that with increase in consolidation (or decrease in consolidation rate constant) the average granule porosity decreases.

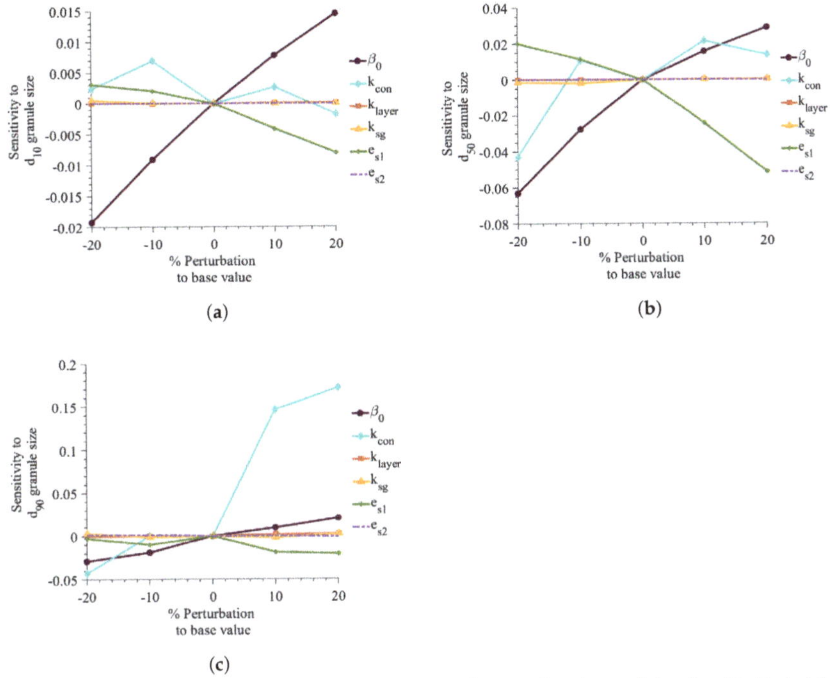

Figure 6. Effect of changes in adjustable parameter values on d_{10}, d_{50}, and d_{90}. Sensitivity to (**a**) d_{10}, (**b**) d_{50}, (**c**) d_{90}.

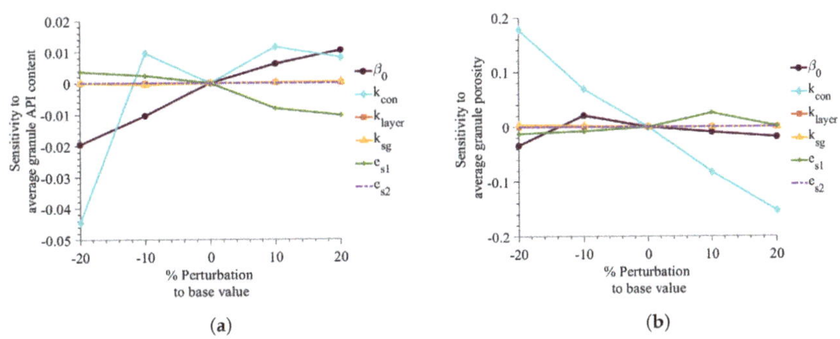

Figure 7. Effect of changes in adjustable parameter values on porosity and API content. (**a**) Sensitivity to average API content, (**b**) Sensitivity to granule average porosity

3. Results and Discussions

3.1. Optimization and Parameter Estimation

Experimental data from Muthancheri and Ramachandran [22] is used to estimate the unknown parameters in the model. The cumulative granule size fraction and content uniformity measurements were used for the estimation. For the multi-objective optimization for minimizing the error for both size and content uniformity, a Pareto optimal solution techniques was used. The method provides an optimal solution when one objective function cannot decrease without increasing the other objective function [40].

The tuned parameters are aggregation constant (β_0), coefficient of restitution of API (e_{s1}) and excipient (e_{s2}), consolidation rate constant (k_{con}), excipient layering rate constant (k_{layer}), and rate of surface growth (k_{sg}). Out of the six experimental runs, four were used for parameter estimation, and two were used to validate the calibrated model. The optimized values of the variables are provided in the Table 2.

Table 2. Optimized parameters.

Parameters	Notation	Value
Aggregation rate constant	β_0	4.34×10^{-10}
Ibuprofen coefficient of restitution	$e_{s1,ibu}$	0.162
APAP coefficient of restitution	$e_{s1,apap}$	0.103
MCC101 coefficient of restitution	e_{s2}	0.07
Consolidation rate constant	k_{con}	1.67×10^{-3}
Excipient layering rate constant	k_{layer}	2.01×10^{-8}
Surface growth rate constant	k_{sg}	4.59×10^{-11}

3.2. Model Training and Validation

The model was trained and validated using the experiment data from Muthancheri and Ramachandran [22]. In the aforementioned study, two formulations were considered such that there exists a wettability differential between the two components in a formulation. The two cases were (1) ibuprofen-USP and micro-crystalline cellulose and (2) micronized acetaminophen (APAP) and micro-crystalline cellulose. The 40% and 60% cumulative size distributions of formulations were used for model training (Figures 8 and 9). The 50% cumulative size distributions of formulations were used for model validation (Figure 10). This validation helps to concur the assumptions and theories considered in the model. The predicted values had a strong relationship with the experimental values. The overall accuracy of the model was estimated to be 0.89 (Overall accuracy = 1 − sum of square error).

An increase in API percentage composition from 40% to 60% experimentally resulted in an overall increase in granule size or a shift in the distribution curve to the right for both case-I and -II formulation. The model was able to accurately predict this increase in granule size (and distribution) at a high API percentage composition. This dependency of the API composition was reflected in the model through the composition dependent aggregation kernel and nucleation probability.

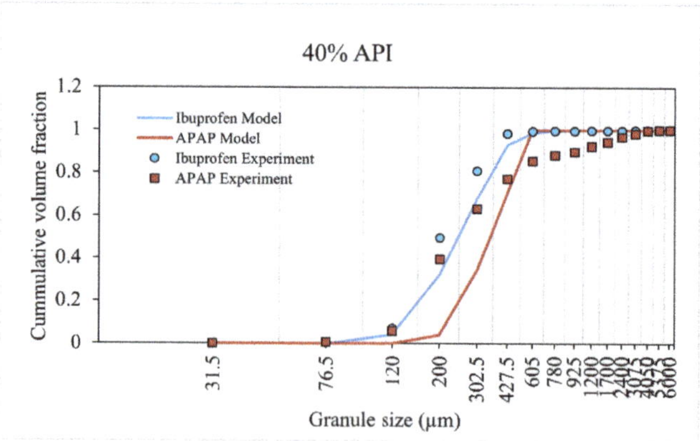

Figure 8. Model prediction for 40% API formulation.

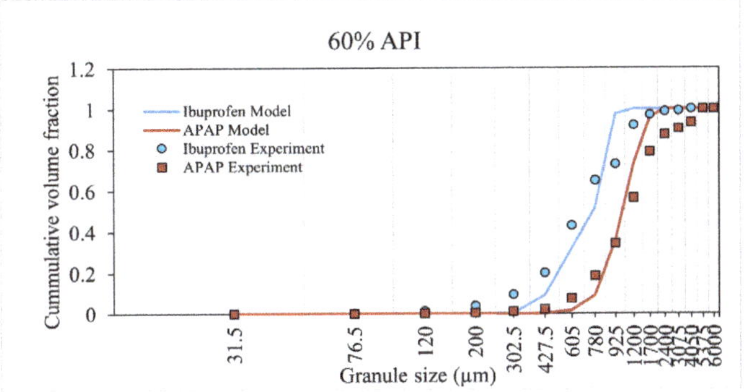

Figure 9. Model prediction for 60% API formulation.

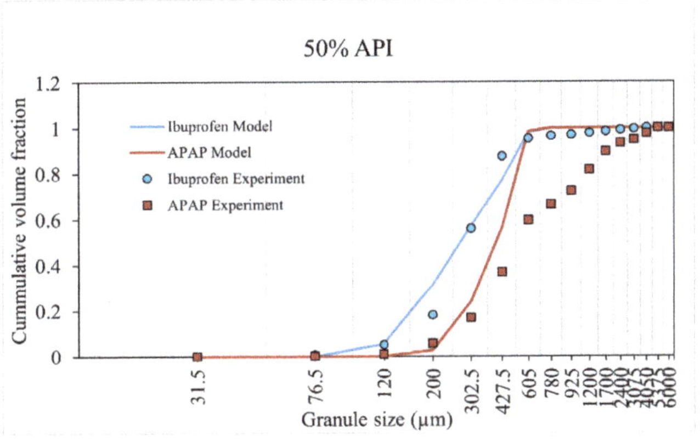

Figure 10. Model validation using 50% API formulation.

The content uniformity of the granules were evaluated using the demixing potential (DP) introduced by Thiel and Nguyen [41] to quantify the distribution of a solid component as a function of particle size. DP can be calculated using the following equation:

$$\mathrm{DP\%} = \frac{100}{\bar{x}}\sqrt{\sum w(x-\bar{x})^2} \tag{42}$$

where x is the API content in a particular size range, $\bar{x}$ is the average API content, and w is the weight fraction of granule in each size range. The quantity is similar to the relative standard deviation used for non-uniformity in mixing by Oka et al. [42]. The larger the value of de-mixing potential, the larger the extent of deviation from the mean of the API across granule size classes. Figure 11 shows the ability of the presented hybrid-modeling framework to predict the change in de-mixing potential with an increase in percentage API. The model predicted a decrease in the extend of de-mixing with increase in percentage API for ibuprofen formulation and an increase in the extent of de-mixing with increase in percentage API for APAP formulation.

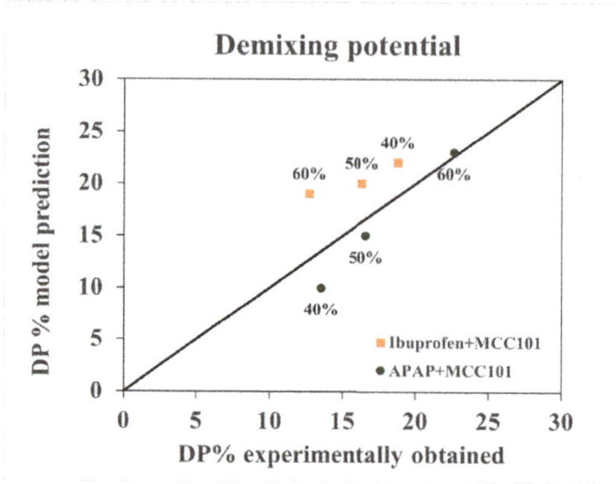

Figure 11. Comparison between model prediction and experimentally obtained de-mixing potential.

3.3. *Model Applications*

3.3.1. Effect of Change in Formulation on Dynamic Granule Formation

Muthancheri and Ramachandran [22] studied the effect of increase in hydrophobic API percentage on the granule size. It was shown that in an insoluble system (all constituent materials were insoluble in the binder liquid), an increase in hydrophobic component results in an overall increase in granule size.

Figure 12 shows that when the percentage API is varied from 30% to 60% the average particle size was found to increase. Here, the liquid-to-solid ratio was kept the same as that during the model optimization (0.6 liquid-to-solid ratio).

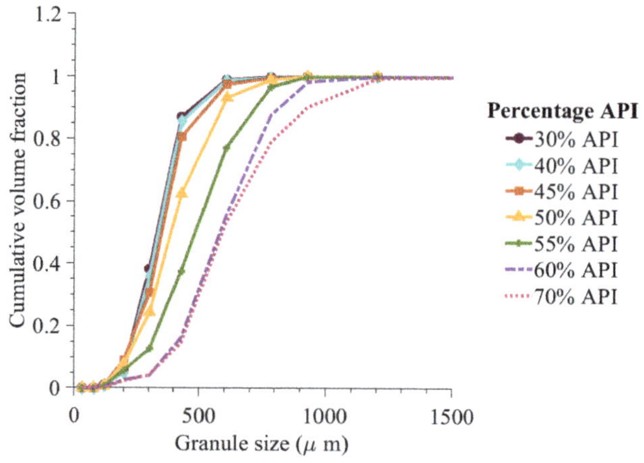

Figure 12. Cumulative volume fraction prediction with change in percentage composition of API and 0.6 liquid-to-solid ratio.

The effect of formulation or hydrophobicity in the dynamic granule formation can be observed in Figure 13. At 40% hydrophobic content, the probability of solid-spread nuclei formation was very low, resulting in a high initial immersion nuclei and consequent granule formation. At 50% hydrophobic component, the probability of solid-spread nuclei is modeled to be 0.67 (calculated from the classification model [30]). This resulted in a small secondary peak at the start of liquid addition (Figure 13b). At a high hydrophobic content of 60%, the high probability of solid-spread nuclei formation led to an increase in granule growth and larger final granules (Figure 13c).

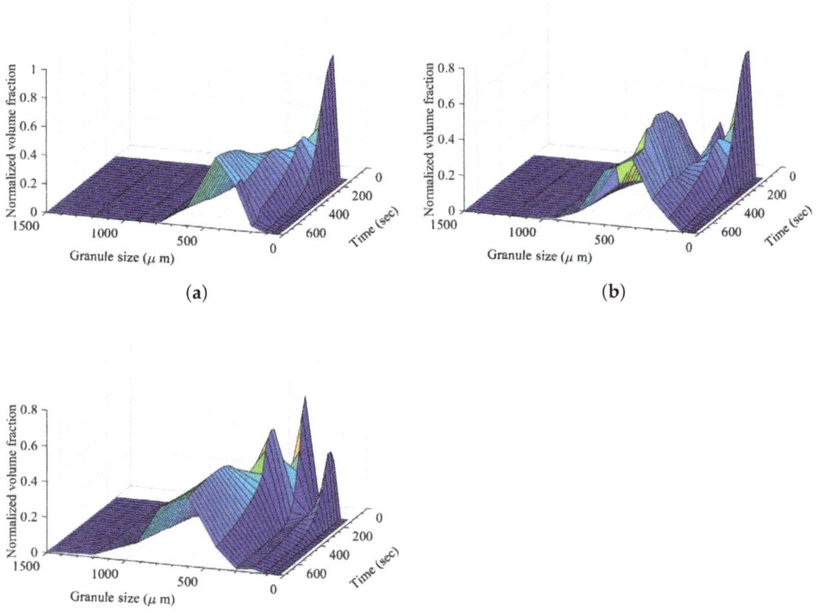

(**a**) (**b**) (**c**)

Figure 13. Granule size distribution with increase in granulation time at a varying degree of hydrophobic content (ibuprofen). (**a**) Granule size distribution with increase in granulation time at 40% hydrophobic component (ibuprofen), (**b**) granule size distribution with increase in granulation time at 50% hydrophobic component (ibuprofen), and (**c**) granule size distribution with increase in granulation time at 60% hydrophobic component (ibuprofen).

At 50% composition of ibuprofen in the formulation made of ibuprofen and MCC-101, the classification model predicted a probability of immersion nucleation to be 0.67. This scenario is further evaluated in Figure 14 in terms of dynamic evolution of granule size as granulation progress. It can be seen that due to wet massing, the solid-spread nuclei combine together to form larger granules and then combine with the immersion nuclei to form a wider distribution of granules.

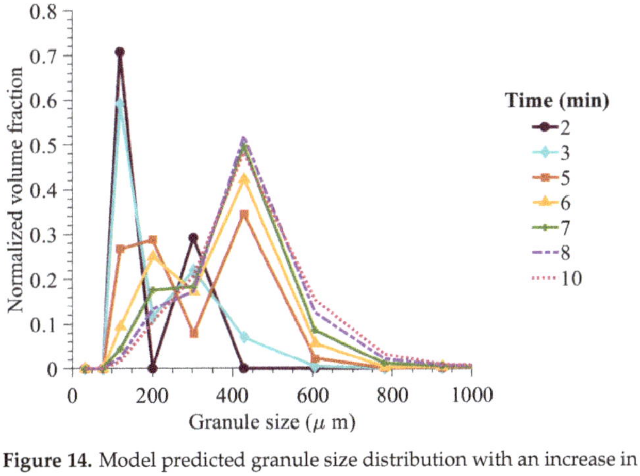

Figure 14. Model predicted granule size distribution with an increase in wet massing time (50% API content).

3.3.2. Effect of Change in Formulation on Granule API Content

Muthancheri and Ramachandran [22] study showed that as the API content increased from 40% to 60%, the granule content non-uniformity decreased. Figure 15 shows a similar trend. As the hydrophobic API content increased from low to high, the Δq, which is the difference between the powder blend API content and average granule API content, became near zero. Smaller granules of high API content were also found during the experiment study, validating the high API content prediction at the start of granulation process.

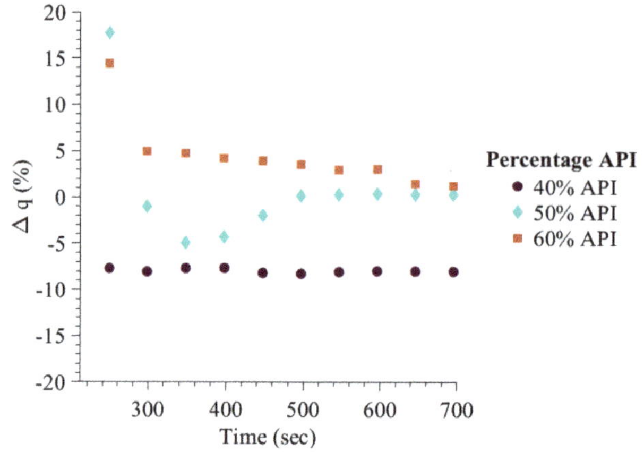

Figure 15. Model predicted granule API content with increase in wet massing time.

3.3.3. Effect of Change in Formulation on Average Granule Porosity

Next, the average granule porosity prediction with change in the percentage API is evaluated. The average porosity prediction is within the porosity range reported previously in the high shear wet granulation model study [43]. The envelop density of the granules obtained in the Muthancheri and Ramachandran [22] was carried out to verify the trend of porosity predicted in Figure 16a.

The envelop density (Figure 16b) was measured using a graphite powder quasi-fluid, known as Dryflo (Micromeritics, Norcross, GA, USA) [44]. The porosity of granule is inversely related to the envelop density. Figure 16b shows the influence of increase in API percentage on the envelop density. Based on the envelop density trend, it is concluded that the decrease in average porosity predicted using the developed model is fairly accurate.

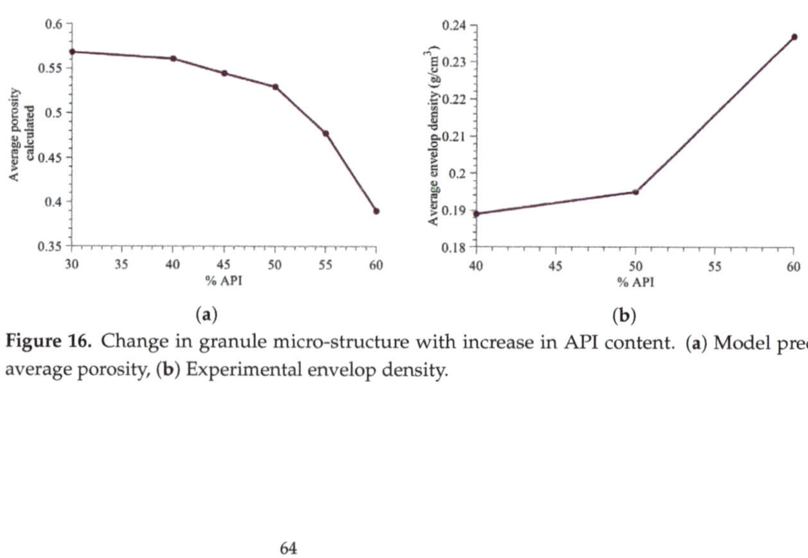

Figure 16. Change in granule micro-structure with increase in API content. (**a**) Model predicted average porosity, (**b**) Experimental envelop density.

3.3.4. Effect of Change in Formulation on Average Liquid Fraction

The change in average liquid fraction with an increase in API percentage is shown in Figure 17. The range of liquid fraction is similar to that reported in Chaudhury and Ramachandran [43]. The increase in liquid content is primarily due to increase in solid-spread nucleation, which results in surface growth. Surface growth of solid-spread nuclei increases the API content as well as liquid content on the granules. The liquid contant also increases due to decrease in hydrophilic excipient. Hydrophilic excipient reduced the available liquid due to the transfer of liquid from the external surface to inside the particles.

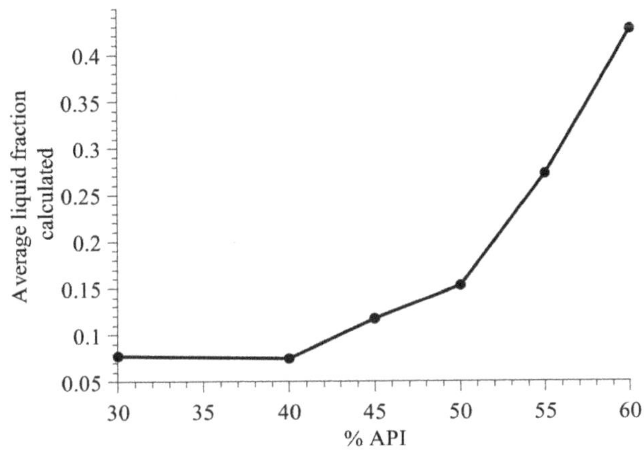

Figure 17. Model predicted average liquid fraction.

4. Conclusions

In this study, a hybrid model (Random Forest-PBM) is developed to describe the bi-component high shear wet granulation process. The model incorporates immersion and solid-spread nucleation based on the change in percentage API. The probability of each nucleation mechanism to occur, for a given formulation, is obtained from the random forest model. The probability is incorporated into the PBM framework such that the rate equations are impacted by the availability of immersion and solid-spread nuclei. It was found that the aggregation and consolidation rate are more sensitive to the granule critical quality attribute predictions. The model predictions are qualitatively in agreement with profiles obtained in the literature [22,43]. The discussed methodology and presented model could be used for predicting various aspects of the granulation process and controlling the transient behavior during the process. As an example, we have provided the average particle size, porosity, liquid content, and demixing-potential of the granules with change in API percentage. The developed model is an improvement to the existing mechanistic modeling framework, such that it incorporates the effect of hydrophobicity to track the granule critical quality attributes.

Author Contributions: Conceptualization, I.M. and R.R.; methodology, I.M.; validation, I.M.; formal analysis, I.M.; investigation, I.M.; resources, R.R.; writing—original draft preparation, I.M.; writing—review and editing, I.M. and R.R.; visualization, I.M.; supervision, R.R.; project administration, I.M.; funding acquisition, R.R. All authors have read and agreed to the published version of the manuscript.

Funding: This research was funded by National Science Foundation CAREER program through grant no: 1350152.

Institutional Review Board Statement: Not applicable.

Informed Consent Statement: Not applicable.

Data Availability Statement: Not applicable.

Acknowledgments: The authors would like to acknowledge the Department of Chemical & Biochemical Engineering, Rutgers University, for funding in the form of a teaching assistantship to I. Muthancheri, as well as the National Science Foundation CAREER program through grant no: 1350152.

Conflicts of Interest: The authors declare no conflict of interest.

Appendix A

Equation (A1) represents the breakage rate equation.

$$R_{brk}(s_1, s_2, p, t) = R_{brk}^{form}(s_1, s_2, p, t) - R_{brk}^{dep}(s_1, s_2, p, t) \tag{A1}$$

where R_{brk}^{form} and R_{brk}^{dep} are the rates of formation of smaller particles and rate of depletion of larger particles.

$$R_{brk}^{form}(s_1, s_2, p, t) = \int_{s_1}^{s_{1,max}} \int_{s_2}^{s_{2,max}} \int_{p}^{p_{max}} b(s_1, s_2, p, s_1', s_2', p') K_{brk}(s_1, s_2, p) \times F(s_1', s_2', p', t) ds_1' ds_2' dp' \tag{A2}$$

$$R_{brk}^{dep}(s_1, s_2, p, t) = K_{brk}(s_1, s_2, p) F(s_1, s_2, p, t) \tag{A3}$$

where K_{brk} is the breakage rate constant and b is the probability distribution function of daughter particles. For the purposes of this study, a uniform probability distribution was assumed for all possible daughter particles [32].

References

1. Suresh, P.; Sreedhar, I.; Vaidhiswaran, R.; Venugopal, A. A comprehensive review on process and engineering aspects of pharmaceutical wet granulation. *Chem. Eng. J.* **2017**, *328*, 785–815. [CrossRef]
2. Gernaey, K.V.; Cervera-Padrell, A.E.; Woodley, J.M. A perspective on PSE in pharmaceutical process development and innovation. *Comput. Chem. Eng.* **2012**, *42*, 15–29. [CrossRef]
3. Ramkrishna, D.; Singh, M.R. Population balance modeling: Current status and future prospects. *Annu. Rev. Chem. Biomol. Eng.* **2014**, *5*, 123–146. [CrossRef]
4. Liu, L.; Litster, J. Population balance modelling of granulation with a physically based coalescence kernel. *Chem. Eng. Sci.* **2002**, *57*, 2183–2191. [CrossRef]
5. Chaudhury, A.; Wu, H.; Khan, M.; Ramachandran, R. A mechanistic population balance model for granulation processes: Effect of process and formulation parameters. *Chem. Eng. Sci.* **2014**, *107*, 76–92. [CrossRef]
6. Poon, J.M.H.; Immanuel, C.D.; Doyle, F.J., III; Litster, J.D. A three-dimensional population balance model of granulation with a mechanistic representation of the nucleation and aggregation phenomena. *Chem. Eng. Sci.* **2008**, *63*, 1315–1329.
7. Chaudhury, A.; Kapadia, A.; Prakash, A.V.; Barrasso, D.; Ramachandran, R. An extended cell-average technique for a multi-dimensional population balance of granulation describing aggregation and breakage. *Adv. Powder Technol.* **2013**, *24*, 962–971. [CrossRef]
8. Ramachandran, R.; Barton, P.I. Effective parameter estimation within a multi-dimensional population balance model framework. *Chem. Eng. Sci.* **2010**, *65*, 4884–4893. [CrossRef]
9. Barrasso, D.; Ramachandran, R. Multi-scale modeling of granulation processes: Bi-directional coupling of PBM with DEM via collision frequencies. *Chem. Eng. Res. Des.* **2015**, *93*, 304–317. [CrossRef]
10. Matsoukas, T.; Lee, K.; Kim, T. Mixing of components in two-component aggregation. *AIChE J.* **2006**, *52*, 3088–3099. [CrossRef]
11. Matsoukas, T.; Kim, T.; Lee, K. Bicomponent aggregation with composition-dependent rates and the approach to well-mixed state. *Chem. Eng. Sci.* **2009**, *64*, 787–799. [CrossRef]
12. Marshall, C.L., Jr.; Rajniak, P.; Matsoukas, T. Multi-component population balance modeling of granulation with continuous addition of binder. *Powder Technol.* **2013**, *236*, 211–220. [CrossRef]
13. Kapur, P.; Kapur, P.; Fuerstenau, D. An auto-layering model for the granulation of iron ore fines. *Int. J. Miner. Process.* **1993**, *39*, 239–250. [CrossRef]
14. Ingram, G.D.; Cameron, I.T. Formulation and comparison of alternative multiscale models for drum granulation. *Comput. Aided Chem. Eng.* **2005**, *20*, 481–486.
15. Barrasso, D.; Ramachandran, R. Qualitative assessment of a multi-scale, compartmental PBM-DEM model of a continuous twin-screw wet granulation process. *J. Pharm. Innov.* **2016**, *11*, 231–249. [CrossRef]
16. Goodson, M.; Kraft, M.; Forrest, S.; Bridgwater, J. A multi-dimensional population balance model for agglomeration. In Proceedings of the PARTEC 2004—International Congress for Particle Technology, Nuremburg, Germany, 16–18 March 2004.

17. Hounslow, M.; Oullion, M.; Reynolds, G. Kinetic models for granule nucleation by the immersion mechanism. *Powder Technol.* **2009**, *189*, 177–189. [CrossRef]
18. Hapgood, K.P.; Litster, J.D.; Smith, R. Nucleation regime map for liquid bound granules. *AIChE J.* **2003**, *49*, 350–361.
19. Bellinghausen, S.; Gavi, E.; Jerke, L.; Ghosh, P.K.; Salman, A.D.; Litster, J.D. Nuclei size distribution modelling in wet granulation. *Chem. Eng. Sci. X* **2019**, *4*, 100038. [CrossRef]
20. Immanuel, C.D.; Doyle, F.J., III. Mechanistic modelling of aggregation phenomena in population balances of granulation. *IFAC Proc. Vol.* **2005**, *38*, 416–421. [CrossRef]
21. Farber, L.; Al-aaraj, D.K.; Smith, R.; Gentzler, M. Formation and internal microstructure of granules from wetting and non-wetting efavirenz/iron oxide blends. *Chem. Eng. Sci.* **2020**, *227*, 115909. [CrossRef]
22. Muthancheri, I.; Ramachandran, R. Mechanistic understanding of granule growth behavior in bi-component wet granulation processes with wettability differentials. *Powder Technol.* **2020**, *367*, 841–859. [CrossRef]
23. Washino, K.; Tan, H.; Hounslow, M.; Salman, A. Meso-scale coupling model of DEM and CIP for nucleation processes in wet granulation. *Chem. Eng. Sci.* **2013**, *86*, 25–37. [CrossRef]
24. Washino, K.; Tan, H.; Hounslow, M.; Salman, A. A new capillary force model implemented in micro-scale CFD–DEM coupling for wet granulation. *Chem. Eng. Sci.* **2013**, *93*, 197–205. [CrossRef]
25. Hapgood, K.; Farber, L.; Michaels, J. Agglomeration of hydrophobic powders via solid spreading nucleation. *Powder Technol.* **2009**, *188*, 248–254. [CrossRef]
26. Chen, Y.; Ierapetritou, M. A framework of hybrid model development with identification of plant-model mismatch. *AIChE J.* **2020**, *66*, e16996. [CrossRef]
27. Azarpour, A.; Borhani, T.N.; Alwi, S.R.W.; Manan, Z.A.; Mutalib, M.I.A. A generic hybrid model development for process analysis of industrial fixed-bed catalytic reactors. *Chem. Eng. Res. Des.* **2017**, *117*, 149–167. [CrossRef]
28. Sampat, C.; Ramachandran, R. Identification of Granule Growth Regimes in High Shear Wet Granulation Processes Using a Physics-Constrained Neural Network. *Processes* **2021**, *9*, 737. [CrossRef]
29. Barrasso, D.; Tamrakar, A.; Ramachandran, R. A reduced order PBM–ANN model of a multi-scale PBM–DEM description of a wet granulation process. *Chem. Eng. Sci.* **2014**, *119*, 319–329. [CrossRef]
30. Muthancheri, I.; Oka, S.; Ramachandran, R. Analysis and prediction of nucleation mechanisms in a bi-component powder bed with wettability differentials. *Powder Technol.* **2021**, *390*, 209–218. [CrossRef]
31. Braumann, A.; Goodson, M.J.; Kraft, M.; Mort, P.R. Modelling and validation of granulation with heterogeneous binder dispersion and chemical reaction. *Chem. Eng. Sci.* **2007**, *62*, 4717–4728. [CrossRef]
32. Barrasso, D.; Ramachandran, R. A comparison of model order reduction techniques for a four-dimensional population balance model describing multi-component wet granulation processes. *Chem. Eng. Sci.* **2012**, *80*, 380–392. [CrossRef]
33. Jeong, J.I.; Choi, M. A bimodal particle dynamics model considering coagulation, coalescence and surface growth, and its application to the growth of titania aggregates. *J. Colloid Interface Sci.* **2005**, *281*, 351–359. [CrossRef]
34. Planchette, C.; Biance, A.; Lorenceau, E. Transition of liquid marble impacts onto solid surfaces. *EPL (Europhys. Lett.)* **2012**, *97*, 14003. [CrossRef]
35. Balakin, B.V.; Kutsenko, K.V.; Lavrukhin, A.A.; Kosinski, P. The collision efficiency of liquid bridge agglomeration. *Chem. Eng. Sci.* **2015**, *137*, 590–600. [CrossRef]
36. Chaudhury, A.; Armenante, M.E.; Ramachandran, R. Compartment based population balance modeling of a high shear wet granulation process using data analytics. *Chem. Eng. Res. Des.* **2015**, *95*, 211–228. [CrossRef]
37. Barrasso, D.; El Hagrasy, A.; Litster, J.D.; Ramachandran, R. Multi-dimensional population balance model development and validation for a twin screw granulation process. *Powder Technol.* **2015**, *270*, 612–621. [CrossRef]
38. Muthancheri, I.; Chaturbedi, A.; Bétard, A.; Ramachandran, R. A compartment based population balance model for the prediction of steady and induction granule growth behavior in high shear wet granulation. *Adv. Powder Technol.* **2021**, *32*, 2085–2096. [CrossRef]
39. Courant, R.; Friedrichs, K.; Lewy, H. On the partial difference equations of mathematical physics. *IBM J. Res. Dev.* **1967**, *11*, 215–234. [CrossRef]
40. Gunantara, N. A review of multi-objective optimization: Methods and its applications. *Cogent Eng.* **2018**, *5*, 1502242. [CrossRef]
41. Thiel, W.; Nguyen, L. Fluidized bed granulation of an ordered powder mixture. *J. Pharm. Pharmacol.* **1982**, *34*, 692–699. [CrossRef] [PubMed]
42. Oka, S.; Kašpar, O.; Tokárová, V.; Sowrirajan, K.; Wu, H.; Khan, M.; Muzzio, F.; Štěpánek, F.; Ramachandran, R. A quantitative study of the effect of process parameters on key granule characteristics in a high shear wet granulation process involving a two component pharmaceutical blend. *Adv. Powder Technol.* **2015**, *26*, 315–322. [CrossRef]
43. Chaudhury, A.; Ramachandran, R. Integrated population balance model development and validation of a granulation process. *Part. Sci. Technol.* **2013**, *31*, 407–418. [CrossRef]
44. Ansari, M.A.; Stepanek, F. The effect of granule microstructure on dissolution rate. *Powder Technol.* **2008**, *181*, 104–114. [CrossRef]

pharmaceutics

MDPI

Article

Conceptualisation of an Efficient Particle-Based Simulation of a Twin-Screw Granulator

John P. Morrissey, Kevin J. Hanley and Jin Y. Ooi *

School of Engineering, Institute for Infrastructure & Environment, The University of Edinburgh, Edinburgh EH9 3JL, UK; J.Morrissey@ed.ac.uk (J.P.M.); K.Hanley@ed.ac.uk (K.J.H.)
* Correspondence: j.ooi@ed.ac.uk

Abstract: Discrete Element Method (DEM) simulations have the potential to provide particle-scale understanding of twin-screw granulators. This is difficult to obtain experimentally because of the closed, tightly confined geometry. An essential prerequisite for successful DEM modelling of a twin-screw granulator is making the simulations tractable, i.e., reducing the significant computational cost while retaining the key physics. Four methods are evaluated in this paper to achieve this goal: (i) develop reduced-scale periodic simulations to reduce the number of particles; (ii) further reduce this number by scaling particle sizes appropriately; (iii) adopt an adhesive, elasto-plastic contact model to capture the effect of the liquid binder rather than fluid coupling; (iv) identify the subset of model parameters that are influential for calibration. All DEM simulations considered a GEA ConsiGma™ 1 twin-screw granulator with a 60° rearward configuration for kneading elements. Periodic simulations yielded similar results to a full-scale simulation at significantly reduced computational cost. If the level of cohesion in the contact model is calibrated using laboratory testing, valid results can be obtained without fluid coupling. Friction between granules and the internal surfaces of the granulator is a very influential parameter because the response of this system is dominated by interactions with the geometry.

Keywords: powder agglomeration; Discrete Element Method; cohesion; wet granulation; twin-screw granulation

Citation: Morrissey, J.P.; Hanley, K.J.; Ooi, J.Y. Conceptualisation of an Efficient Particle-Based Simulation of a Twin-Screw Granulator. *Pharmaceutics* **2021**, *13*, 2136. https://doi.org/10.3390/pharmaceutics13122136

Academic Editor: Colin Hare

Received: 29 October 2021
Accepted: 6 December 2021
Published: 12 December 2021

1. Introduction

Wet granulation is a process used to create larger stable agglomerates (granules) from fine powders. This has many desirable outcomes such as improving flowability, compactibility and homogeneity. Granulation is commonly employed in the food, pharmaceutical, detergent and fertilizer industries. Despite its widespread adoption, it is often inefficiently operated [1,2], with high recycle ratios in continuous processes and high rejection rates in batch processes [3]. Wet granulation is the most common type of granulation and, in pharmaceutical applications, it is a critical step in tablet manufacturing that affects the uniformity and compactibility of the final dosage form.

Traditionally, batch granulation was the favoured granulation approach in the pharmaceutical industry due to the challenges associated with continuous processing such as the changeover cost and inability to monitor product quality reliably. Since the introduction of Quality by Design (QbD) and Process Analytical Technology (PAT) by the FDA in 2003 [4] and due to recent advances in continuous monitoring, there has been a move from traditional batch towards continuous processing. Although tremendous efforts have been made to gain scientific insight into the granulation process [5–12], a fundamental understanding of wet granulation is still lacking due to the complexity of the mechanisms involved, i.e., governing rate processes of wetting and nucleation, aggregation and consolidation (or consolidation and coalescence), breakage and attrition, and layering [1,9].

Twin-screw granulation developed from early work on single- [13] and twin-screw extruders [14] in the late 1980s. It has become popular within the pharmaceutical industry in the past decade due to the many advantages it offers over high shear and fluidised bed granulators including limited or no scale-up requirement, lower space requirements, continuous operation with monitoring and higher throughput. Since a patent was awarded to Ghebre-Sellassie et al. [15], there has been increased adoption of this method of granulation in industry.

A twin-screw granulator (TSG) consists of three main components: the powder feed, liquid addition mechanism and two intermeshed screws, which may be either co- or counter-rotating, enclosed in a barrel. The powder is fed into the screw barrels at one end and exits the open end of the barrel. This is in contrast to twin-screw extruders where the granulated product is forced through a die or plate at the end of the barrel. Twin-screw granulators are often described by their length-to-diameter (L/D) ratio [16] with extruder lengths in the range of 12–50 L/D common in industry [17]. Screws are required to tightly fit within the bore of the barrel to create a closely confined flow path for materials. Researchers have reported that the confinement offered by the low clearance between the screws and barrel ensures a similar shear history for all particles, which helps produce more consistent granules than batch granulation processes [17,18]. Screws typically operate part-filled, with solid fraction dependent on screw geometry, material feed rate and screw speed [16]. The screws are modular in nature and comprise elements of different types to give the required configuration.

In general, there are four element types: conveying elements, kneading elements, chopping elements and comb mixer elements (not considered in this paper). Multiple elements are combined to form a block; multiple alternating blocks form the complete screw. Kneading blocks impart high mechanical energy to the wetted material, producing high shear forces, compaction and distributive mixing. Further details on screw elements, profiles and configuration can be found in the literature [16–22].

Depending on the angle of offset, kneading blocks can produce forwarding or reversing flow [23,24]. Reversing kneading blocks force material back against the direction of flow, which leads to areas of high pressure and compaction. Although this allows strong granules to be formed, there is a high likelihood of blockages [25]. Chopping elements are shortened kneading elements whose purpose is to break up oversized agglomerates at the end of the screw.

Twin-screw granulation has been studied in much detail in recent years [16,17]. Several experimental studies have investigated the effects of key process variables [26–37], screw configurations [24,25,38–42] and formulation variables [26,36,40], and have resulted in a regime map [43] of the twin-screw granulator. However, these experimental studies cannot provide any insight into the micro-scale phenomena that govern granulation processes. Particle-scale simulations have the potential to provide this insight. The Discrete Element Method (DEM) has become the dominant particle-scale simulation tool in the last 20 years. DEM has been applied to investigate various wet granulation processes, e.g., Mishra et al. [44] and Liu et al. [45] investigated granulation in a rotating drum; Goldschmidt et al. [46] and Kafui et al. [47,48] simulated fluidised bed granulation; Gantt and Gatzke [49], Hassanpour et al. [50], Nakamura et al. [51], Watson et al. [52], Tamrakar et al. [53] and Börner et al. [54] studied high shear granulation. Researchers have also explored granulation processes by coupling DEM and population balance modelling [55–59]. However, there are few prior DEM studies of twin-screw granulation. Dhenge et al. [29] studied the effect of binder amount and binder viscosity at varying powder feed rates on the granulation behaviour and final granule properties. Due to the large computational cost, their simulations use conveying screws only within a 16 mm periodic domain. They also do not include cohesion between particles in the simulations. Only a qualitative visual comparison was made to the experimental results. Recently Zheng et al. [60] have used DEM to study residence time distributions for dry, cohesionless, mono-sized elastic spheres of various sizes for a fixed configuration of conveying elements with two blocks of kneading elements at various operating RPMs. Zheng et al. [61] extended the study further to include the effect of particle shape in the TSG. Spherical particles were observed to have slightly lower mean residence times than particles of other shapes, which were largely the same, regardless of shape. The influence of the many DEM parameters on the residence time has not been investigated in these studies, which only considered dry cohesionless materials. Kumar et al. [62] used mono-disperse periodic DEM simulations to study liquid distributions within the mixing zone of a GEA ConsiGma 25 TSG; however, this zone was only one kneading element thick (8× particle diameters). While liquid transfer and associated agglomeration could be observed, the limited domain means no longitudinal travel or transfer from conveying elements were considered in this model, providing a somewhat isolated overview of the dynamics within a TSG.

The objectives of this research study are threefold:

- To develop and verify computationally efficient reduced-scale DEM models that closely match full-scale simulations of the entire granulator;
- To carry out a comprehensive sensitivity study to explore how operating parameters and DEM model parameters affect the performance of a TSG, and thereby identify the subset of influential parameters requiring calibration (noting that this study does not attempt to perform a full calibration of the granular material against a real solid);
- To demonstrate that the effect of liquid binder can be captured using an adhesive, elasto-plastic contact model.

In order to create a numerical model that is capable of producing numerical predictions that can be validated against experimental results, it is necessary to first verify that the conceptual model is operating in the correct manner, similar to its real-life counterpart. As such, the operation of the TSG model is verified by comparing the particle dynamics, flow characteristics and residence time distributions against previously reported sources. Validation is concerned with establishing the predictive capability of the computational model in relation to the real system. Computational results should be compared with independent experimental results, and a computational model will be considered validated once the discrepancy between the simulation and experimental result of interest falls within pre-determined bounds. This discrepancy must be evaluated against the degree of uncertainty in both the validation experiment and the multiscale model. Validation of the model outputs against careful experimental measurements such as granule porosities and granule size distributions is not part of this study.

2. Twin-Screw Granulator Model

The test apparatus chosen is a ConsiGma™ 1/25 continuous tableting line (CTL) manufactured by GEA. The ConsiGma 1/25 granulator consists of two co-rotating screws, each housed within a 25 mm diameter barrel. The dimensions of the TSG and screws are summarised in Tables 1 and 2, respectively. A key figure is the free barrel fraction of 0.4731, i.e., around 47% of the total volume of the granulator barrels is free space for particles to occupy. The ConsiGma 1/25 granulator uses three types of screw element: conveying, kneading and chopping. CAD models of these individual elements were constructed in 3D, as shown in Figure 1. All elements consist of the same cross-section. Conveying elements of three different lengths were formed with a helical extrusion, and kneading and chopping elements with simple extrusions. The key difference between kneading and chopping elements is the depth, with kneading elements being 6.25 mm deep and chopping elements being just 4.167 mm deep.

Table 1. Key ConsiGma 1/25 Dimensions.

Property	Value
Barrel Diameter, D [mm]	25
Channels [-]	2
Channel Depth [mm]	2.58
Centre line Distance [mm]	19.31
Barrel–Screw Clearance [mm]	0.440
Screw–Screw Clearance [mm]	0.175
Area of Double Intermeshed Barrels [mm^2]	920.0
Free Barrel Area [mm^2]	435.26
Free Barrel Fraction	0.4731

Table 2. Key ConsiGma Screw Properties.

Property	Value
Screw Radius—Outer, R [mm]	12.06
Screw Radius—Inner, r [mm]	6.90
Screw Diameter Ratio	1.75
Screw Transition Radius, R + r [mm]	18.96
Screw Length, L [mm]	500
Screw Lead, S [mm]	25
Screw Section Perimeter [mm]	59.56
Screw Section Area [mm^2]	242.37
Screw Transition Angle	38.15°
Screw Tip Angle	13.72°

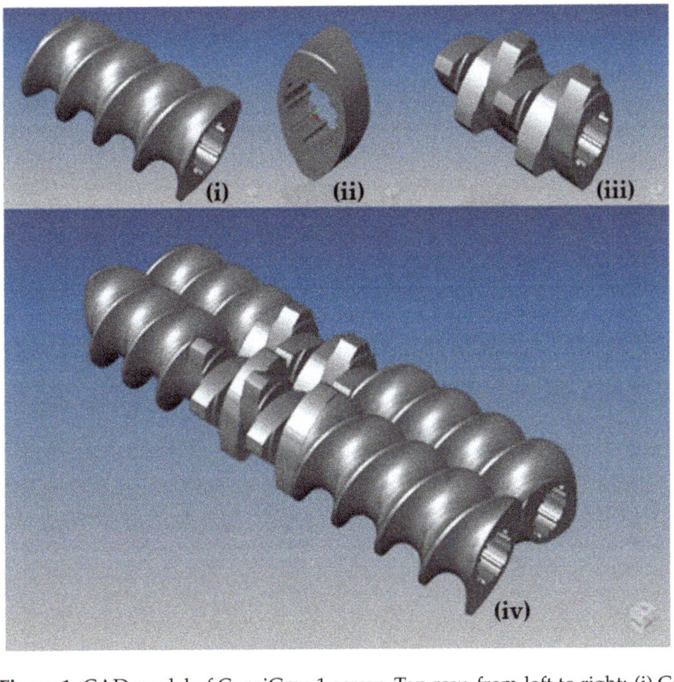

Figure 1. CAD model of ConsiGma 1 screw. Top row, from left to right: (i) Conveying element of length 37.5 mm (ii) Kneading element (iii) 6 × 60 R block of kneading elements; Bottom row: (iv) 6 × 60 R block of kneading elements with adjoining conveying elements.

2.1. DEM Model Configuration

DEM simulations of the ConsiGma TSG were carried out with the commercial code EDEM [63]. The CAD model described in the previous section was imported into EDEM (Figure 2). A high level of precision was used during import to ensure that no unnatural edges existed in the intermeshing region where screw clearances are very tight. Twin-screw granulators are operated at steady state: the total mass entering the system is equal to the mass leaving the system. In the simulations, three mass flow sensors were included along the barrel to assess when the granulator reached a steady state during operation.

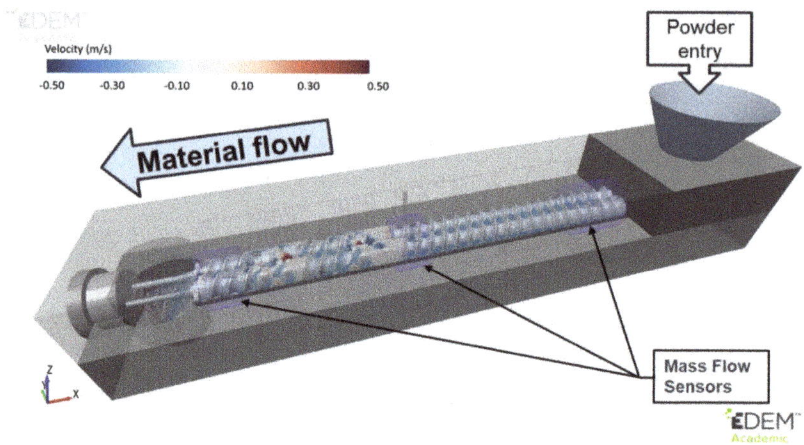

Figure 2. Full−scale DEM model of ConsiGma 1/25.

The stainless-steel barrel and screws were modelled as rigid bodies in EDEM. Appropriate rotational dynamics were added for the chosen screw speed. The screw configuration used in the simulations is that of a typical experimental setup, summarised in Figure 3, which consists of two blocks of six kneading elements with a 60° offset in a reversing configuration, separated by a 1.5D conveying element. Two chopping elements were located at the screw exit to break up any oversized agglomerates that may have formed.

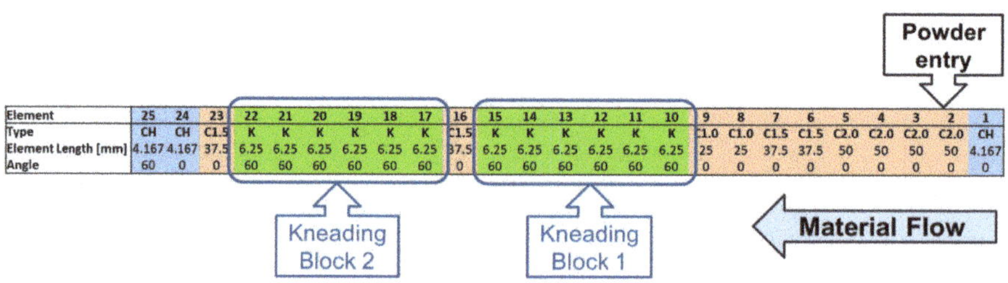

Figure 3. Typical experimental screw configuration, as used in the full-scale DEM simulations. CH represents a chopping element, C1.5 represents a conveying element of length 1.5 times the screw lead and K denotes a kneading element.

Both cohesionless and cohesive systems were studied. Cohesionless systems are first introduced to study the effect of parameters such as friction and restitution, in recognition of the fact that most DEM simulations of twin-screw granulators still adopt simple cohesionless models. The subsequent inclusion of cohesion allows its influence to be distinguished from the other parameters.

In the absence of cohesion, a conventional Hertz-Mindlin contact model was used. For cohesive interparticle interactions, the Edinburgh Elasto-Plastic Adhesion (EEPA) model [64,65] (Figure 4) was used: an adhesive, elasto-plastic contact model that captures the key characteristic behaviour for agglomerates. This non-linear model is used to capture the role of the liquid binder in the TSG. It is based on the physical phenomena observed in adhesive contact between micron-sized particles or small agglomerates [66]. The EEPA model accounts for both the elastic–plastic contact deformation and the contact-area-dependent adhesion. Key parameters of this model are the constant pull-off force (f_0), the slope exponent (n) and the contact plasticity ratio, which relates the unloading/reloading stiffness k_2 to the loading stiffness k_1. The EEPA model has been successfully used to model the behaviour of wet iron ore fines [64,67,68], detergent powders [69,70], powder mixing [71] and cohesive soils [72]. The self-cleaning nature of intermeshing, co-rotating screws means that the cohesionless Hertz-Mindlin contact model was used for all particle–geometry contacts.

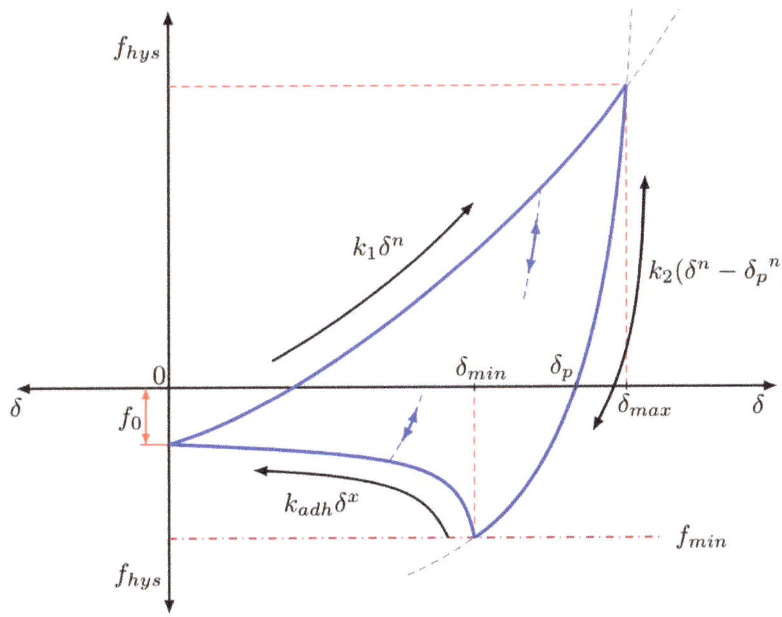

Figure 4. Force–displacement behaviour of the Edinburgh Elasto-Plastic Adhesion model.

Breakage of the primary particles was not considered in the model. However, agglomerates formed by cohesion were able to break and reform, similar to real agglomeration behaviour. Liquid migration was not considered in the model to reduce the computational cost. Values of key input parameters used for the cohesionless simulations are given in Table 3. These parameters form the reference case for the sensitivity study presented later. The chosen bulk density is that of a typical pharmaceutical solid: paracetamol. The remaining parameters are chosen as typical median values that are representative of a cohesive powder. The "standard deviation" of 0.05 in Table 3 defines the width of the normal Gaussian distribution in particle diameter, about a mean value of 400 µm. This distribution is truncated at an upper and lower limit (0.5x–1.5x), relative to the mean. The effect of cohesion in the TSG is considered through a set of cohesion parameters that result in high levels of cohesion (given later in Table 4). These parameters have not been calibrated for any specific material or level of liquid binder and are simply to explore the effect of cohesion in the system and its relative importance.

Table 3. Reference values of the DEM input parameters.

Property	Particles	Geometry
Poisson's ratio, ν	0.25	0.25
Shear modulus, G (Pa)	1.7×10^7	1×10^8
Coefficient of restitution, χ	0.5	0.5
Static friction, μ_s	0.5	0.5
Rolling friction, μ_r	0.01	0.01
Density (kg/m^3)	1263	7850
Mean particle radius (m)	0.0002	
Standard deviation	0.05	
Normalised truncation limits	(0.5, 1.5)	
Gravitational acceleration (m/s^2)	9.81	

Table 4. Variations from the reference case defined in Table 3, where χ, μ_s and μ_r apply to both particle–particle and particle–geometry interactions.

	Reference	Variations
Coefficient of restitution, χ	0.5	0.1, 0.9
Static friction, μ_s	0.5	0.1
Rolling friction, μ_r	0.01	0.2
Relative Standard Deviation	0.05	0.2
Cohesion	Cohesionless	High ($\gamma = 15\ J/m^2, f_0 = -0.001$ N)

The main processing parameters in twin-screw granulation include the liquid-to-solid (L/S) ratio, the feed rate, the screw speed, the screw configuration and the barrel temperature. The combination of screw speed and feed rate leads to the development of a specific fill level or porosity on each element type, with a lower porosity typically found on kneading elements. The stress state developed is largely determined by the screw speed in combination with the geometric clearances in relation to particle size. A higher powder feed rate increases the degree of compaction and densification of the powder in the TSG barrel. The feed rate chosen for the simulations is a typical value of approximately 14.4 kg/h. A single screw speed of 600 RPM was chosen. The level of cohesion in the simulations was used to capture the effect of liquid in the system. The level of cohesion could be calibrated against flowability measurements from experimental characterisation tests, e.g., direct shear or uniaxial tests, for the various L/S ratios used. Thermal effects and barrel temperature were not considered in these DEM simulations.

2.2. *Numerical Instabilities*

There are various instabilities that can develop in a numerical model, which require careful consideration to avoid unwanted effects on the simulation results. Numerical instabilities in DEM simulations can arise from various sources, but the most common source of errors is the integration timestep. The sensitivity of the results to the integration timestep was checked independently for 1%, 10% and 20% of the critical Rayleigh timestep and no significant variation in results was noted. As such, all simulations were carried out with a timestep of 5×10^{-7} s, which is 10% of the critical Rayleigh timestep for 400 μm particles or 20% for 200 μm particles. This timestep is sufficiently low to ensure that the contacts are identified accurately in this highly dynamic environment. Although a reduced shear modulus has been chosen to reduce the computational time, it is kept large enough to ensure numerical accuracy [73,74].

Given the tight tolerances at play in a twin-screw granulator and the generally large L/D ratios for the barrel, the accuracy of geometry discretization as well as the choice of elements is also of importance here. The curvature of the screw and barrel needs to be sufficiently captured to ensure particles are not getting "pinched" in gaps that are artefacts of the discretistation of the CAD geometry. It is also important to ensure that small, well-shaped meshes are used and to avoid any stretched triangles, especially in the barrel, as this could cause some strange effects.

Finally, adding large amounts of cohesion to the system can cause many problems such as creating excessive overlaps at collisions to forming a single agglomerate containing all particles in the simulation. It should also be noted that even if the amount of cohesion used is not excessive, it can still lead to the build-up of material in the kneading elements that would cause jamming of the real granulator. In this numerical situation, rather than the screws stopping, the particles will just be forced through the barrel walls.

2.3. *Full-Scale vs. Reduced Domain Models*

In this study a comparison is made between a full-sized domain model of the granulator and a more computationally efficient reduced domain model to investigate whether it is possible to use computationally efficient models without affecting the correctness and reliability of the simulation results.

2.3.1. Full-Size Computational Domain

This simulation captures the continuous flow through the granulator where transitions between different element types lead to different phenomena. The full length of the TSG has been modelled from inlet feed to outlet. It was observed that due to the consistent behaviour of particles on conveying elements (which serve only a transport purpose before liquid injection), the particle generation point could be moved to mid-way along the granulator to element number 7 on Figure 3 (immediately before the liquid inlet points on the physical TSG) without any observable difference in results. The computational cost of simulating all of these elements is high and moving the particle generation location reduced the required run-time by approximately 40%. It is important to note that the measured residence time for the shortened granulator would need to be adjusted for experimental comparisons where the granulator length cannot be reduced.

2.3.2. Reduced Domain Models

The computational cost of running the full simulations is considerable: from start-up to steady state, and then operation at steady state for several seconds, requires millions of particles over a typical 10–20 s simulation. To reduce this cost, the twin-screw granulator can be broken down into different sub-domains of interest, such as a block of conveying elements, a block of kneading elements or a block of conveying and kneading elements.

A reduced domain model can be created to focus on the specific sub-domain of interest and can be created with either periodic boundaries at each end of the domain in the direction of flow (termed *periodic*) or with particle generation at one end of the screw and the standard domain from which particles will exit at the other (termed *non-periodic*). These reduced domain models are numerically efficient as the number of particles is considerably reduced. This can allow greater fidelity in the models through the use of smaller particles or the implementation of more complex physics for the same computational cost as the full-size granulator. Selecting the elements for the reduced domain model requires consideration of the key zones in the full granulator. The mixing zones containing the kneading elements are significantly more important to the granulation process than the conveying elements. As such, these mixing zones are the focus of the reduced models in this study.

At this point it is worth considering some of the limitations of the two types of reduced domain models. A periodic sub-domain model cannot consist only of kneading elements since these elements are mixing elements rather than transport elements, and without sufficient transport elements there would be little or no forward flow in the periodic sub-domain model. This would lead to residence times that would tend towards infinity as there is no longitudinal flow being provided. The length of the periodic sub-domain model must be such that that the same screw position is matched on the start and the end screw profiles to prevent loss of particles and ensure a clean and error-free recycling of the particles. This can make periodic sub-domain models longer than non-periodic sub-domain models, which do not need to recycle particles. Periodic sub-domain models will have a fixed number of particles, which means they cannot enforce a fixed flow rate, whereas non-periodic sub-domain models utilise dynamic particle generation and can be used to study the granular behaviour under fixed flow rates. Periodic sub-domain models cannot be used to study the developing granule size distribution as formed granules would be recycled through the initial flow boundary instead of virgin feed. In this type of study some models with continuous generation would be required.

2.3.3. Implemented Reduced Domain Model

An equivalent periodic model that comprises a full block of kneading elements (six elements) and the adjoining conveying elements was developed to investigate the complex behaviour within a kneading block. This is shown in Figure 5. The extent of the simulation domain, which is periodic in the x-direction, is outlined in red. The kneading elements in the centre of the system are outlined in green, with one kneading element zone, which is the same thickness as a kneading element, highlighted. The extent of the adjoining conveying elements for analysis purposes is marked by the blue zones. The difference between the simulation domain (red) and the analysis domain (blue) creates buffer zones where the particles are not considered for analysis, as they may be affected by some backflow or particles oscillating at the periodic boundary. The average mass of particles from the full-sized simulation, for the full duration at steady state in the same domain, was generated statically in the free space around the screws as the starting point for the periodic system. The simulation was then run for a short period of time to reach a steady state—approximately the time taken for a single particle to circulate through the system (0.5–0.75 s)—to allow particles to distribute naturally along the elements. All data analysis was carried out after this initial start-up time. This setup produces a very good qualitative match to the flow patterns observed in the full-scale simulations.

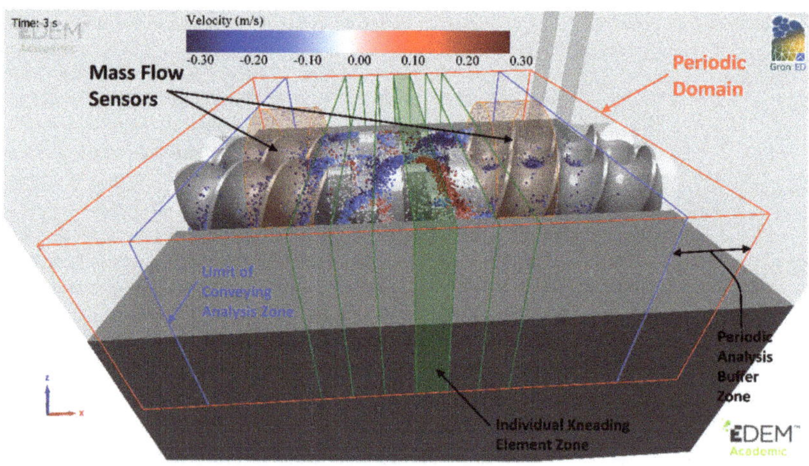

Figure 5. Periodic simulation zones at steady state.

2.3.4. Comparison to Full-Scale Simulation

The full-scale DEM simulation was run for 2 s after steady state operation had been achieved. Particle streamlines were extracted for a random selection of particles and are shown in Figure 6. These streamlines and resulting residence times correlate well with those measured experimentally using PEPT (positron emission particle tracking) [7,75].

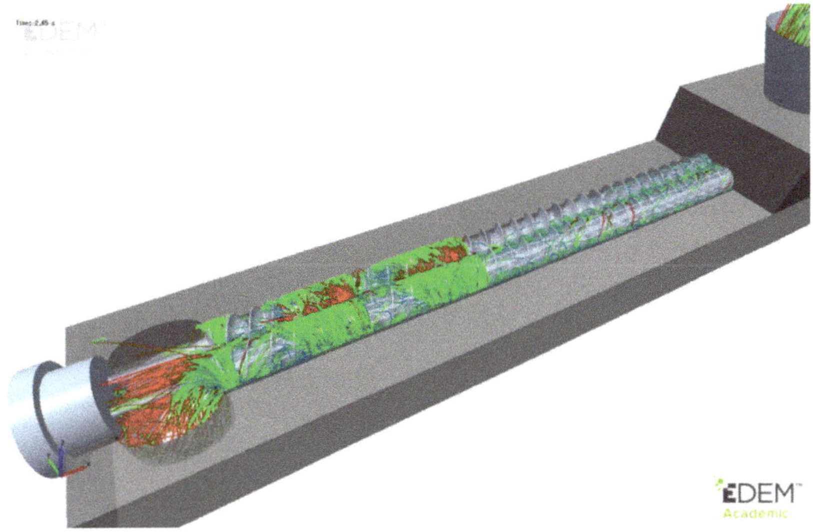

Figure 6. Particle streamlines for full granulator.

Cross-sections of the temporally (5 screw revolutions) and spatially averaged solid fraction are shown in Figure 7 for the two main element types: conveying and kneading elements. The figure shows that there are three main transport locations for the conveying element: between the screws in the intermeshing zone and at the bottom of each screw bore. The kneading elements show a different pattern due to their flat nature and tend to only push particles around the plane of the element. The DEM simulations predict more material in one bore of the conveying elements, where the driving screw is located, than the other. This matches previous experimental observations [76].

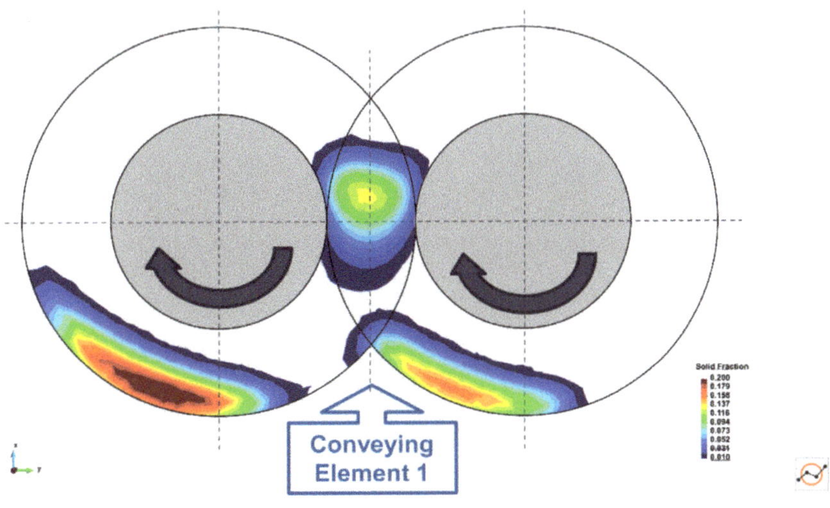

(**a**) Conveying element

Figure 7. *Cont.*

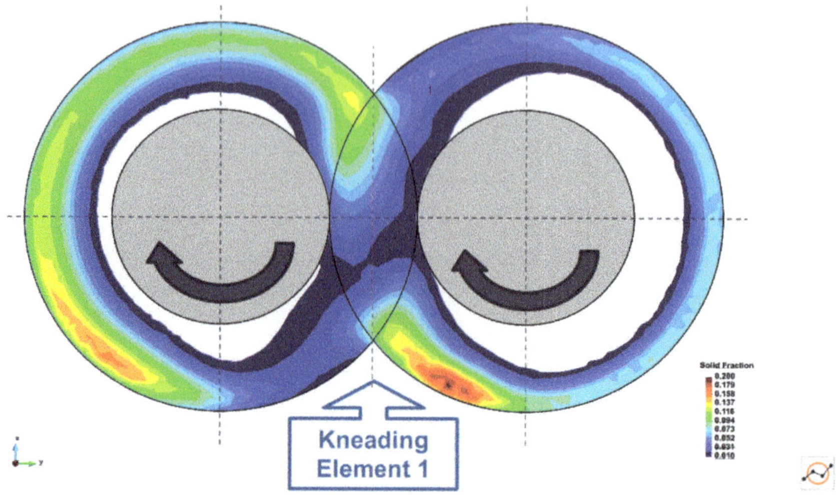

(**b**) Kneading element

Figure 7. Transport locations in different element types.

The Residence Time Distribution (RTD) for the full TSG simulation was extracted, which shows a mean total residence time of approximately 3 s. This is in the region of what has been measured experimentally for a free-flowing material in similar equipment [7,27,28,39]. RTDs for the full-size DEM simulation and the periodic simulation in the same domain, over a 1 s period at steady state, are compared in Figure 8. The periodic simulation is capturing the same behaviour as the full-size simulation. For the total residence time, the periodic results include a spike at 0 s, which is an artefact due to the re-circulation of particles through the periodic domain and the time they spend in the buffer zone at each end. The agreement is also excellent for the RTDs by element type.

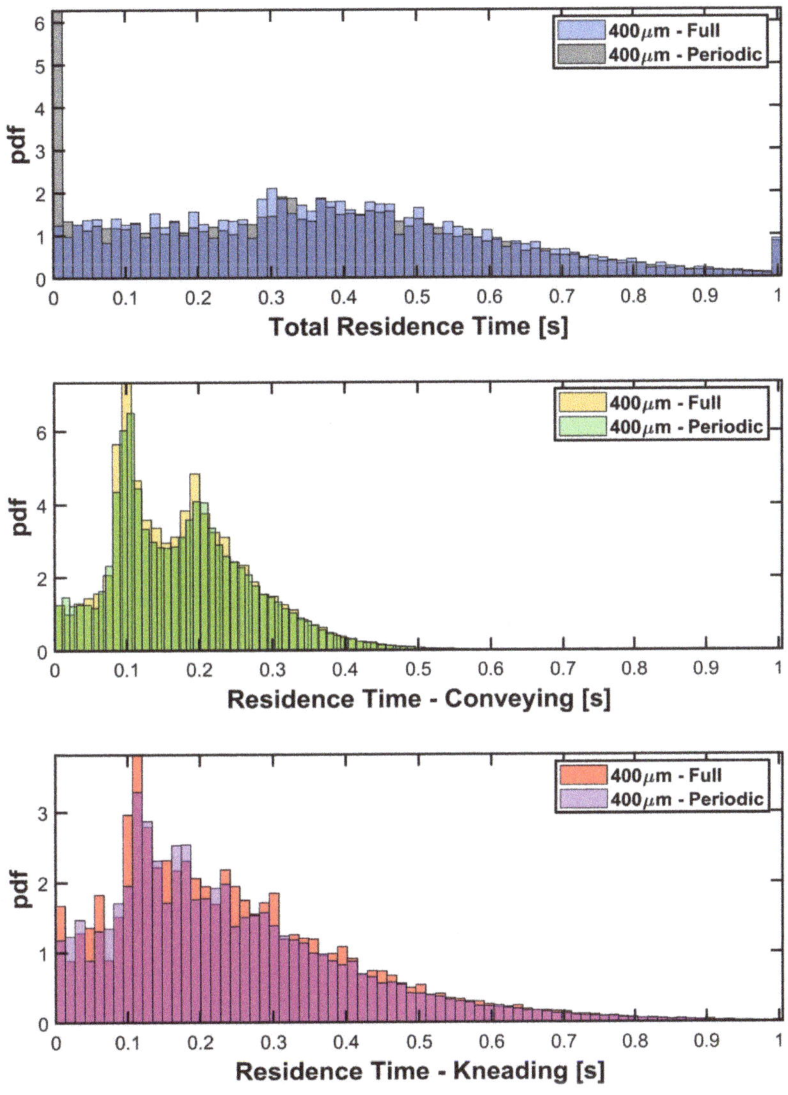

Figure 8. Comparison of the Residence Time Distributions (RTDs) for the full model and periodic simulation.

Figure 9 shows a comparison between various temporally-averaged measured quantities (residence time, solid fraction, average velocity magnitude and average longitudinal (X) velocity) per element for the two model types. The results from both model types are in excellent agreement, verifying that the periodic system is capable of reproducing the same phenomena observed in the full-size simulation and can safely be used in the sensitivity study to investigate the effect of the various DEM parameters. Both models show that there is a strong link between the residence time and solid fraction for each element as the largest residence time per unit length is found where the highest solid fractions exist. It should be noted that the measured region of the conveying elements is longer (approx. 3.5*x*) than the kneading elements. Figure 9 also shows the velocity magnitude and longitudinal velocity comparisons for the two models. The axial velocities show a strong connection to the respective solid fractions and residence time values, whereas this is less clear for the velocity magnitudes, which are similar across all kneading elements. The velocity magnitude, resulting from the longitudinal and in-plane movement, is, however, much higher for kneading elements than for conveying elements due to the rotational path enforced by the geometry and is, therefore, heavily influenced by the screw RPM.

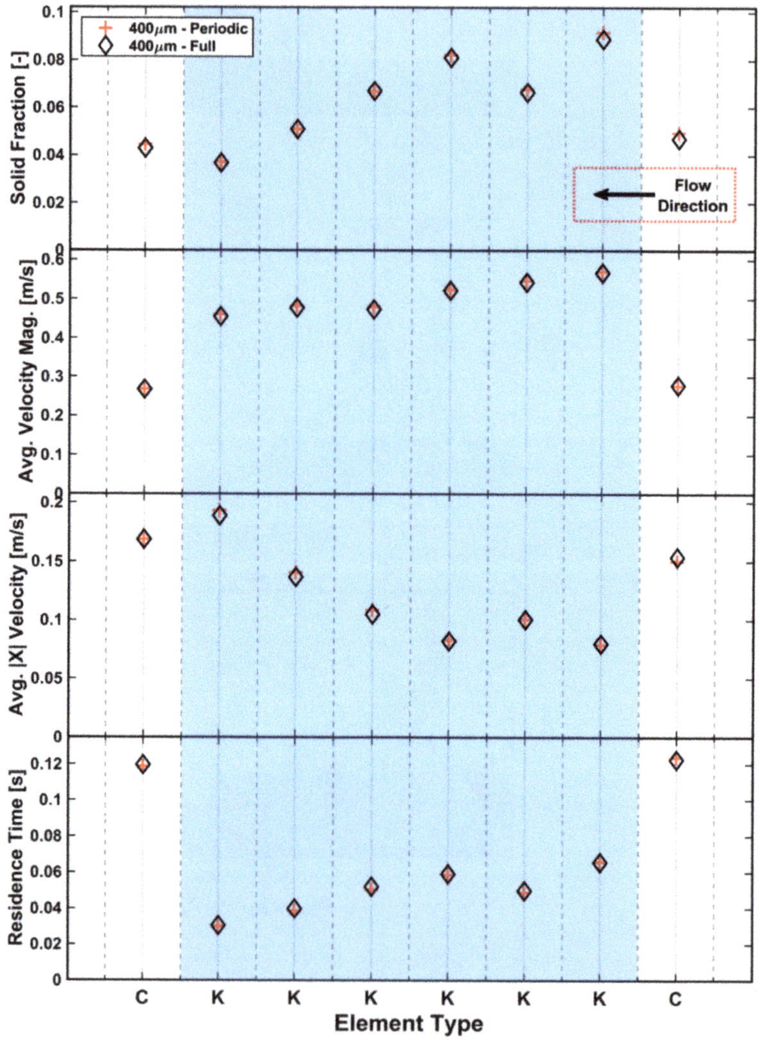

Figure 9. Comparison between periodic and full-scale simulations for various measured quantities.

2.4. *Influence of Particle Size*

The mean particle diameter was varied from a reference value of 400 μm to values in the inclusive range between 200 μm and 1500 μm, all with the same truncated normal size distribution with the same standard deviation. The total mass of particles generated in the domain is the same for all particle sizes, being obtained from the steady-state mass in the same elements in the full-size model. The effects of the change in mean particle size on the solid fraction, velocity and residence time are shown in Figure 10. Figure 11 shows the snapshots at the same steady-state time instant for all simulations.

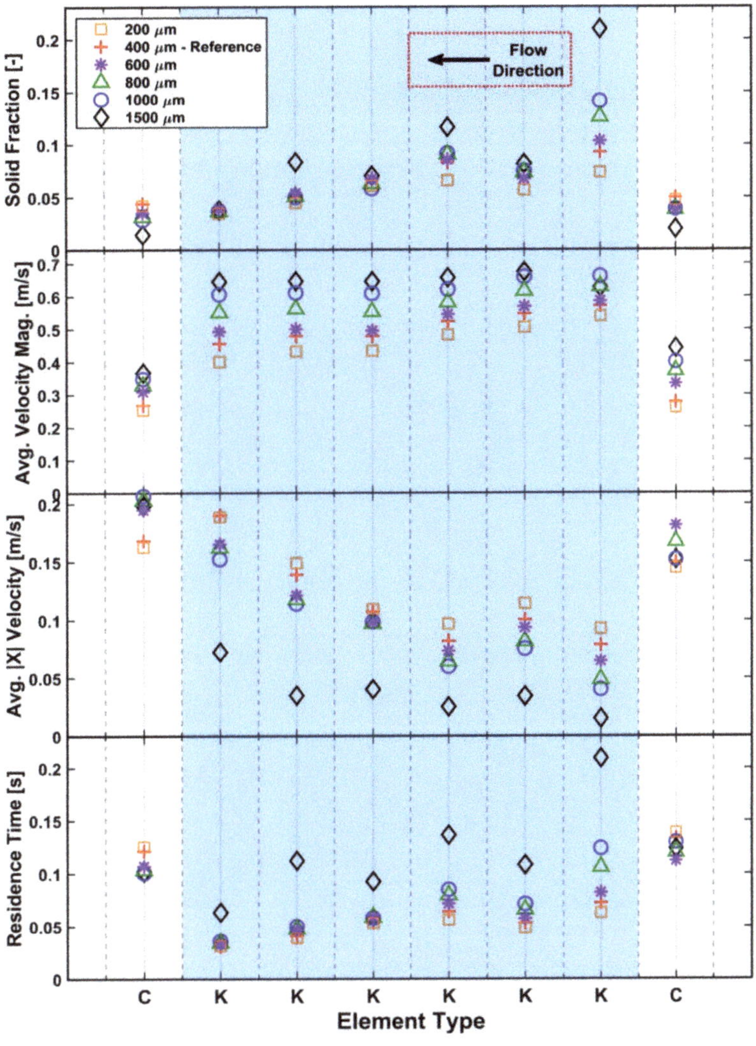

Figure 10. Effect of particle size on solid fraction, velocity and residence time.

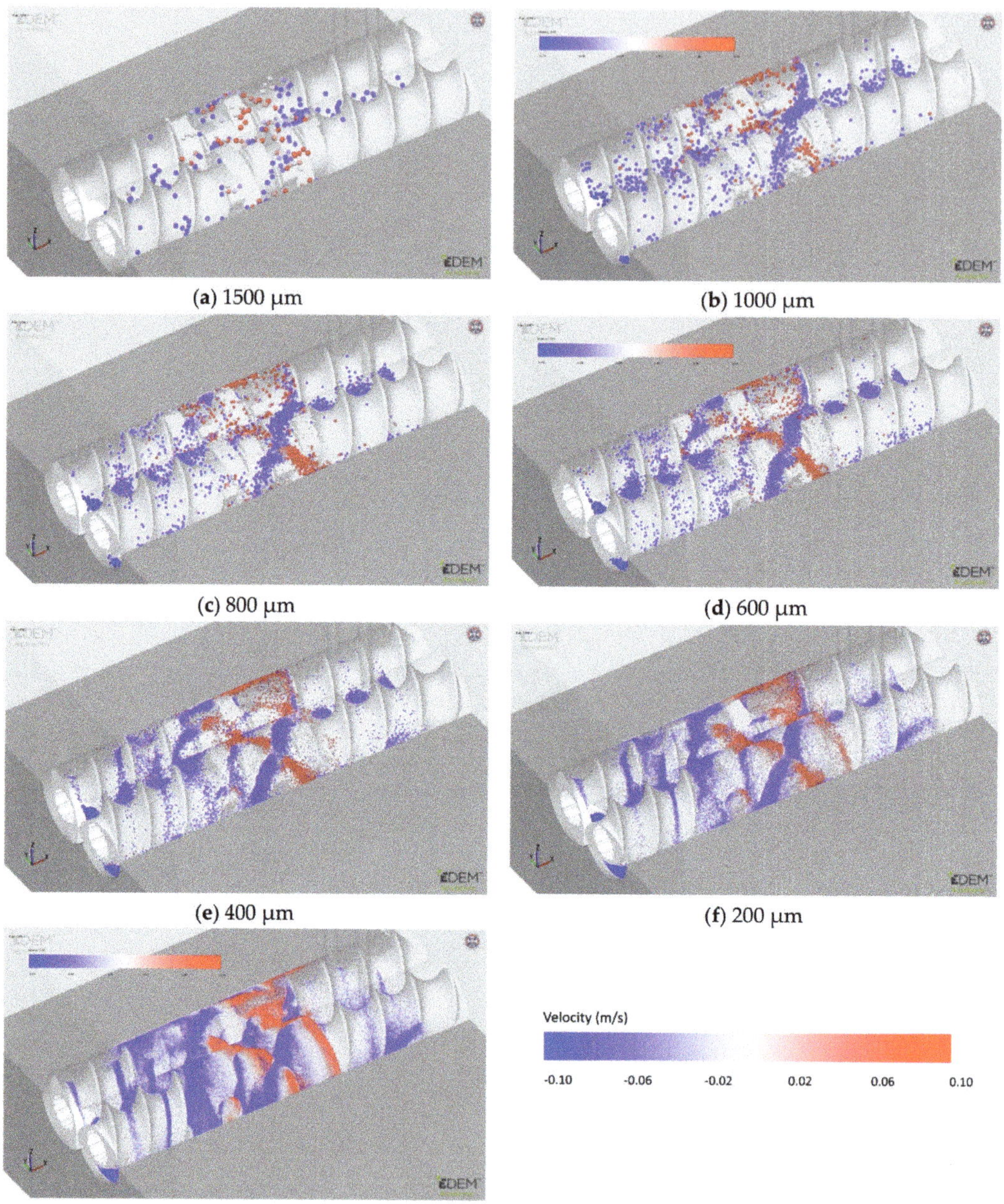

Figure 11. Particle snapshots at steady state as the mean particle diameter is varied. The particle colour indicates longitudinal velocity in the direction parallel to the screws; a negative velocity indicates the movement of particles towards the outlet of the granulator.

The solid fraction results on conveying elements fall into three separate classes for small (<500 μm), medium (500–1000 μm) and large (>1000 μm) particles. Large particles lead to the lowest solid fraction on conveying elements due to the geometric constraints of the screw geometry: larger spheres cannot occupy the space around the screws as efficiently as smaller spheres. The kneading block of six elements appears to act as two blocks of three elements, with significantly different behaviours observed on the first three and the last three kneading elements. In the last three elements, the observed solid fractions are largely the same for all particle sizes, apart from the large 1500 µm particles. The first three kneading elements show significant variation with the larger particles generally having a higher solid fraction.

For the longitudinal (X) velocity of the particles along the barrel, the largest particles have the lowest longitudinal velocity on the kneading elements, suggesting that these particles experience more obstructions by geometrical constraints at the transitions between elements, hampering their path through the granulator. All particle sizes follow the same general trend where there is a significant velocity decrease at the transfer from conveying to kneading elements. However, the magnitude of the retardation is heavily dependent on particle size, with the larger particles reduced to the lowest velocity. The 1500 µm particles do not follow the general trend of significantly increasing longitudinal velocity as the particles progress through the kneading block, with the velocity remaining almost constant until the last kneading element before an increase is noted. All other sizes show a clearer velocity increase along the kneading block and have returned to the original screw translational velocity by the time they have left the kneading block.

Figure 10 also shows a consistent trend of increasing velocity magnitude with increasing size, with an almost constant value for each element type. Interestingly, there is a large variation in the velocity magnitudes for differently sized particles in the conveying elements, with larger particles having larger velocities. This appears to be a result of the number of particles in the system. As the size of the particles increases, the number of particles in the system decreases, which can lead to decreasing levels of particle–particle collisions in the system. With only several large particles able to sit on a conveying element at once, there are relatively few interparticle contacts, leading to lower energy dissipation. This results in the larger particles being more active in the screw elements.

A similar trend of high velocity magnitude for larger particles is also observed in the kneading elements. Within the kneading zone, the increased velocities can also be attributed to the reduction in particle numbers. In the kneading zone, particle collisions play a significant role in helping particles progress along the screw. Without the particle collisions to help "nudge" particles onto the next element, the larger particles have a tendency to remain circulating in-plane, being pushed around by the flat lobes of the kneading elements. There is also a geometric constraint at play in the kneading zone, with the larger particles more likely to collide with the next kneading element and remain on the current element than smaller particles, which are less likely to be constrained in this manner. These factors contribute to the reduced longitudinal velocity in the larger particles, which is also reflected in the observed residence times that are highest for the largest particles.

Pradhan et al. [77] found from experiments and geometrical analysis on static 3D CAD models that particle breakage was related to the maximum void space on an element, with 20% of particles on mixing elements broken at just 60% of the maximum size and all particles suffering some breakage at less than the maximum size. This geometrical breakage constraint was recently implemented in a new population balance model kernel for particle breakage [78] and should be considered when choosing particle sizes for simulations so as to not adversely affect the simulation results. The simulation results in the current study suggest the existence of a critical ratio of particle diameter to the screw void space (effectively the difference between the screw's root and outer radii) for unhindered particle flow that is significantly less than the breakage limit identified by Pradhan et al. [77]. For the configuration used in this study, the maximum dimensions within the void spaces are approximately 5.5 mm and 6.25 mm for conveying and kneading elements, respectively, suggesting a diameter-to-void limit of approximately 10–20% for unhindered particle flow. This limit may also be affected by both the offset angle and direction of the kneading/mixing elements, which could reduce the maximum dimensions within the void spaces below the figures quoted here.

The significance of the particle size in relation to the screw–barrel clearance can also be observed in Figure 10. This critical clearance is 440 μm (Tables 1 and 2) in this model, which means approximately half of the particle sizes studied will not pass through this outer region. With the exception of kneading element 1, the particle behaviour that is dominated by collisions in the central intermeshing region, there is little difference between the particle sizes that fit through this outer gap (200 μm, 400 μm) and those that are too large (1500 μm particles excluded) to pass through. This suggests that the screw–barrel clearance is not a significant concern when choosing particle size.

Finally, it is worth considering the qualitative particle dynamics when investigating particle size effects. Figure 11 compares the snapshots at the same time instant for all simulations. For the 1500 μm particles, there is no clear or obvious pattern in the particle dynamics due to the low number of particles, with just under 1200 existing at this particle size. However, a flow pattern starts to appear when the particle size is decreased to 1000 μm (approximately 4100 particles). At 800 μm and 600 μm, this "*x-shaped*" crossover pattern becomes increasingly well defined. Further reductions in particle size do not significantly enhance this flow pattern. This was checked for particles as small as 100 μm (Figure 11g). The 400 μm reference case is a good compromise between computational efficiency and capturing the particle dynamics of the system, and 400 μm has been selected as the reference size for the study. Omitted from the comparison in Figure 10 were 100 μm particles due to the large number of particles (approx. 4.1 M in total) and limited simulation time limiting the amount of available data.

3. Sensitivity Study of DEM Input Parameters

The results of a sensitivity study on the mean particle diameter are presented in Figures 10 and 11. For the remaining parameters, the sensitivity study employs a univariate approach to quantify the effect of a DEM parameter relative to the reference case defined in Table 3. In cases where the reference value is an intermediate value, both extreme high and low values were tested; when the reference value is already an extreme value, an opposite extreme value was tested without an intermediate value. Variations from the reference case are given in Table 4.

For the cohesive case, the contact plasticity ratio, tangential stiffness multiplier (ζ_{tm}), slope exponent (n) and tension exponent (x) were fixed at 0.8, 0.667, 1.5 and 20, respectively. The contact plasticity ratio was arbitrarily chosen as a value suitable for a highly compressible cohesive powder and the adhesion energy is chosen as 15 J/m^2 and $f_0 = -0.001$ N. Both n and ζ_{tm} are consistent with the non-linear Hertz model used for the cohesionless simulations.

3.1. Particle Size Distribution

Two different size distributions with the same mean (400 μm) were considered, with both distributions truncated at 200 μm and 600 μm. Adopting the broader distribution yields far more large and small particles in a simulation than the narrower distribution. Figure 12 shows the effect of the size distribution on the various measured quantities for each element. In general, the spread of the size distribution has little effect. The exception is the first kneading element at which there is approximately a 10% difference in the solid fractions. The transition from the conveying elements to the first kneading element is a highly chaotic region, with much backflow from other kneading elements meeting the forward flow from the conveying. This can lead to large fluctuations at this element when there is a large variation in particle sizes. A broader size distribution leads to a slightly reduced average longitudinal velocity in the kneading elements, while the conveying elements, which are typically plug flow due to the geometric constraint, have almost identical velocities despite the variation in size distribution. This correlates well with the solid fraction: the slightly reduced longitudinal velocity increases the solid fraction and, hence, the observed residence times at the same mass flow rate. The average velocity magnitude, which includes the in-plane particle movement, remains almost identical for both distributions.

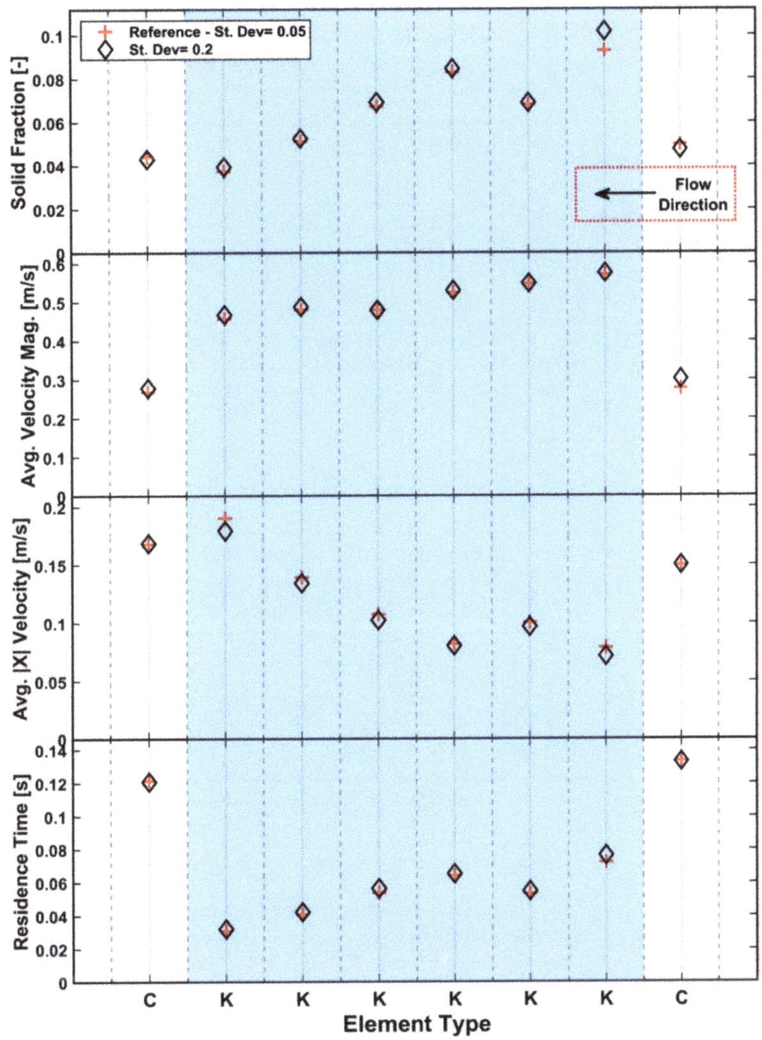

Figure 12. Effect of particle size distribution on solid fraction, velocity and residence time.

3.2. Coefficient of Restitution

The effect of damping in the simulations has been considered in terms of both particle–particle and particle–geometry collisions to establish which has the greater influence on the results. Figure 13 shows the effect of damping on the various quantities computed. Reducing the coefficient of restitution, χ, (increasing damping) from the reference value of 0.5 to 0.1 for either interparticle or particle–geometry contacts has little effect on the averaged solid fraction. Increasing χ to 0.9 for interparticle contacts similarly has little effect.

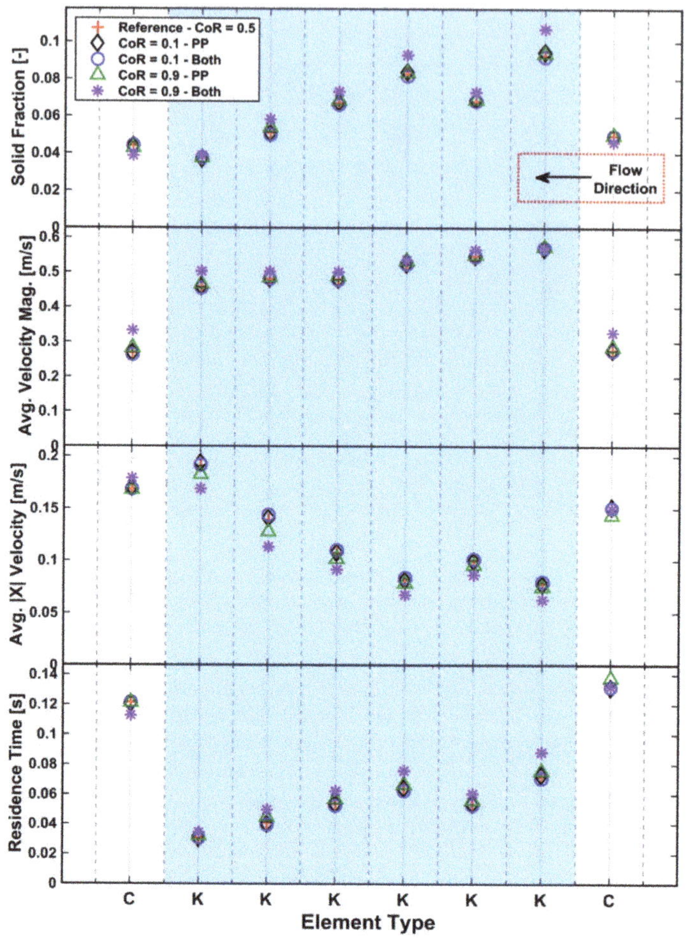

Figure 13. Effect of coefficient of restitution on solid fraction, velocity and residence time.

However, increasing χ for both the particle–geometry contacts and the interparticle contacts leads to a significant effect on the average solid fraction for both kneading and conveying elements. Low particle–geometry damping leads to more chaotic behaviour as less energy is dissipated. The increased chaotic behaviour appears to be preventing particles from freely moving axially through the granulator, shown by the reduced axial velocity for kneading elements in Figure 13, which in turn leads to higher residence times for the kneading elements. While the longitudinal velocity is decreased when the particle–geometry coefficient of restitution is increased, the opposite trend is seen in the average velocity magnitude. This shows that there is significantly increased in-plane velocities, which leads to the larger velocity magnitude. Provided that a sensible value is chosen, the TSG model is relatively insensitive to the amount of damping between particles or geometries, with the particle–geometry coefficient of restitution being the more influential. This effect is likely to be relative to the screw speed, with higher screw speeds likely to exaggerate the effect.

3.3. Static & Rolling Friction Coefficients

The static friction coefficient, μ_s, is often a key parameter as it makes a significant contribution to the generation of shear strength in DEM simulations. The reference case used a relatively high value of $\mu_s = 0.5$. Increasing μ_s beyond this value will not cause very large changes as the effect of particle friction tends to saturate. This limited study investigated the effect of friction by decreasing μ_s to 0.1 for both particle–particle ($\mu_{s,pp}$) and particle–geometry ($\mu_{s,pg}$) contacts, individually and in tandem. When the averaged solid fraction for the four cases is studied (Figure 14), the results fall into two distinct groups distinguished by the $\mu_{s,pg}$ values. A lower friction coefficient leads to significantly lower solid fractions across all elements, particularly for the kneading elements. The trend is less well defined for the first and second kneading elements due to the nearby transition from conveying to kneading. The division into two groups defined by $\mu_{s,pg}$ is even more apparent in the longitudinal velocity (also in Figure 14). Larger $\mu_{s,pg}$ leads to much lower longitudinal velocities as the increased friction leads to increased shear along the barrel surface, retarding flow. This also increases the residence time for the higher $\mu_{s,pg}$. As particles' longitudinal velocity reduce and they spend longer on each element, the in-plane velocity also increases due to the longer time spent being rotated by the element. This leads to an increase in the velocity magnitude, especially for the kneading element at which the longitudinal velocity is the lowest. In general, $\mu_{s,pp}$ has minimal influence in the TSG, which appears to be a system dominated by geometry interactions and geometry dynamics.

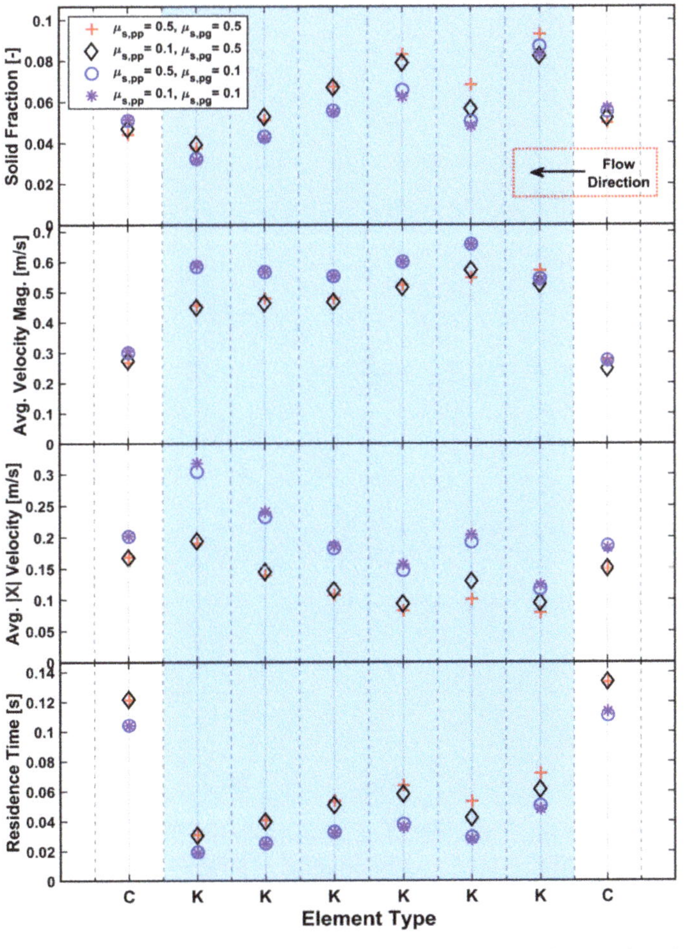

Figure 14. Effect of static friction coefficient on solid fraction, velocity and residence time.

Rolling friction is a commonly used DEM approach to incorporate the effect of particle shape without the computational expense of simulating non-spherical particles. It works by applying an additional torque to resist particle rolling. The reference case used a minimal value of 0.01 as the rolling friction coefficient. In the same manner as for static friction, the opposite extreme is explored for particle–particle ($\mu_{r,pp}$) and particle–geometry ($\mu_{r,pg}$) coefficients to assess the effect on the TSG model.

Figure 15 shows that rolling friction has a minimal effect on the average solid fraction on most elements, except for the first and second kneading elements. More particles are held up on these elements when either rolling friction coefficient increases. Increasing rolling friction leads to reduced longitudinal velocities across all elements. Increasing $\mu_{r,pp}$ and $\mu_{r,pg}$ has a similar effect, although the former seems to have a greater significance: the longitudinal velocity reduced more for particle–particle contacts than for particle–geometry contacts. The combination of both leads to the largest reduction in velocity. The average residence time (Figure 15) is strongly linked to the average longitudinal velocity and displays the same trends, with $\mu_{r,pp}$ having the most significant effect.

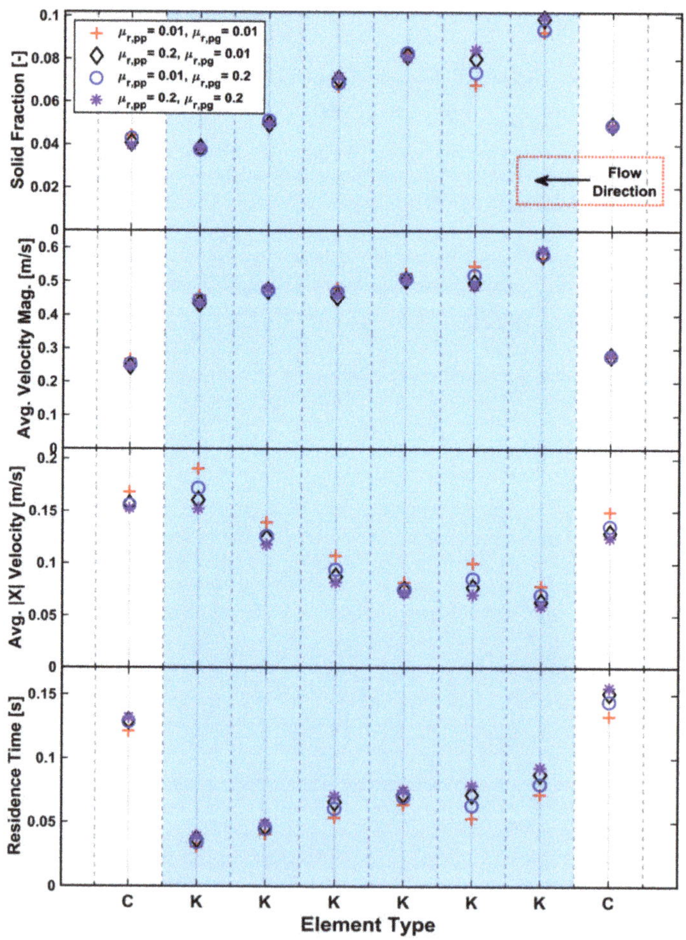

Figure 15. Effect of rolling friction coefficient on solid fraction, velocity and residence time.

3.4. Effect of Cohesion

The introduction of cohesion into the system is required to capture the behaviour of the real material, which will have significant levels of cohesion arising from the added liquid binder. The additional cohesion leads to dramatically different behaviour (Figure 16) in comparison to the previously explored cohesionless material as agglomerates begin to form in the granulator. There is a significant increase in the measured solid fraction for the first two kneading elements; thereafter, the solid fraction reduces to be much closer to the cohesionless reference case. The conveying elements show a significant decrease in solid fraction, which is a consequence of the mass balance requirement of a periodic system. With the initial mass fixed (based on the cohesionless steady-state mass), a significant increase in solid fraction in one area must be compensated by decreases in solid fraction elsewhere. In this case, cohesion has significantly decreased the mass flow rate on the conveying element from the expected 14.4 kg/h.

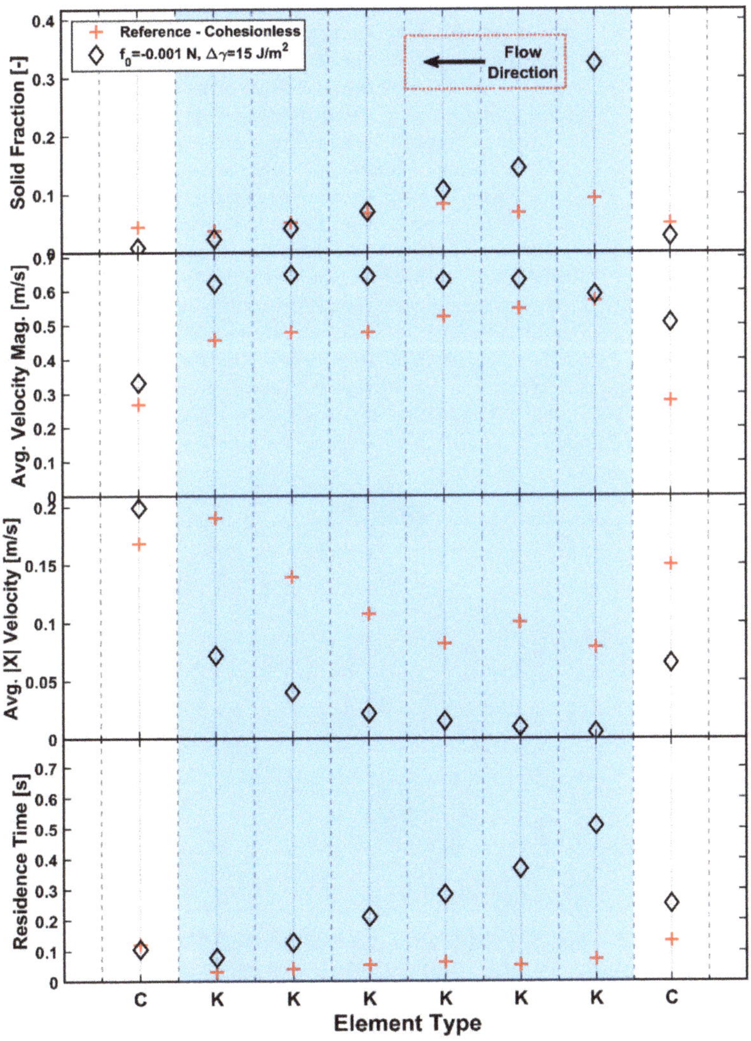

Figure 16. Effect of level of cohesion on solid fraction, velocity and residence time.

The average longitudinal velocities also decrease when cohesion is introduced, as it is now more difficult for particles to flow forward individually. The effect of cohesion on the longitudinal velocity may also be enhanced as a side-effect of the periodic system. As the mass balance requirements have led to a drop in the mass flow rate on the conveying input section, there is now less forward momentum pushing particles through the kneading zone, which leads to the significant drop in longitudinal velocity. As a result of the reduction in longitudinal velocities, the average residence times (Figure 16) increase significantly with increasing cohesion.

In terms of solid distribution on the elements, which is displayed in Figure 7 for the reference case of 400 μm cohesionless particles, a preliminary study on 200 μm particles (not included in this paper) indicates that, for the conveying elements, the introduction of cohesion reduces the amount of material transported along the bottom of the barrel. With cohesion, a greater proportion of the material is caught on the intermeshing region between the screws due to the formation of agglomerates. The kneading elements have the same flow patterns with and without cohesion. However, the solid fraction increases on average across the section, particularly at the outer extremes of the element and at the intermeshing zone due to the greater amount of material build-up on these elements.

As previously mentioned, in order to study the formation of granules, a non-periodic reduced domain model is required to prevent recycling of formed granules. This process is shown in Figure 17 where virgin material is generated at the rightmost side of the domain. In this model some small agglomerates are formed on the intermeshing region of the screws and transported towards the kneading elements. The kneading elements create a build-up of material and lead to significant consolidation in this zone, helping to create dense agglomerates. As these dense agglomerates leave the kneading zone, there is a slight chopping action of the screw element in the intermeshing region, which helps break up the dense agglomerates and create some independent agglomerates. These agglomerates are still quite large at this point and would most likely be significantly reduced in size by the chopping elements (not included in this reduced domain model) normally included at the TSG outlet. The inclusion of cohesion in the model is highly significant in capturing the natural formation of granules in a TSG.

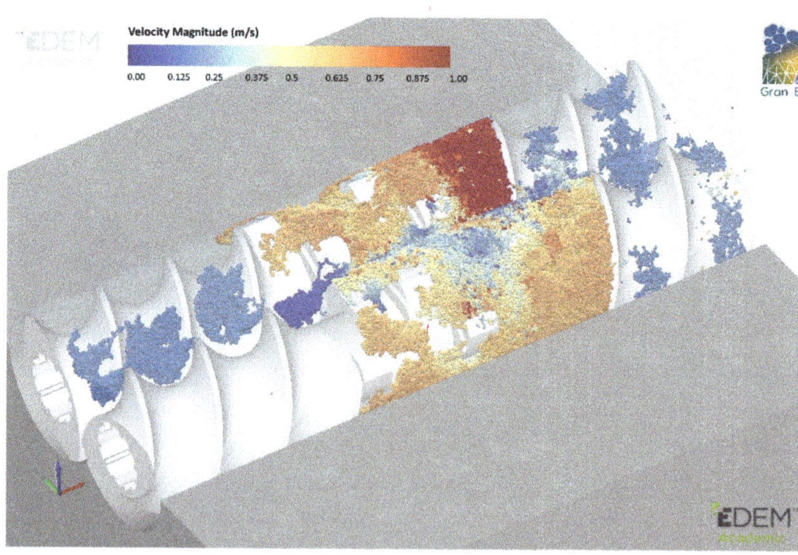

Figure 17. Agglomerate formation in the non-periodic reduced domain model (direction of flow: right to left).

Capturing the formation of agglomerates within the TSG is a key aspect of the model, and this is accomplished though the inclusion of the cohesive forces, which represent the liquid binder in the real granulator. No changes in primary particle sizes occur in the model; instead, primary particles agglomerate because of the cohesive forces.

4. Discussion

In this study, the reference case was the full-scale, cohesionless system, which had a mass flow rate set at 14.4 kg/h. The initial mass of the periodic system was calculated from the steady-state operation of the full-size model. The computationally efficient periodic system is then used extensively to assess how the DEM input parameters would influence the behaviour in the granulator. The results in Section 3 have considered how the particle dynamics in the various elements were affected in terms of velocity, solid fraction and residence time. However, due to the periodic system having a fixed mass, it is possible to measure the change in mass flow rate on the conveying elements at the inlet and outlet of the periodic domain. Any significant changes to the bulk behaviour will result in variations in the observed mass flow rates and a difference from the full-scale model value, which can be used to assess the magnitude of a parameter effect. The larger the difference is, the more influential that parameter is.

To consider this effect and the sensitivity of the mass flow rate to the parameters, the mass flow rates were calculated at sensors (MFS_1 and MFS_2) at each end of the periodic system and plotted for each case in Figure 18. Steady state exists in the system when the averaged mass flow rate is approximately equal for both the inlet and outlet sensor.

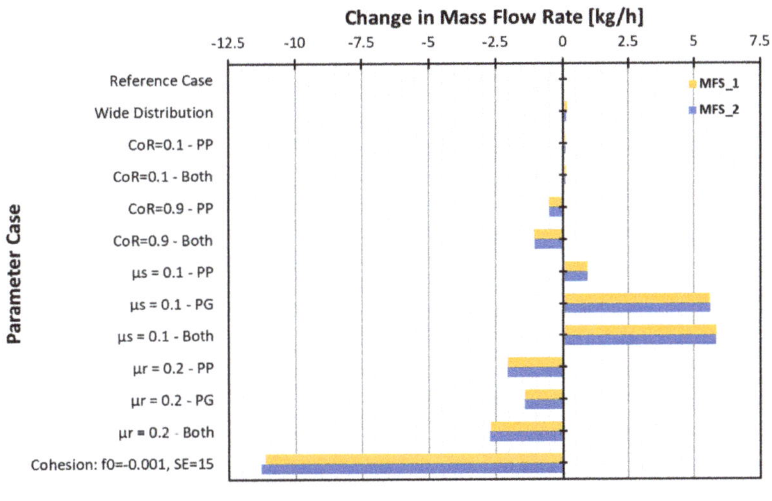

Figure 18. Observed changes in mass flow rate in the periodic system.

For a changing size distribution and low restitution, there is almost no change from the expected 14.4–14.6 kg/h mass flow rate. The decrease in damping at $\chi = 0.9$ leads to slightly higher solid fractions being observed for the kneading elements, which had the effect of slightly reducing the mass flow to 13.7–14.1 kg/h.

The reduction in static friction (μ_s) led to reduced solid fractions, which manifested as a significant increase in the mass flow rate of approximately 5 kg/h to just over 20 kg/h. Increasing rolling friction, μ_r, led to an increase in solid fraction for some elements, which led to a reduction in the mass flow rate to approximately 12–13.3 kg/h depending on the combination of parameters used. The effect of cohesion is the most dramatic, with the mass flow rates reduced by just over 10 kg/h for the cohesive case. This is due to the significant agglomeration and build-up of solids on the kneading elements, which reduces the circulating mass in the periodic system. The dominant effect of cohesion explains why it was considered separately from other parameters, allowing the relative importance of each to be quantified without being outweighed by the more dominant parameters.

5. Conclusions

Discrete Element Method (DEM) simulations were used in this study of the particle dynamics that occur in a twin-screw granulator comprising both conveying and kneading elements. Both full-scale and reduced periodic models were used.

The DEM simulations for a cohesionless system provide a mean residence time—a key characteristic of a TSG—that is in line with expectations when compared to experimental results with various different granulators and cohesive materials. The residence time increased, as expected, when an adhesive, elasto-plastic contact model was used to capture the effect of liquid binder and the resulting agglomeration in the granulator.

The significant computational expense of the full-scale simulations was reduced through (carefully defined) periodic simulations. Results show that the residence time distributions across elements in a periodic simulation are almost identical to their counterparts in the full-scale simulation. These reduced domain models allow the key particle dynamics to be captured at a much lower computational cost, making larger studies more feasible in the future. These models, combined with a DEM contact model that includes cohesion, are an effective way to simulate and investigate the wet granulation process.

A sensitivity study showed that the level of cohesion is a key determinant of granule size and must be carefully calibrated using experiments. An important finding is that the size of the fundamental particles in the simulation may be larger than in reality while still capturing the correct dynamics for the TSG. The results suggest the existence of a critical ratio of particle diameter to the screw void space (effectively the difference between the screw's root and outer radii). Once the particle size is below a certain threshold, the velocity, solid fraction and residence time are largely unaffected by particle size, with the exception of the first kneading element where there is the transition from the conveying elements. This is a highly chaotic region that appears to be heavily influenced by particle size. This effect was also noted when a broader particle size distribution was used; the increased numbers of larger particles lead to slightly increased solid fractions and residence times on the first kneading element only. The effect of the screw–barrel clearance is also found to be negligible with particles that are both smaller and larger than this critical size behaving the same in most locations in the granulator.

The effect of restitution coefficient (damping) was relatively insignificant for materials not considered to be elastic materials. The particle–geometry damping is more influential than the particle–particle damping, which is a feature of a system where geometries are moving at very high speeds, leading to high relative impact velocities. Similar trends were seen for static friction: reducing the particle–geometry static friction leads to significant changes in the solid fraction and residence times. Rolling friction has a lesser effect than static friction, in general. Increased cohesion leads to increased solid fractions and residence times on kneading elements. The results suggest that the TSG is a system that is dominated by geometry interactions, with velocities very much controlled by the screw speed and pitch.

Author Contributions: Conceptualization, J.P.M., K.J.H. and J.Y.O.; methodology, J.P.M., K.J.H. and J.Y.O.; software, J.P.M.; validation, J.P.M., K.J.H. and J.Y.O.; formal analysis, J.P.M.; visualization, J.P.M.; investigation, J.P.M.; data curation, J.P.M.; writing—original draft preparation, J.P.M.; writing—review and editing, J.P.M., K.J.H. and J.Y.O.; supervision, J.Y.O.; project administration, J.Y.O.; resources, J.Y.O.; funding acquisition, J.Y.O. All authors have read and agreed to the published version of the manuscript.

Funding: This research was funded by the UK's Centre for Process Innovation (CPI) as part of the project "Models for Manufacturing of Particulate Products" (MMPP).

Institutional Review Board Statement: Not applicable.

Informed Consent Statement: Not applicable.

Data Availability Statement: The data presented in this study are openly available on Edinburgh DataShare at https://doi.org/10.7488/ds/3251. Data citation: Morrissey, John P; Hanley, Kevin; Ooi, Jin Y. (2021). Data from "Conceptualisation of an efficient particle-based simulation of a twin-screw granulator", [dataset]. University of Edinburgh. School of Engineering. Institute for Infrastructure & Environment. https://doi.org/10.7488/ds/3251.

Acknowledgments: The authors acknowledge the contribution made by the entire Models for Manufacturing of Particulate Products (MMPP) consortium of eight partners (AstraZeneca, Pfizer Inc., Johnson Matthey, Proctor & Gamble, The University of Sheffield, The University of Edinburgh, Altair EDEM (DEM Solutions Ltd.) and PSE Ltd.), supported by the CPI. In particular, we thank Gavin Reynolds and Sean Clifford from AstraZeneca for their assistance.

Conflicts of Interest: The authors declare no conflict of interest. The funders had no role in the design of the study; in the collection, analyses, or interpretation of data; in the writing of the manuscript, or in the decision to publish the results.

References

1. Cameron, I.T.; Wang, F.Y.; Immanuel, C.D.; Stepanek, F. Process Systems Modelling and Applications in Granulation: A Review. *Chem. Eng. Sci.* **2005**, *60*, 3723–3750. [CrossRef]
2. Barrasso, D.; Ramachandran, R. Multi-Scale Modeling of Granulation Processes: Bi-Directional Coupling of PBM with DEM via Collision Frequencies. *Chem. Eng. Res. Des.* **2015**, *93*, 304–317. [CrossRef]
3. Wang, F.Y.; Cameron, I.T. Review and Future Directions in the Modelling and Control of Continuous Drum Granulation. *Powder Technol.* **2002**, *124*, 238–253. [CrossRef]
4. Department of Health and Human Services. *Pharmaceutical CGMPs for the 21st Century—A Risk-Based Approach*; FDA: Silver Spring, MD, USA, 2004.
5. Litster, J.D. Scaleup of Wet Granulation Processes: Science Not Art. *Powder Technol.* **2003**, *130*, 35–40. [CrossRef]
6. Liu, L.; Litster, J.D.; Iveson, S.; Ennis, B. Coalescence of Deformable Granules in Wet Granulation Processes. *AIChE J.* **2000**, *46*, 529–539. [CrossRef]
7. Lee, K.T.; Ingram, A.; Rowson, N.A. Twin Screw Wet Granulation: The Study of a Continuous Twin Screw Granulator Using Positron Emission Particle Tracking (PEPT) Technique. *Eur. J. Pharm. Biopharm.* **2012**, *81*, 666–673. [CrossRef]
8. Li, H.; Thompson, M.R.; O'Donnell, K.P. Understanding Wet Granulation in the Kneading Block of Twin Screw Extruders. *Chem. Eng. Sci.* **2014**, *113*, 11–21. [CrossRef]
9. Iveson, S.M.; Litster, J.D.; Hapgood, K.P.; Ennis, B.J. Nucleation, Growth and Breakage Phenomena in Agitated Wet Granulation Processes: A Review. *Powder Technol.* **2001**, *117*, 3–39. [CrossRef]
10. Kayrak-Talay, D.; Dale, S.; Wassgren, C.R.; Litster, J.D. Quality by Design for Wet Granulation in Pharmaceutical Processing: Assessing Models for a Priori Design and Scaling. *Powder Technol.* **2013**, *240*, 7–18. [CrossRef]
11. El Hagrasy, A.S.; Litster, J.D. Granulation Rate Processes in the Kneading Elements of a Twin Screw Granulator. *AIChE J.* **2013**, *59*, 4100–4115. [CrossRef]
12. Hapgood, K.P.; Lveson, S.M.; Litster, J.D.; Liu, L.X. Granulation Rate Processes. In *Handbook of Powder Technology*; Salman, A.D., Hounslow, M.J., Seville, J.P.K., Eds.; Elsevier Science B.V.: Amsterdam, The Netherlands, 2007; Volume 11, pp. 897–977. ISBN 9780444518712.
13. Gamlen, M.J.; Eardley, C. Continuous Extrusion Using a Raker Perkins MP50 (Multipurpose) Extruder. *Drug Dev. Ind. Pharm.* **1986**, *12*, 1701–1713. [CrossRef]
14. Lindberg, N.-O.; Tufvesson, C.; Holm, P.; Olbjer, L. Extrusion of an Effervescent Granulation with a Twin Screw Extruder, Baker Perkins MPF 50 D. Influence on Intragranular Porosity and Liquid Saturation. *Drug Dev. Ind. Pharm.* **1988**, *14*, 1791–1798. [CrossRef]
15. Ghebre-Sellassie, I.; Mollan, M.J.; Pathak, N.; Lodaya, M.; Fessehaie, M. Continuous Production of Pharmaceutical Granulation 2000. U.S. Patent No. 6,499,984, 31 December 2002.
16. Seem, T.C.; Rowson, N.A.; Ingram, A.; Huang, Z.; Yu, S.; de Matas, M.; Gabbott, I.; Reynolds, G.K. Twin Screw Granulation—A Literature Review. *Powder Technol.* **2015**, *276*, 89–102. [CrossRef]
17. Thompson, M.R. Twin Screw Granulation—Review of Current Progress. *Drug Dev. Ind. Pharm.* **2015**, *41*, 1223–1231. [CrossRef]
18. Keleb, E.I.; Vermeire, A.; Vervaet, C.; Remon, J.P. Extrusion Granulation and High Shear Granulation of Different Grades of Lactose and Highly Dosed Drugs: A Comparative Study. *Drug Dev. Ind. Pharm.* **2004**, *30*, 679–691. [CrossRef]
19. Keleb, E.I.; Vermeire, A.; Vervaet, C.; Remon, J.P. Twin Screw Granulation as a Simple and Efficient Tool for Continuous Wet Granulation. *Int. J. Pharm.* **2004**, *273*, 183–194. [CrossRef]
20. Rauwendaal, C. The Geometry of Self-Cleaning Twin-Screw Extruders. *Adv. Polym. Technol.* **1996**, *15*, 127–133. [CrossRef]
21. Rauwendaal, C. Twin Screw Extruders. In *Polymer Extrusion*; Carl Hanser Verlag GmbH & Co. KG: München, Germny, 2014; pp. 697–761.

22. Rauwendaal, C. Extruder Screw Design. In *Polymer Extrusion*; Carl Hanser Verlag GmbH & Co. KG: München, Germny, 2014; pp. 509–652.
23. Barrasso, D.; el Hagrasy, A.; Litster, J.D.; Ramachandran, R. Multi-Dimensional Population Balance Model Development and Validation for a Twin Screw Granulation Process. *Powder Technol.* **2015**, *270*, 612–621. [CrossRef]
24. Djuric, D. *Continuous Granulation with a Twin-Screw Extruder*; Cuvillier Verlag: Göttingen, Germany, 2008; ISBN 3867276439.
25. Djuric, D.; Kleinebudde, P. Impact of Screw Elements on Continuous Granulation with a Twin-Screw Extruder. *J. Pharm. Sci.* **2008**, *97*, 4934–4942. [CrossRef]
26. Dhenge, R.M.; Cartwright, J.J.; Hounslow, M.J.; Salman, A.D. Twin Screw Wet Granulation: Effects of Properties of Granulation Liquid. *Powder Technol.* **2012**, *229*, 126–136. [CrossRef]
27. Dhenge, R.M.; Cartwright, J.J.; Doughty, D.G.; Hounslow, M.J.; Salman, A.D. Twin Screw Wet Granulation: Effect of Powder Feed Rate. *Adv. Powder Technol.* **2011**, *22*, 162–166. [CrossRef]
28. Dhenge, R.M.; Fyles, R.S.; Cartwright, J.J.; Doughty, D.G.; Hounslow, M.J.; Salman, A.D. Twin Screw Wet Granulation: Granule Properties. *Chem. Eng. J.* **2010**, *164*, 322–329. [CrossRef]
29. Dhenge, R.M.; Washino, K.; Cartwright, J.J.; Hounslow, M.J.; Salman, A.D. Twin Screw Granulation Using Conveying Screws: Effects of Viscosity of Granulation Liquids and Flow of Powders. *Powder Technol.* **2013**, *238*, 77–90. [CrossRef]
30. Lute, S.v.; Dhenge, R.M.; Hounslow, M.J.; Salman, A.D. Twin Screw Granulation: Understanding the Mechanism of Granule Formation along the Barrel Length. *Chem. Eng. Res. Des.* **2016**, *110*, 43–53. [CrossRef]
31. Lute, S.v.; Dhenge, R.M.; Salman, A.D. Twin Screw Granulation: An Investigation of the Effect of Barrel Fill Level. *Pharmaceutics* **2018**, *10*, 67. [CrossRef]
32. Lute, S.v.; Dhenge, R.M.; Salman, A.D. Twin Screw Granulation: Effects of Properties of Primary Powders. *Pharmaceutics* **2018**, *10*, 68. [CrossRef]
33. Vercruysse, J.; Córdoba Díaz, D.; Peeters, E.; Fonteyne, M.; Delaet, U.; van Assche, I.; de Beer, T.; Remon, J.P.; Vervaet, C. Continuous Twin Screw Granulation: Influence of Process Variables on Granule and Tablet Quality. *Eur. J. Pharm. Biopharm.* **2012**, *82*, 205–211. [CrossRef] [PubMed]
34. Vercruysse, J.; Delaet, U.; van Assche, I.; Cappuyns, P.; Arata, F.; Caporicci, G.; de Beer, T.; Remon, J.P.; Vervaet, C. Stability and Repeatability of a Continuous Twin Screw Granulation and Drying System. *Eur. J. Pharm. Biopharm.* **2013**, *85*, 1031–1038. [CrossRef] [PubMed]
35. Fonteyne, M.; Vercruysse, J.; de Leersnyder, F.; Besseling, R.; Gerich, A.; Oostra, W.; Remon, J.P.; Vervaet, C.; de Beer, T. Blend Uniformity Evaluation during Continuous Mixing in a Twin Screw Granulator by In-Line NIR Using a Moving F-Test. *Anal. Chim. Acta* **2016**, *935*, 213–223. [CrossRef] [PubMed]
36. Kyttä, K.M.; Lakio, S.; Wikström, H.; Sulemanji, A.; Fransson, M.; Ketolainen, J.; Tajarobi, P. Comparison between Twin-Screw and High-Shear Granulation—The Effect of Filler and Active Pharmaceutical Ingredient on the Granule and Tablet Properties. *Powder Technol.* **2020**, *376*, 187–198. [CrossRef]
37. Ryckaert, A.; Stauffer, F.; Funke, A.; Djuric, D.; Vanhoorne, V.; Vervaet, C.; de Beer, T. Evaluation of Torque as an In-Process Control for Granule Size during Twin-Screw Wet Granulation. *Int. J. Pharm.* **2021**, *602*, 120642. [CrossRef]
38. Vercruysse, J.; Toiviainen, M.; Fonteyne, M.; Helkimo, N.; Ketolainen, J.; Juuti, M.; Delaet, U.; van Assche, I.; Remon, J.P.; Vervaet, C.; et al. Visualization and Understanding of the Granulation Liquid Mixing and Distribution during Continuous Twin Screw Granulation Using NIR Chemical Imaging. *Eur. J. Pharm. Biopharm.* **2014**, *86*, 383–392. [CrossRef]
39. Kumar, A.; Vercruysse, J.; Toiviainen, M.; Panouillot, P.E.; Juuti, M.; Vanhoorne, V.; Vervaet, C.; Remon, J.P.; Gernaey, K.v.; de Beer, T.; et al. Mixing and Transport during Pharmaceutical Twin-Screw Wet Granulation: Experimental Analysis via Chemical Imaging. *Eur. J. Pharm. Biopharm.* **2014**, *87*, 279–289. [CrossRef]
40. el Hagrasy, A.S.; Hennenkamp, J.R.; Burke, M.D.; Cartwright, J.J.; Litster, J.D. Twin Screw Wet Granulation: Influence of Formulation Parameters on Granule Properties and Growth Behavior. *Powder Technol.* **2013**, *238*, 108–115. [CrossRef]
41. Li, J.; Pradhan, S.U.; Wassgren, C.R. Granule Transformation in a Twin Screw Granulator: Effects of Conveying, Kneading, and Distributive Mixing Elements. *Powder Technol.* **2019**, *346*, 363–372. [CrossRef]
42. Liu, H.; Ricart, B.; Stanton, C.; Smith-Goettler, B.; Verdi, L.; O'Connor, T.; Lee, S.; Yoon, S. Design Space Determination and Process Optimization in At-Scale Continuous Twin Screw Wet Granulation. *Comput. Chem. Eng.* **2019**, *125*, 271–286. [CrossRef]
43. Da Tu, W.; Ingram, A.; Seville, J. Regime Map Development for Continuous Twin Screw Granulation. *Chem. Eng. Sci.* **2013**, *87*, 315–326. [CrossRef]
44. Mishra, B.K.; Thornton, C.; Bhimji, D. A Preliminary Numerical Investigation of Agglomeration in a Rotary Drum. *Miner. Eng.* **2002**, *15*, 27–33. [CrossRef]
45. Liu, P.Y.; Yang, R.Y.; Yu, A.B. Particle Scale Investigation of Flow and Mixing of Wet Particles in Rotating Drums. *Chem. Eng. Sci.* **2013**, *86*, 99–107. [CrossRef]
46. Goldschmidt, M.J.v.; Weijers, G.G.C.; Boerefijn, R.; Kuipers, J.A.M. Discrete Element Modelling of Fluidised Bed Spray Granulation. *Powder Technol.* **2003**, *138*, 39–45. [CrossRef]
47. Kafui, D.K.; Thornton, C. Fully-3D DEM Simulation of Fluidised Bed Spray Granulation Using an Exploratory Surface Energy-Based Spray Zone Concept. *Powder Technol.* **2008**, *184*, 177–188. [CrossRef]
48. Kafui, K.D.; Thornton, C.; Adams, M.J. Discrete Particle-Continuum Fluid Modelling of Gas-Solid Fluidised Beds. *Chem. Eng. Sci.* **2002**, *57*, 2395–2410. [CrossRef]

49. Gantt, J.A.; Gatzke, E.P. High-Shear Granulation Modeling Using a Discrete Element Simulation Approach. *Powder Technol.* **2005**, *156*, 195–212. [CrossRef]
50. Hassanpour, A.; Pasha, M.; Susana, L.; Rahmanian, N.; Santomaso, A.C.; Ghadiri, M. Analysis of Seeded Granulation in High Shear Granulators by Discrete Element Method. *Powder Technol.* **2013**, *238*, 50–55. [CrossRef]
51. Nakamura, H.; Fujii, H.; Watano, S. Scale-up of High Shear Mixer-Granulator Based on Discrete Element Analysis. *Powder Technol.* **2013**, *236*, 149–156. [CrossRef]
52. Watson, N.J.; Povey, M.J.W.; Reynolds, G.K.; Ding, Y.; Xu, B.H. Development of a Discrete Element Model with Moving Realistic Geometry to Simulate Particle Motion in a Mi-Pro Granulator. *Comput. Chem. Eng.* **2016**, *93*, 234–247. [CrossRef]
53. Tamrakar, A.; Chen, S.W.; Ramachandran, R. A DEM Model-Based Study to Quantitatively Compare the Effect of Wet and Dry Binder Addition in High-Shear Wet Granulation Processes. *Chem. Eng. Res. Des.* **2019**, *142*, 307–326. [CrossRef]
54. Börner, M.; Michaelis, M.; Siegmann, E.; Radeke, C.; Schmidt, U. Impact of Impeller Design on High-Shear Wet Granulation. *Powder Technol.* **2016**, *295*, 261–271. [CrossRef]
55. Sen, M.; Barrasso, D.; Singh, R.; Ramachandran, R. A Multi-Scale Hybrid CFD-DEM-PBM Description of a Fluid-Bed Granulation Process. *Processes* **2014**, *2*, 89–111. [CrossRef]
56. Barrasso, D.; Eppinger, T.; Pereira, F.E.; Aglave, R.; Debus, K.; Bermingham, S.K.; Ramachandran, R. A Multi-Scale, Mechanistic Model of a Wet Granulation Process Using a Novel Bi-Directional PBM-DEM Coupling Algorithm. *Chem. Eng. Sci.* **2015**, *123*, 500–513. [CrossRef]
57. Barrasso, D.; Ramachandran, R. Qualitative Assessment of a Multi-Scale, Compartmental PBM-DEM Model of a Continuous Twin-Screw Wet Granulation Process. *J. Pharm. Innov.* **2016**, *11*, 231–249. [CrossRef]
58. Kulju, T.; Paavola, M.; Spittka, H.; Keiski, R.L.; Juuso, E.; Leiviskä, K.; Muurinen, E. Modeling Continuous High-Shear Wet Granulation with DEM-PB. *Chem. Eng. Sci.* **2016**, *142*, 190–200. [CrossRef]
59. Wang, L.G.; Morrissey, J.P.; Barrasso, D.; Slade, D.; Clifford, S.; Reynolds, G.; Ooi, J.Y.; Litster, J.D. Model Driven Design for Twin Screw Granulation Using Mechanistic-Based Population Balance Model. *Int. J. Pharm.* **2021**, *607*, 120939. [CrossRef] [PubMed]
60. Zheng, C.; Zhang, L.; Govender, N.; Wu, C.Y. DEM Analysis of Residence Time Distribution during Twin Screw Granulation. *Powder Technol.* **2021**, *377*, 924–938. [CrossRef]
61. Zheng, C.; Govender, N.; Zhang, L.; Wu, C.-Y. GPU-Enhanced DEM Analysis of Flow Behaviour of Irregularly Shaped Particles in a Full-Scale Twin Screw Granulator. *Particuology* **2021**, *61*, 30–40. [CrossRef]
62. Kumar, A.; Radl, S.; Gernaey, K.v.; de Beer, T.; Nopens, I. Particle-Scale Modeling to Understand Liquid Distribution in Twin-Screw Wet Granulation. *Pharmaceutics* **2021**, *13*, 928. [CrossRef]
63. Altair Engineering Inc. Altair EDEM, 2020.3 2021. Available online: https://www.altair.com/edem/(accessed on 7 July 2021).
64. Morrissey, J.P. Discrete Element Modelling of Iron Ore Fines to Include the Effects of Moisture and Fines. 2013. Available online: https://era.ed.ac.uk/handle/1842/8270 (accessed on 7 July 2021).
65. Thakur, S.C.; Morrissey, J.P.; Sun, J.; Chen, J.F.; Ooi, J.Y. Micromechanical Analysis of Cohesive Granular Materials Using Discrete Element Method with an Adhesive Elasto-Plastic Contact Model. *Granul. Matter* **2014**, *16*, 383–400. [CrossRef]
66. Jones, R. From Single Particle AFM Studies of Adhesion and Friction to Bulk Flow: Forging the Links. *Granul. Matter* **2003**, *4*, 191–204. [CrossRef]
67. Morrissey, J.P.; Ooi, J.Y.; Chen, J.F.F.; Tano, K.T.; Horrigmoe, G. Measurement and Prediction of Compression and Shear Behavior of Wet Iron Ore Fines. In Proceedings of the 7th World Congress on Particle Technology (WCPT7), Beijing, China, 19–22 May 2014; p. 8.
68. Morrissey, J.P.; Ooi, J.Y.; Chen, J.F. Effect of Solid Cohesion and Friction on Silo Discharge. In Proceedings of the 7th World Congress on Particle Technology (WCPT7), Beijing, China, 19–22 May 2014; p. 8.
69. Thakur, S.C.; Ahmadian, H.; Sun, J.; Ooi, J.Y. An Experimental and Numerical Study of Packing, Compression, and Caking Behaviour of Detergent Powders. *Particuology* **2014**, *12*, 2–12. [CrossRef]
70. Thakur, S.C.; Ooi, J.Y.; Ahmadian, H. Scaling of Discrete Element Model Parameters for Cohesionless and Cohesive Solid. *Powder Technol.* **2015**, *293*, 130–137. [CrossRef]
71. Pantaleev, S.; Yordanova, S.; Janda, A.; Marigo, M.; Ooi, J.Y. An Experimentally Validated DEM Study of Powder Mixing in a Paddle Blade Mixer. *Powder Technol.* **2017**, *311*, 287–302. [CrossRef]
72. Janda, A.; Ooi, J.Y. DEM Modeling of Cone Penetration and Unconfined Compression in Cohesive Solids. *Powder Technol.* **2016**, *293*, 60–68. [CrossRef]
73. Härtl, J.; Ooi, J.Y. Experiments and Simulations of Direct Shear Tests: Porosity, Contact Friction and Bulk Friction. *Granul. Matter* **2008**, *10*, 263–271. [CrossRef]
74. Lommen, S.; Schott, D.; Lodewijks, G. DEM Speedup: Stiffness Effects on Behavior of Bulk Material. *Particuology* **2014**, *12*, 107–112. [CrossRef]
75. Lee, K.T. Continuous Granulation of Pharmaceutical Powder Using a Twin Screw Granulator. 2013. Available online: https://etheses.bham.ac.uk/id/eprint/4002/ (accessed on 7 July 2021).
76. Chan Seem, T.; Rowson, N.A.; Gabbott, I.; de Matas, M.; Reynolds, G.K.; Ingram, A. Asymmetric Distribution in Twin Screw Granulation. *Eur. J. Pharm. Biopharm.* **2015**, *106*, 50–58. [CrossRef]

77. Pradhan, S.U.; Sen, M.; Li, J.; Litster, J.D.; Wassgren, C.R. Granule Breakage in Twin Screw Granulation: Effect of Material Properties and Screw Element Geometry. *Powder Technol.* **2017**, *315*, 290–299. [CrossRef]
78. Wang, L.G.; Pradhan, S.U.; Wassgren, C.; Barrasso, D.; Slade, D.; Litster, J.D. A Breakage Kernel for Use in Population Balance Modelling of Twin Screw Granulation. *Powder Technol.* **2020**, *363*, 525–540. [CrossRef]

pharmaceutics

MDPI

Article

Impact of Powder Properties on the Rheological Behavior of Excipients

Pauline H. M. Janssen [1,*], Sébastien Depaifve [2], Aurélien Neveu [2], Filip Francqui [2] and Bastiaan H. J. Dickhoff [1]

1 DFE Pharma, Klever Str. 187, 47574 Goch, Germany; bastiaan.dickhoff@dfepharma.com
2 Granutools, Rue Jean-Lambert Defrêne, 107, 4340 Awans, Belgium; sebastien.depaifve@granutools.com (S.D.); aurelien.neveu@granutools.com (A.N.); filip.francqui@granutools.com (F.F.)
* Correspondence: pauline.janssen@dfepharma.com; Tel.: +31-6-211-545-79

Abstract: With the emergence of quality by design in the pharmaceutical industry, it becomes imperative to gain a deeper mechanistic understanding of factors impacting the flow of a formulation into tableting dies. Many flow characterization techniques are present, but so far only a few have shown to mimic the die filling process successfully. One of the challenges in mimicking the die filling process is the impact of rheological powder behavior as a result of differences in flow field in the feeding frame. In the current study, the rheological behavior was investigated for a wide range of excipients with a wide range of material properties. A new parameter for rheological behavior was introduced, which is a measure for the change in dynamic cohesive index upon changes in flow field. Particle size distribution was identified as a main contributing factor to the rheological behavior of powders. The presence of fines between larger particles turned out to reduce the rheological index, which the authors explain by improved particle separation at more dynamic flow fields. This study also revealed that obtained insights on rheological behavior can be used to optimize agitator settings in a tableting machine.

Keywords: powder flow; die filling; powder rheology; dynamic cohesive index; excipients; tableting; continuous manufacturing; quality by design

Citation: Janssen, P.H.M.; Depaifve, S.; Neveu, A.; Francqui, F.; Dickhoff, B.H.J. Impact of Powder Properties on the Rheological Behavior of Excipients. *Pharmaceutics* **2021**, *13*, 1198. https://doi.org/10.3390/pharmaceutics13081198

Academic Editor: Colin Hare

Received: 30 June 2021
Accepted: 2 August 2021
Published: 4 August 2021

1. Introduction

Powders are used in many industries and in a broad range of processes and applications. Often understanding powder behavior is crucial to properly design the process and equipment [1]. In a continuous manufacturing line for example, consistent and continuous flow through the system is a critical requirement for finished product quality [2]. Flow is also relevant for manufacturing efficiency for batch processes. It determines for example whether bins can be used or hand scooping is required, to what extend product is scraped at the beginning or end of a run, and the allowable production rate of products [3].

In pharmaceutical processing, insufficient flow can lead to product quality failures, due to large weight or dosage variations [4]. Weight variations in the final dosage form can occur when the flowability of the final formulation limits the separation of small quantities of powder from the larger mass of powder. Because of powder flow limitations in this step, a slow-speed process that works well may not work at all when rates are increased [3]. The production capacity of a tablet press is directly determined by the rotation frequency of the die table and limited by powder flow into the dies [5,6].

Predicting how powder will flow into the dies is however not easy, as flowability is a complex, multidimensional property. Flowability is the result of material physical properties and the equipment used for handling, storing or processing the material [3]. The forces that influence material flowability depend on the flow field and stress state as well as on external factors, like equipment design and material, temperature, and humidity. This also explains why flow behavior (and ranking) of powders can be different between different measurement methods and applications.

Many different characterization techniques to measure flow have been developed and correlated to the powder flow behavior in different processing units [1,7]. Pharmaceutical compendial flow characterization methods include angle of repose [8,9], compressibility index [10,11], flow through an orifice [12], rotating drum [13,14], powder rheometers [15], and several variants of the shear cell tester [16]. The different flow characterization methods quantify powder flow differently, due to the differences in flow field and degree of stresses applied during measurement [17]. Stavrou et al. [18] for example showed differences in powder flow behavior when different stress levels were employed.

Although many different characterization techniques have been developed, care must be taken to select the most suitable characterization technique for a specific approach [1]. Simple flow measurements tend to fail in the prediction of powder flow into tableting dies, due to a lack of simulation of the right flow field and stress state. So far, only a few characterization techniques have been developed with specific focus to mimic the die filling process. Wu et al. [19] for example used transparent dies and moving feeding shoes to study the powder flow of different metallurgical powder components in air and vacuum. He showed that powder characteristics, shoe speed, and die geometry play an important role in the die filling process. Mendez et al. [20] used a fixed frame and a moving die disc system to examine the effect of blend composition, shoe properties and die parameters on uniformity of die filling. Mehrotra et al. [21] simulated the die filling process for cohesive materials. He concluded that cohesive powder take longer to fill dies and hence could be a potential cause of tablet weight variability.

Described research has shown to predict the tablet die filling process from a die shoe well. A common practical challenge of the described studies however, is that they all studied the die filling process with simplified systems. The effect of different flow fields in the feeding frame has not been considered. This factor can be very important in the prediction of die filling, especially when powder fluidity is influenced by the presence of paddle feeders, wipers, or agitator arms [3].

With the emergence of quality by design (QbD), it becomes imperative to gain a deeper mechanistic understanding of how different materials respond to differences in flow field [22]. The aim of this research is to investigate the effect of rheological behavior during powder flow into tablet dies. A wide range of excipients with a broad range of material properties was evaluated. A new parameter for rheological behavior is introduced, which is a measure for the change in dynamic cohesive index upon changes in stress state and flow field. The objective of this study is to identify which material properties do have an impact on this rheological index parameter. In addition, the fidelity of the rheological index was validated by correlating it to tableting performance in a rotary tablet press with agitators.

2. Materials and Methods

2.1. Materials

Anhydrous lactose (Lactopress® anhydrous, SuperTab® 21 AN, SuperTab® 22 AN, SuperTab® 24 AN), sieved lactose monohydrate (Pharmatose® 80 M), milled lactose monohydrate (Pharmatose® 150 M, Pharmatose® 200 M, Pharmatose® 450 M), modified lactose monohydrate (SuperTab® 11 SD, SuperTab® 14 SD, SuperTab® 50 ODT, SuperTab® 30 GR), microcrystalline cellulose (Pharmacel® 101, Pharmacel® 102), and superdisintegrants (Primojel®, Primellose®) were obtained from DFE Pharma (Goch, Germany).

2.2. Material Characterization

An overview of the characterization techniques, with corresponding physical properties and abbreviations is provided in Table 1.

Table 1. List of material characterization techniques, the material properties they measure and corresponding abbreviations of the measured material properties

Characterization Technique	Physical Property	Abbreviation	Range of Values	Unit
Visual observation by scanning electron microscopy	Shape	Shape	-	-
Particle size distribution (PSD) by laser diffraction	10% cumulative undersize of volumetric PSD	×10	3.0–77.6	µm
	50% cumulative undersize of volumetric PSD	×50	18.3–243	µm
	90% cumulative undersize of volumetric PSD	×90	49.4–406	µm
	Span of the volumetric PSD*	Span	1.22–2.83	-
Karl fisher titration	Total moisture content	KF	0.1–5.8	%w/w
Thermogravimetric balance	Loss on drying	LOD	0.0–9.3	%w/w
Brunauer–Emmett–Teller analysis with Krypton	Specific surface area	SSA	0.1–5.1	m^2/g
Graduated cylinder	Bulk density	BD	0.35–0.79	g/mL
	Tapped density	TD	0.49–0.98	g/mL
	Hausner ratio	HR	1.15–1.60	-
Ring shear cell tester	Flow function coefficient at 4 kPa pre-consolidation pressure	ffc	2.4–17.3	-
Electric charge analyzer	Initial charge density	q0	−1.5–0.2	nC/g
	Final charge density	qf	−1.5–0.2	nC/g
	Tribo-charging density variation	Δq	−5.0–0.1	nC/g
Gas pycnometer	True density	TrD	1.52–1.58	g/mL
Heckel testing by compaction simulation	Yield pressure at 0.01 mm/s—slow	PyS	78–229	MPa
	Yield pressure at 300 mm/s—fast	PyF	80–236	MPa
	Strain Rate Sensitivity	SRS	0–48	%
Rotating drum	Rheological index	RI	−0.54–0.86	rpm^{-1}

2.2.1. Shape

Scanning electron microscopy (SEM) images were recorded using a Phenom ProX scanning electron microscope (Thermo Fischer Scientific, Waltham, MA, USA). Prior to the measurements, samples were coated with a gold layer with a thickness of 4 nm. Images were recorded at an acceleration voltage of 10 kV. The shape of particles is defined by visual observation.

2.2.2. Laser Diffraction

Particle size distributions (PSD) were determined (n = 3) by dry laser diffraction (Helos/KR, Sympatec, Clausthal-Zellerfeld, Germany). A dry dispersion unit with a feed rate of 50% and an air pressure of 0.5 bar was used. The particle size was reported as a volume equivalent sphere diameter. The 10%, 50%, and 90% cumulative undersize of the volumetric distribution was described as ×10, ×50, and ×90 respectively. The span of the volumetric particle size indicates the width of the particle size distribution and was calculated from the indicated parameters with the equation

$$\text{Span} = (\times 90 - \times 10)/\times 50 \tag{1}$$

2.2.3. Moisture Content

The total moisture content (KF) was determined (n = 2) via Karl-Fisher titration. The loss on drying (LOD) is determined (n = 2) with a Sartorius MA150 Q moisture analyzer (Sartorius AG, Göttingen, Germany). Samples are measured according to the Ph. Eur. Methods.

2.2.4. Specific Surface Area

Specific surface area (SSA) is determined (n = 2) with a Tristar II physisorption instrument (Micromeritics, Norcross, USA) based on a static volumetric technology. Samples are degassed at 40 °C for 2 h under nitrogen flow before analysis. Krypton is used as adsorption gas to analyze samples of 1–2 g. Isotherm data were elaborated with Tristar II 3020 2.02 software (Micromeritics, Norcross, GA, USA) using a Brunauer–Emmett–Teller analysis (BET) model.

2.2.5. Bulk and Tapped Density

Bulk and tapped density were measured (n = 2) according to Ph. Eur. Method 1. 100 g of powder was poured into a 250 mL graduated cylinder mounted on an automatic tapping device STAV 2003 stampfvolumeter (Engelsmann, Ludwigshafen am Rhein, Germany). The Hausner ratio (HR) was calculated as the quotient of the tapped density (TD) and the bulk density (BD)

$$\mathrm{HR} = \mathrm{TD}/\mathrm{BD} \tag{2}$$

2.2.6. Ring Shear Testing

A ring shear tester (RST-XS, Dietmar Schulze, Wolfenbuttel, Germany) was used to measure (n = 2) the flow function coefficient (ffc). The ffc is defined as the ratio of the consolidation stress and the unconfined yield strength. Powders were measured at a pre-consolidation stress (σ_{pre}) of 4 kPa and normal stresses of 1, 2, and 3 kPa were used for shear to failure.

2.2.7. Charge Density

Triboelectric charging of powders was measured (n = 3) with a GranuCharge (GranuTools, Awans, Belgium). The initial charge was measured by introducing the powder inside the Faraday cup connected to an electrometer. The initial charge density (q_0) was calculated by dividing the net charge by the mass of the powder sample. After this measurement, powders were fed into V-shaped stainless-steel 316 L tubing using a vibratory feeder. At the end of the tubing system, samples were collected inside a Faraday cup connected to an electrometer to determine the final charge density (q_f). The tribo-charging density variation (Δq) was calculated as the difference between final (q_f) and initial charge density (q_0)

$$\Delta q = q_f - q_0 \tag{3}$$

2.2.8. True Density and Heckel Testing

The true density (TrD) of samples was determined (n = 2) with an AccuPyc II 1340 Helium pycnometer (Micromeritics, Norcross, USA). The equilibrium rate was 0.02 psig and the number of purges 5.

A compaction simulator (Phoenix, Brierley Hill, UK) with a V-shaped compaction profile and 10 mm round flat faced punches was used to perform Heckel analysis (n = 3). The punch speed for the slow evaluation was set to 0.01 mm/s and the fast at 300 mm/s. Data were analyzed by the compaction analysis software program to generate values for yield pressure (P_y) using the Heckel equation with the relative density of the compact (D), the applied pressure (P), and the gradient of the line in the linear region (k)

$$\ln(1/(1-\mathrm{D})) = \mathrm{kP} + \mathrm{A} \tag{4}$$

The strain rate sensitivity (SRS) is calculated by comparing the yield pressure at high speed (PyF) and slow speed (PyS)

$$\%SRS = (P_yF - P_yS)/(P_yS) \times 100 \quad (5)$$

2.3. Preparation of the Blends

A fine lactose grade (Pharmatose® 450 M) is blended with a coarse sieved lactose (Pharmatose® 80 M) in a Turbula blender T2 at 96 rpm for 8 min. Amounts of fines were increased in steps of 5% w/w between 0% and 40% w/w fines, and in steps of 20% w/w between 40% and 100% w/w fines.

2.4. Rotating Drum Method

The rheological behavior of different excipient grades and blends is evaluated ($n = 2$ (Note that blends with 5% w/w, 10% w/w, 15% w/w, 25% w/w, and 40% w/w fines were measured once (n = 1) instead of in duplicate. Replicates on other batches ensure robustness and repeatability of measurements)) with the rotating drum method GranuDrum (GranuTools, Awans, Belgium). Rotational speed is increased from 2–20 rpm in steps of 2 and from 20–60 rpm in steps of 5 rpm. For each rotating speed, 40 snapshots of the powder bed separated by 1 s are taken by a CCD camera. The position of the powder/air interface in these snapshots is detected by an edge detection algorithm. The average interface position is used to compute the flowing angle. The standard deviation from the temporal fluctuations of the interface is used to compute the dynamic cohesive index. A cohesive powder leads to an intermitted flow while a non-cohesive powder leads to a regular flow. Therefore, a dynamic cohesive index close to zero corresponds to a non-cohesive powder. When the powder cohesiveness increases, the cohesive index increases accordingly. The cohesive index measure has been shown in previous work to be relevant to evaluate the powder cohesiveness [23,24].

A rheological index parameter (RI) is defined as the slope of a linear fit of the dynamic cohesive index as function of rotational speed. Positive rheological index indicates shear thickening, while negative rheological index indicates shear thinning. Powder that exhibits shear thickening behavior will have increased cohesiveness at higher stresses, which is associated with a decrease of flowability. Shear thinning indicates opposite behavior.

2.5. Multivariate Analyses

Partial least squares (PLS) type MVA models were developed with Simca-P 16 software (Umetris, Umeå, Sweden). Models were developed by regressing the material property descriptors in Table 1 (X) versus the Rheological Index parameter (Y). The importance of X-variables is evaluated by a variable influence on projection (VIP). Values for the material property descriptors that are used to create PLS models are indicated in Supplementary Table S1.

2.6. Tableting

For the formulations that are tableted, 99.5% w/w filler is blended with 0.5 % w/w MgSt for 2 min in a Turbula blender T2 at 96 rpm. Blends are compressed on a RoTab rotary tableting press at 25 rpm. 9 mm flat beveled punches (iHolland) are used and compaction force is set to 10 kN. The filling depth is set to obtain tablets of 250 mg at 10 rpm agitator (optfiller) speed. The agitator speed is increased from 10–45 rpm in steps of 5 rpm to get different amounts of agitation without changing any further settings.

2.7. Tablet Testing

Tablets are analyzed on weight by using an automated tablet tester (Sotax AT50). Twenty tablets are analyzed and the average and standard deviation is reported.

3. Results and Discussion

3.1. Raw Material Characterization

A range of physical properties of the materials that are evaluated in this study are shown in Table 2. SEM pictures that are used to identify the shape of particles are provided in Supplementary Figure S1. Physical properties shown are expected to be relevant for flow behavior, although a broader set of material properties was used to characterize the materials and to investigate correlations with the Rheological Index in the PCA. Additional parameters that are measured for all materials are provided in Supplementary Table S1.

Table 2. Physical properties such as type, shape and particle size distribution, density and flow properties for the set of excipients that is used in this study

Grade	Abbreviation	Type	Shape	×10 (µm)	×50 (µm)	×90 (µm)	Span	Bulk Density (g/mL)	Hausner Ratio (-)	ffc @4 kPa (-)
Lactopress® anhydrous	LP anh	Anhydrous lactose	Shards	16.5	133	323	2.30	0.69	1.28	7.5
SuperTab® 21 AN	21 AN	Anhydrous lactose	Shards	24.1	180	387	2.02	0.72	1.27	7.7
SuperTab® 22 AN	22 AN	Anhydrous lactose	Shards	47.0	203	359	1.54	0.68	1.17	15
SuperTab® 24 AN	24 AN	(Granulated) anhydrous lactose	Granular	37.0	121	298	2.15	0.54	1.25	13
Pharmatose® 80 M	80 M	Lactose monohydrate (sieved)	Tomahawk	76.6	242	406	1.36	0.79	1.19	13
Pharmatose® 150 M	150 M	Lactose monohydrate (milled)	Tomahawk/fines	7.4	68.4	189	2.66	0.72	1.36	3.8
Pharmatose® 200 M	200 M	Lactose monohydrate (milled)	Tomahawk/fines	4.4	37.7	111	2.83	0.62	1.58	3.7
Pharmatose® 450 M	450 M	Lactose monohydrate (milled)	Fines	3.0	18.3	49.4	2.54	0.50	1.60	2.4
SuperTab® 30 GR	30 GR	Modified lactose monohydrate	Granular	38.3	126	297	2.05	0.63	1.24	17
SuperTab® 11 SD	11 SD	Modified lactose monohydrate	Spherical	44.0	119	223	1.51	0.63	1.19	17
SuperTab® 14 SD	14 SD	Modified lactose monohydrate	Spherical	47.7	124	227	1.44	0.62	1.15	14
SuperTab® 50 ODT	50 ODT	Modified lactose monohydrate	Spherical	30.9	106	199	1.58	0.71	1.17	13
Pharmacel® 101	MCC101	Microcrystalline cellulose	Spherical/Fibers	20.0	62.2	137	1.89	0.34	1.45	5.9
Pharmacel® 102	MCC102	Microcrystalline cellulose	Spherical/Fibers	29.9	86.9	200	1.95	0.33	1.39	7.0
Pharmacel® sMCC90	sMCC90	Microcrystalline cellulose, co-processed with silicon dioxide	Spherical/Fibers	29.4	102	233	1.99	0.38	1.32	9.0
Primojel®	PJ	Superdisintegrant	Spherical	21.1	42.3	72.6	1.22	0.79	1.21	12
Primellose®	PL	Superdisintegrant	Fibers	24.4	54.4	114	1.65	0.55	1.35	7.4

The used set of materials covers a large variation in excipient type, particle shape, particle size distribution, density, and powder flow parameters. Four grades (4) of anhydrous lactose are evaluated. SuperTab® 21 AN and Lactopress® anhydrous have a relatively large proportion of fines, as indicated by the ×10 of 15 µm. SuperTab® 22 AN contains a smaller proportion of fines and has improved flow properties according to the Hausner ratio and the flow function coefficient. SuperTab® 24 AN has also improved flow properties and a slightly different shape and lower density. Four grades of non-modified lactose monohydrate are evaluated. Pharmatose® 80 M is a sieved tomahawk shaped material, with a relatively large particle size. Pharmatose® 150 M, Pharmatose® 200 M and Pharmatose® 450 M are milled materials. Particle size, density, and flow properties of these grades decrease with increasing number. The four evaluated modified lactose monohydrate grades consist of three spray dried grades with slightly different particle size distribution and density, and one granulated lactose monohydrate. Flow parameters of these grades all indicate very good flowability. More irregular shaped particles that are evaluated are microcrystalline cellulose and Primellose®. Pharmacel® 101 and Pharmacel® 102 are two microcrystalline cellulose grades with spherical morphology consisting of fibers. Particle size of Pharmacel® 101 is smaller than for Pharmacel® 102, which also explains the reduced flow properties. Pharmacel® sMCC90 is very similar to Pharmacel® 102, but has been co-processed with 2% w/w silicon dioxide to increase the specific surface area and improve the flow properties. Primojel® and Primellose® are two superdisintegrants with small particle size and a spherical and fibrous morphology respectively. The good flow indicated by the flow parameters of Primojel®, is in line with the expectation from the spherical morphology.

Particle size distribution, particle shape and density are parameters that are known to have an impact on the different forces acting on particles during powder flow. This can be understood by looking at the different driving and drag forces that act on particles and determine the powder flow. One of the driving forces for flowability is gravity. Gravitational forces are higher for larger particles, and for particles with higher (true) density [25]. Drag forces on the other hand, typically include adhesive and cohesive forces, which produce a tendency for particles to stick to each other and to other surfaces. Adhesive and cohesive forces are composed mainly of van der Waals forces, capillary bridging, and electrostatics [26]. The magnitude of these forces depends on the nature of the material and on the available surface. Adhesive and cohesive forces will be higher for smaller or irregular shaped particles, as the available surface for these particles is higher [23]. Irregular shaped particles also can have a negative impact on flowability due to the increased risk for mechanical interlocking [27,28].

3.2. Cohesive Index as a Function of Rotational Speed—From Static to Dynamic Regimes

The rheological behavior of the different excipients was evaluated with the rotating drum method GranuDrum. GranuDrum measurements allow to follow the evolution of the flow properties as function of the flow field. At each rotational speed value, the flowing angle is computed from the average interface position, and the dynamic cohesive index is computed from the interface fluctuations. The dynamics of the flowing angle of non-cohesive granular materials is well-known and described in literature. In the considered range of rotating speed, non-cohesive materials lead to continuous flow and the shape of the interface is typically flat and easy to analyze [29]. However, in this paper, we also deal with cohesive powders, of which the flow is irregular and more complex [30,31]. In this paper, we therefore do not focus on the flowing angle as a flow parameter, but on the dynamic cohesive index (CI). The value of the cohesive index is close to zero for non-cohesive powders, and increases with the cohesion of the material.

To highlight the influence of the different flow regimes, the cohesive indices at 2 rpm and 60 rpm are compared with other flow parameters Hausner ratio and ffc from shear cell testing. Figure 1 shows the cohesive index obtained by GranuDrum measurements at 2 rpm and 60 rpm rotational speed as function of ffc and Hausner ratio. At rotational speeds of 2 rpm, the powder is close to the quasi-static regime and the flow behavior is dominated by inter-particle forces, often referred to as cohesiveness [23]. The observations and classification of the GranuDrum at this speed are in line with results obtained in the (quasi-)static flow regime in previous work [32]. Within each category of products, flow ranking based on the cohesive index at 2 rpm is in line with the flow ranking according to Hausner ratio and ffc. For anhydrous lactose rotating at 2 rpm, SuperTab® 21 AN and Lactopress® anhydrous have the highest cohesive index (37–43), while SuperTab® 22 AN and SuperTab® 24 AN have low cohesive index (18). This is in line with the order indicated by Hausner ratio and the shear cell testing at 4 kPa pre-consolidation strength with ffc values >10 for SuperTab® 22 AN and SuperTab® 24 AN and around 7.5 for the other two grades. For lactose monohydrate samples a similar trend is observed. Pharmatose® 80 M has a low cohesive index of 22 at 2 rpm rotational speed, which is in line with a low Hausner ratio and high ffc. The milled lactose monohydrate with smaller particle size all have a cohesive index above 50 at low rotational speed, which is in line with higher Hausner ratio and lower ffc. Modified lactose monohydrate grades all have a cohesive index below 20 at low rotational speeds, which is also in line with low Hausner ratio and ffc >10. For irregular shaped excipients like microcrystalline cellulose and Primellose®, the cohesive index is around 30. Primojel® and Pharmacel® sMCC90 have a cohesive index below 20, in line with the lowest Hausner ratio and highest ffc within its category. Also, over the different product groups, the variance in cohesive index at 2 rpm can be explained for over 60% by linear models of both the Hausner ratio and the flow function coefficient at 4 kPa pre-consolidation strength. The high overlap in powder ranking by these measurements is explained by similarities in the stress state and flow field during these measurements.

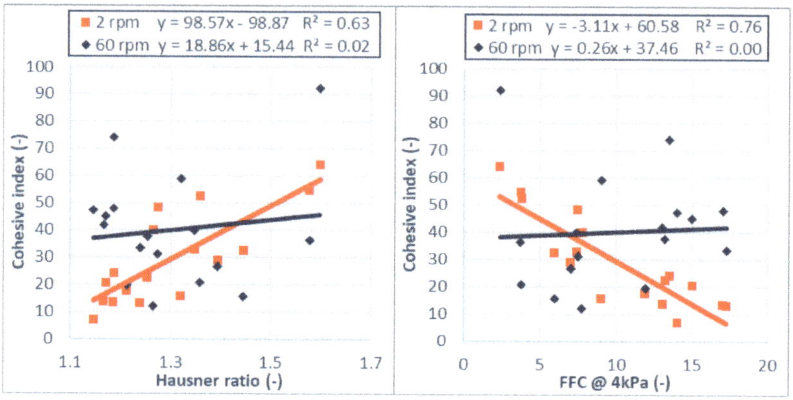

Figure 1. Cohesive index values as function of Hausner ratio (**left**) and flow function coefficient (**right**) at two different rotational speeds. The obtained cohesive index at low rotational speed (2 rpm) correlated with both measurements, while no correlation is observed for the cohesive index at high rotation speed (60 rpm).

At rotational speed of 60 rpm however, no significant correlation with the Hausner ratio or flow function coefficient can be drawn. Fitted linear models for Hausner ratio and flow function coefficient both explain the variance in cohesive index less than 2%. The lack of overlap in powder ranking by these measurements is explained by differences in the flow fields of these measurements. Hausner ratio and shear cell consider a quasi-static flow field [33], while the cohesive index at 60 rpm measures powder flow in a dynamic flow field. Observed differences are in line with Lumay et al. [23], who showed that different flow regimes were influenced by different material properties. In the quasi-static or plastic regime, velocities are small or zero and the space between neighboring particles is low. Powder behavior is dominated by inter-particle contact forces [30]. In the rapid flow regime, particles are moving so fast that friction can be neglected and only short collisions between particles determine the character of flow [34]. The differences in flow classification due to differences in stress state and flow field was also confirmed by other authors [1,3,11,17,23,26,30,34–36].

Cohesive index values as function of rotational speed for the tested materials are presented in Figure 2. First of all, it is important to mention that no irreversible rheological behavior was observed for any of the powders. This was indicated by the absence of a hysteresis between original curve, taken during the increasing rotational speed sequence (2–60 rpm) and the and the reverse curve, taken during the decreasing rotational speed sequence (60–2 rpm). The reverse curve is not shown for clarity reasons. The reversibility in the rheological behavior indicates that no irreversible changes in material properties are induced by the measurement method.

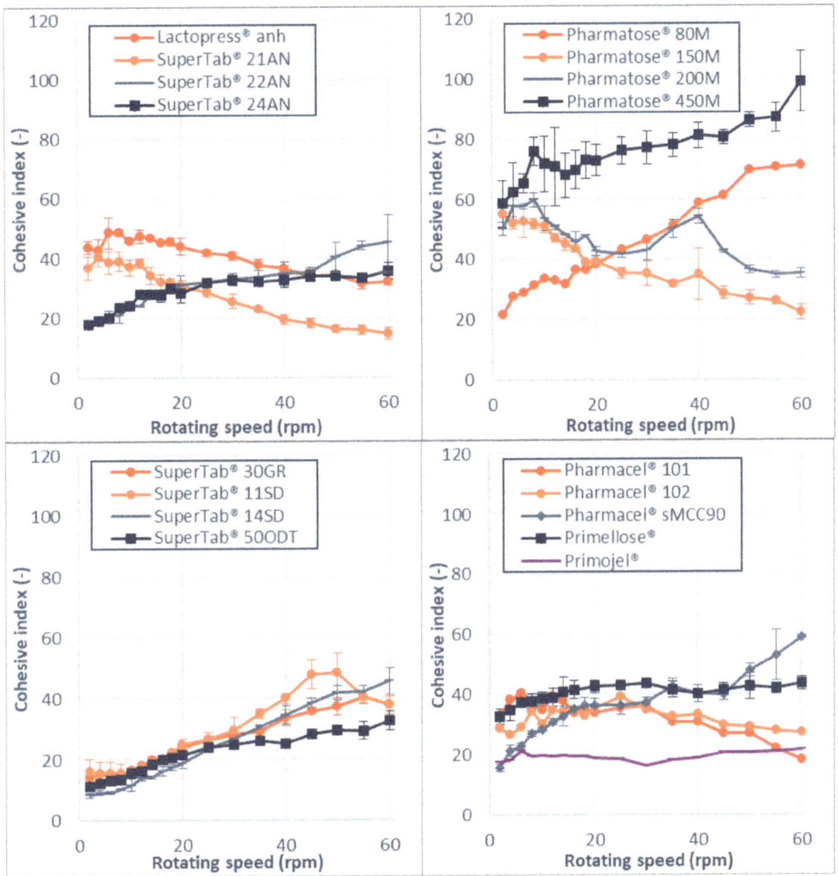

Figure 2. Cohesive index values as function of rotational speed for anhydrous lactose (**top left**), lactose monohydrate (**top right**), modified lactose monohydrate (**bottom left**) and microcrystalline cellulose/superdisintegrants (**bottom right**).

Most powders that are evaluated show shear thickening behavior. This indicates that cohesiveness increases with rotational speed, which is associated with more intermittent and irregular flow. This can be concern for the pharmaceutical industry as more cohesive powders can lead to increased variability and mass flow excursions outside the acceptable target range, as demonstrated by Allenspach et al. [37]. Notable is the shear thinning behavior, which is observed for two grades of anhydrous lactose, two grades of lactose monohydrate and microcrystalline cellulose. The shear thinning behavior can be explained by aeration of the powder at higher rotational speeds, which increases the distance between particles and thereby reduces the cohesive surface interactions [36].

The observed differences in cohesive index and rheological index for sieved (Pharmatose® 80 M) and milled lactose (Pharmatose® 150 M and Pharmatose® 200 M) are in line with findings of Hickey et al. [38] for inhalation grades lactose. These authors reported that an apparent contradiction is present in flow property parameters of milled and sieved powders. Static powder flow measurements (bulk and tapped density, angle of repose) lead to the conclusion that milled powders exhibit poor flow compared to sieved batches. Dynamic rotating drum measurements indicated the reverse, where milled powder flows better than sieved powder.

To evaluate which material properties influence the shear thinning and thickening behavior, a new parameter rheological index (RI) is introduced. The Rheological Index (RI) is defined as the linear slope of dynamic cohesive index as function of rotational speed between 2 rpm and 60 rpm. Figure 3 shows the RI values for the investigated materials. For Lactopress® anhydrous, SuperTab® 21 AN, Pharmatose® 150 M, Pharmatose® 200 M, and Pharmacel® 101 the reduction in cohesiveness due to aeration seems to outweigh the inertial effect, resulting in shear thinning behavior. For Pharmacel® 102, Primojel®, and Primellose® the rheological index is close to zero.

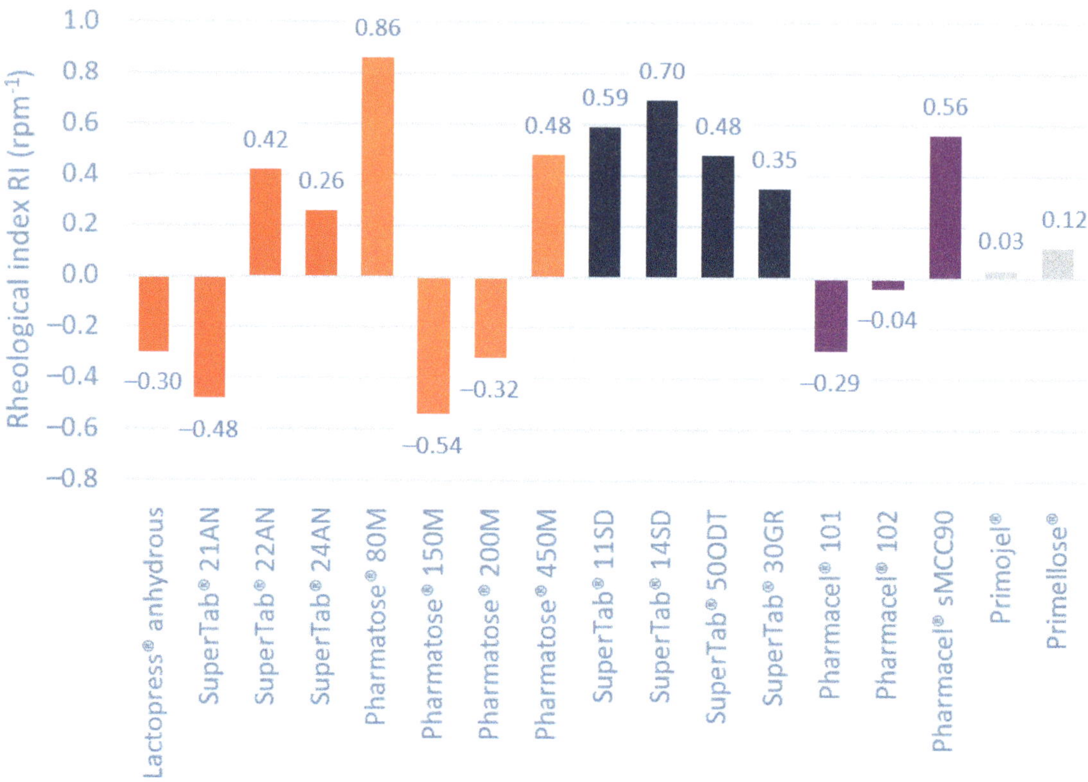

Figure 3. Rheological index (RI) values for the different materials. A positive rheological index indicates shear thickening behavior, while negative rheological index indicates shear thinning behavior. Different colours represent different types of materials.

3.3. *Multivariate Analyses to Reveal Drivers for the Rheological Index*

A partial least squares (PLS) model was created to identify which material properties impact the rheological index parameter (RI) of powders the most. In the current approach, PLS was used as exploratory technique and no optimization of the model was performed. Figure 4 shows the score and loading plot of the PLS model with two principal components. In the PLS loading plot the parameters $\times 10$ and ffc point in the same direction as the RI, indicating a positive correlation. The parameter span is indicated at the exact opposite direction, suggesting a negative correlation. In the PCA loading plot RI is pointing towards the upright direction in quadrant 1. The four modified lactose monohydrate grades and Pharmatose® 80 M are located in the PCA scoring plot in quadrant 1 towards the positive direction of rheological index. Pharmatose® 150 M, Lactopress® anhydrous and SuperTab® 21 AN are located in quadrant 3 towards the negative direction of rheological index.

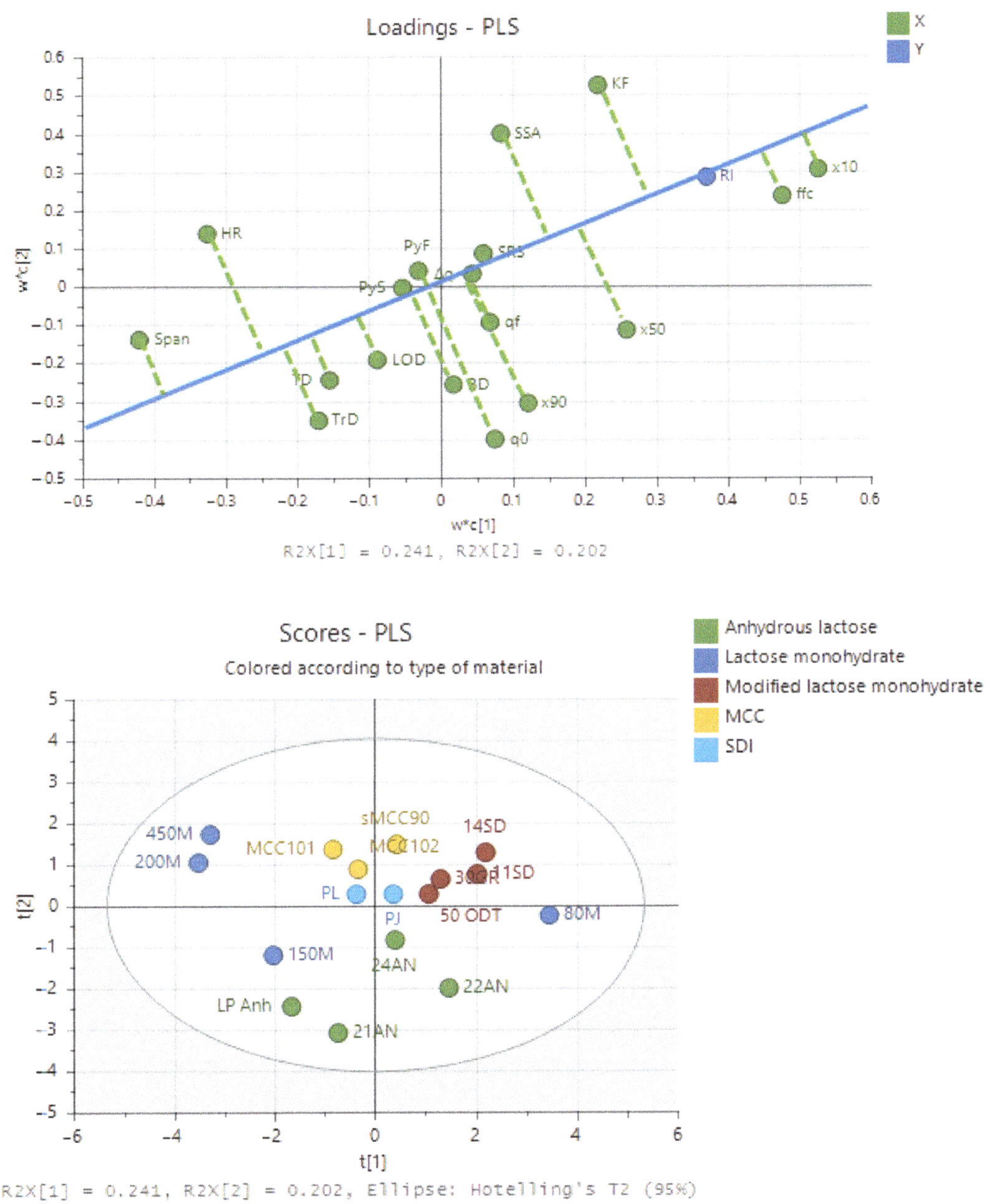

Figure 4. PLS analyses of the dataset with a loading plot (**top**) and score plot (**bottom**).

A variable influence on projection (VIP) plot is constructed to evaluate which material properties have the most significant correlation with rheological behavior. From the VIP plot in Figure 5 it can be observed that the parameters that are the most important in this model towards RI are related to the particle size distribution (×10, span) and parameters related to powder flow (ffc, HR). No relevant contribution is present for true density, yield strengths, SRS, electrostatics, tapped density, and moisture content.

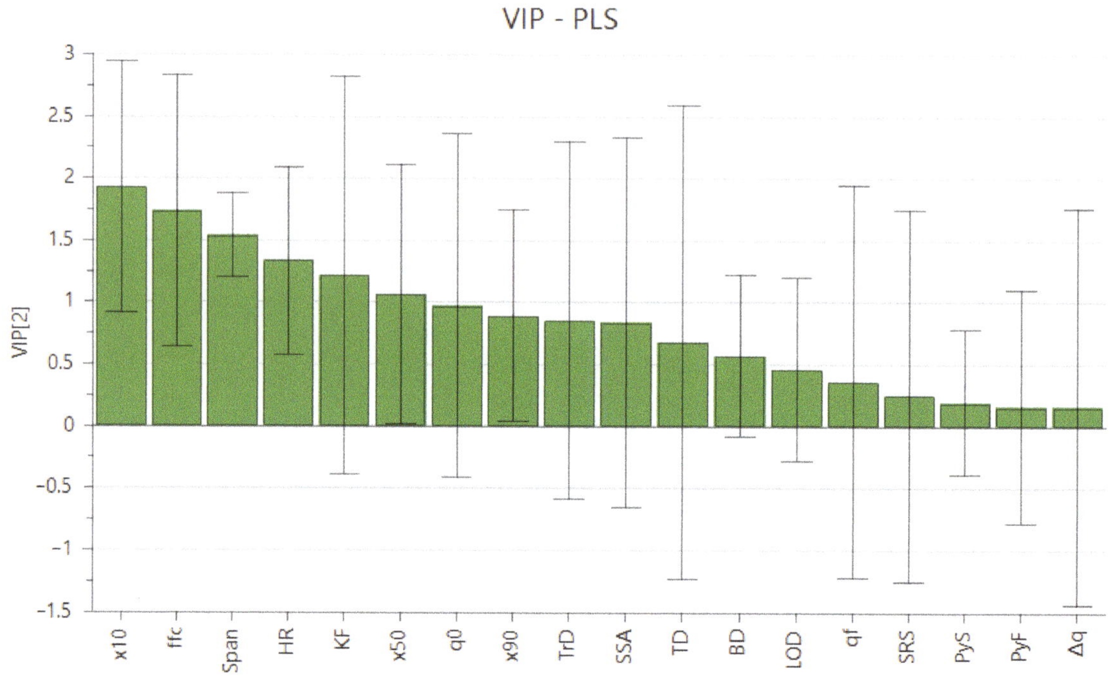

Figure 5. Variable Influence on Projection (VIP) plot indicating the relevance of terms for explaining the rheological index. Error bars indicate the 95% confidence interval.

3.4. Variation of the Amount of Fines

To test the impact of particle size (×10 and span) on the rheological index that was indicated by the PLS, blends with different particle size distributions are further evaluated. Sieved lactose with different amounts of fines added are tested with the rotating drum method. Particle size distribution parameters are indicated in Supplementary Table S2. Figure 6 shows the result of the rotating drum measurements, from which the rheological parameter is derived and summarized in Figure 7. The results of the dynamic flow measurements show that the cohesive index at low rotational speed increases with an increased fraction of fines. This can be understood by looking at the flow regime of the powder. At low rotational speed, the powder is in the quasi-static or plastic regime and the space between neighboring particles is low. Powder behavior is dominated by inter-particle contact forces, like cohesive and adhesive forces [30]. Cohesive and adhesive forces are typically higher with more fine particles present, due to the increased availability of surface and because fines support powder packing [23,38]. At the same time, flowability driving gravitational forces that act on small particles are typically lower, due to the lower mass of these particles.

At higher rotational speed, addition of 5–35% w/w of fines resulted in reduced cohesive index. This is also visible in the reduced rheological index when up to 35% w/w fines are added. Negative rheological index is observed for blends with 10–35% w/w fines. Notable is the tipping point between 35% w/w and 40% w/w of fines. At this point the cohesive index at 2 rpm remains constant, but the rheological index suddenly turns from negative to positive, resulting in significantly larger cohesive index at high rotational speeds. A tipping point in flow behavior around 30% w/w of fines was also observed before by Molerus and Nwylt in quasi-static shear cell flow measurements of binary mixtures of coarse and fine limestone particles [39]. They found that flow behavior at a fines content above 30% w/w was dominated by the behavior of fines. More recently, Pillitery et al. [40] also found a tipping point in compaction of binary granular mixture close to 30–35% w/w of fines. It is expected that, at this level of fines, the coarse particles are completely embedded by the fines, and so flow behavior is governed by interparticle forces between fines. Below this level, contacts between coarse particles dominate flow [41].

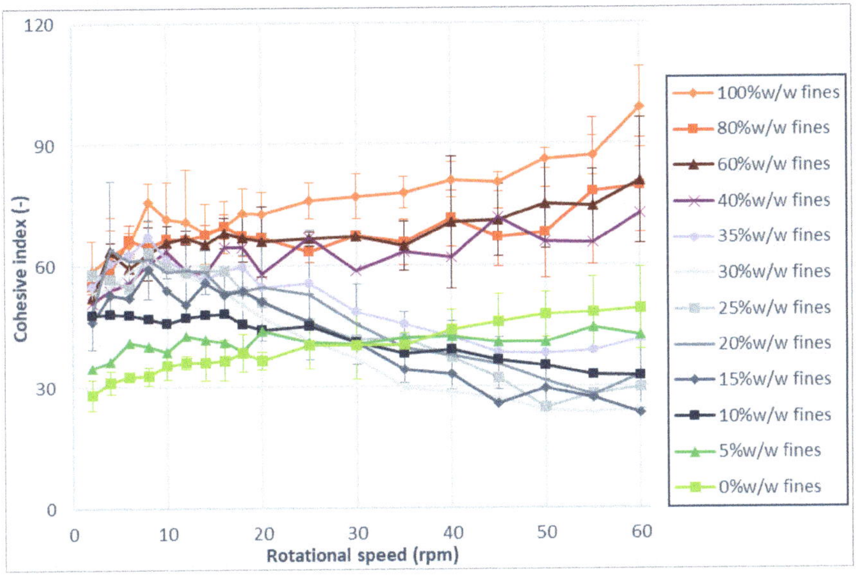

Figure 6. Cohesive index as function of rotational speed for sieved lactose with different amounts of fines added.

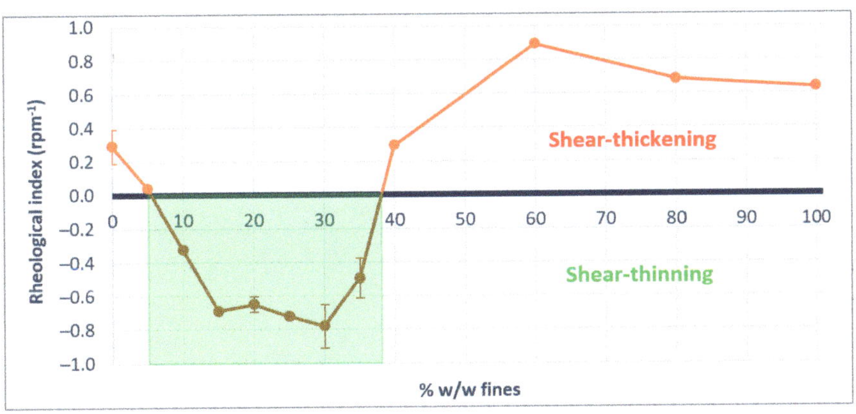

Figure 7. Rheological index parameter as function of the amount of fines added to the sieved lactose. The marked area at 10–35% w/w fines indicates shear thinning behavior.

The observed differences in cohesive index and rheological index for blends with different amounts of fines, are in line with the earlier mentioned findings of Hickey [38] on inhalation lactose. The authors explained that milled powder flows more readily because of the presence of fines. Stretching the findings of Hickey et al. [38], the current studies shows that the rheological behavior is not only different for milled and sieved materials. It reveals that the presence of a small amount of fines in a powder with larger particle size distribution can have a positive effect on the dynamic flow properties and rheological behavior.

An explanation for the reduced cohesive index of blends with 10–35% w/w fines at high rotational speeds is found in the flow field. In a more rapid flow regime, the magnitude of forces acting on the particles are different. On the one hand, the inertial effect is stronger, which could lead to increased cohesive index. On the other hand, friction by cohesive and adhesive forces is typically reduced. This is due to the fact that rolling friction is typically much smaller than static or sliding friction [34]. It is the authors hypothesis that the presence of fines even further reduces the friction when particles are moving. This is explained by a reduction of bridge formation in moving powder when small particles are present between larger particles. This improves particle separation, resulting in more readily powder flow. In this study, we find that blends with 10–35% w/w fines correspond to optimal mixtures where the separation of large grains by interstitial fines improves the rheological properties of the blends. Indeed, the cohesive index of the blends with 10–35% w/w fines reduces with the rotational speed, which can be explained by an aeration of the powders leading to a reduction of the cohesive forces acting on the grains. When the fines content increases over 40% w/w, the flow behavior is dominated by the fine fraction embedding the coarse particles, and a more cohesive powder is observed. The hypothesis is that for blends of coarse and fine material a tipping point for rheological behavior exist at the point where coarse particles are completely embedded by fines. This tipping point is therefore expected to depend on the particle shape and particle size ratio of the fines and the coarse particles.

3.5. Correlating Rheological Behavior to Tableting Performance

To evaluate the practical application of shear thickening and shear thinning, a tableting study was performed based on two excipients that showed these two types of rheological behavior. Direct compression grades were selected with a cohesive index below 50 over the entire range and strong rheological behavior. SuperTab® 11 SD showed strong shear thickening with a rheological index of 0.59 rpm^{-1}, while SuperTab® 21 AN did show shear thinning with a rheological index of −0.48 rpm^{-1} in the GranuDrum. The mass and mass variability of tablets was evaluated as a measure for powder flow into the dies during a tableting process. The agitator speed was increased in steps of 5 rpm between 10–45 rpm to investigate different flow fields and stress states. Figure 8 shows the average mass and the variability in the mass of tablets that was obtained for tableting with different agitator speeds.

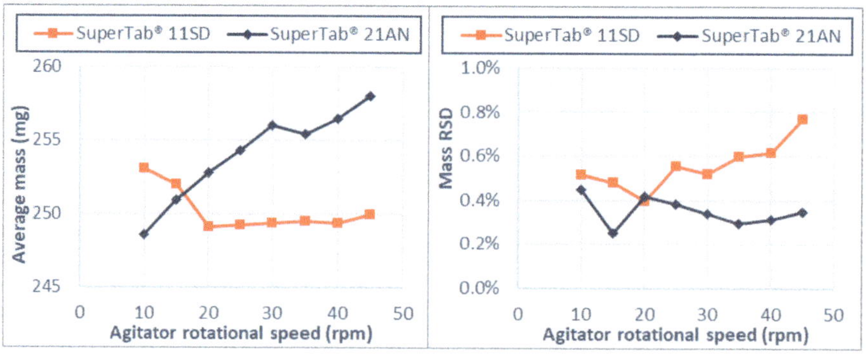

Figure 8. Tableting results (n = 20) on average mass and mass variability (%RSD) of a shear thinning material (SuperTab® 21 AN) and a shear thickening material (SuperTab® 11 SD). The agitator rotational speed was increased from 10–45 rpm in steps of 5 rpm.

Differences in rheological behavior were also visible during flow into the dies in the tableting process. Increased agitator speed resulted in higher mass tablets with less variability for the formulation with SuperTab® 21 AN. This indicates that at low agitator speed, the powder flow into the dies was incomplete and that improvement was obtained by increased flow via adjustment of the flow field. These findings confirm that powders with a negative rheological index exhibit a lower cohesion and better flowability with a more dynamic process. This is especially visible for grades like SuperTab® 21 AN, which have as a significant fraction of fines present and a large bridging propensity in the quasi-static state [32]. For SuperTab® 11 SD on the other hand, increased agitator speed did not result in higher mass tablets. A slight increase in mass RSD is observed for this formulation, indicating more cohesive powder flow. This is in line with the positive rheological index of SuperTab® 11 SD, in line with the sharp particle size distribution and low bridging propensity in the quasi-static state [32]. These results show that rheological index measurements in the lab can be a tool to set-up a tableting process.

4. Conclusions

In this study, the effect of rheological behavior during powder flow into tablet dies is investigated. It was shown that a difference in flow field and stress state can have an effect on the cohesive index and die filling of powders. This effect depends on the raw material properties. A new parameter for rheological behavior (RI) was introduced, which is a measure for the change in dynamic cohesive index upon changes in stress state and flow field. Most powders evaluated show shear thickening behavior. However, shear thinning is observed for some materials as well. Of all physical/chemical parameters tested, the particle size distribution ($\times 10$) was shown to have the largest impact on the rheological behavior of powders. At higher rotational speeds, addition of 5–35% w/w of fines to coarse sieved lactose resulted in a decrease of cohesive index. This is explained by the presence of fines between larger particles, that can improve particle separation thus reducing the cohesive forces acting between grains. The fidelity of the rheological index was validated by correlating it to tableting performance in a rotary tablet press with agitators. It was shown that insights on the rheological index (RI) obtained by rotating drum experiments can be used to optimize agitator settings in a tableting machine.

Supplementary Materials: The following are available online at https://www.mdpi.com/article/10.3390/pharmaceutics13081198/s1. Figure S1. Scanning electron micrographs of the excipients used in this study: (a) Lactopress® anhydrous, (b) SuperTab® 21 AN, (c) SuperTab® 22 AN, (d) SuperTab® 24 AN, (e) Pharmatose® 80 M, (f) Pharmatose® 150 M, (g) Pharmatose® 200 M, (h) Pharmatose® 450 M, (i) SuperTab® 30 GR, (j) SuperTab® 11 SD, (k) SuperTab® 14 SD, (l) SuperTab® 50 ODT, (m) Pharmacel® 101, (n) Pharmacel® 102, (o) Pharmacel® sMCC90, (p) Primojel®, (q) Primellose®. The scale bars represent 100 μm. Table S1: Physical properties for the set of excipients that is used in this study, being total moisture content by Karl–Fisher titration (KF), free moisture content by Loss On Drying (LOD), specific surface area (SSA), tapped density (TD), initial charge density (q_i), final charge density (q_f), tribo-charging density variation (Δq), true density (TrD), yield pressure at 0.01 mm/s (PyS), yield pressure at 300 mm/s (PyF), and strain rate sensitivity (SRS). Table S2: Particle size distribution of the blends of different amounts of fines (Pharmatose® 450 M) and coarse lactose (Pharmatose® 80 M).

Author Contributions: Conceptualization, P.H.M.J.; Methodology, P.H.M.J., S.D. and A.N.; Formal analysis, P.H.M.J.; Investigation, P.H.M.J., S.D. and A.N.; Resources, P.H.M.J., S.D., A.N., B.H.J.D. and F.F.; Data curation, P.H.M.J.; Writing—original draft preparation, P.H.M.J.; Writing—review and editing, P.H.M.J., S.D., A.N. and B.H.J.D.; Visualization, P.H.M.J.; Supervision, B.H.J.D. All authors have read and agreed to the published version of the manuscript.

Funding: This research received no external funding.

Institutional Review Board Statement: Not applicable.

Informed Consent Statement: Not applicable.

Data Availability Statement: Not applicable.

Conflicts of Interest: The authors declare no conflict of interest. The companies had no role in the design of the study; in the collection, analyses, or interpretation of data; in the writing of the manuscript, and in the decision to publish the results.

References

1. Krantz, M.; Zhang, H.; Zhu, J. Characterization of Powder Flow: Static and Dynamic Testing. *Powder Technol.* **2009**, *194*, 239–245. [CrossRef]
2. Pernenkil, L.; Cooney, C.L. A Review on the Continuous Blending of Powders. *Chem. Eng. Sci.* **2006**, *61*, 720–742. [CrossRef]
3. Prescott, J.K.; Barnum, R.A. On Powder Flowability. *Pharm. Technol.* **2000**, *24*, 60–85.
4. Blackshields, C.A.; Crean, A.M. Continuous Powder Feeding for Pharmaceutical Solid Dosage Form Manufacture: A Short Review. *Pharm. Dev. Technol.* **2018**, *23*, 554–560. [CrossRef] [PubMed]
5. Puckhaber, D.; Eichler, S.; Kwade, A.; Finke, J.H. Impact of Particle and Equipment Properties on Residence Time Distribution of Pharmaceutical Excipients in Rotary Tablet Presses. *Pharmaceutics* **2020**, *12*, 283. [CrossRef] [PubMed]
6. Peeters, E.; De Beer, T.; Vervaet, C.; Remon, J.P. Reduction of Tablet Weight Variability by Optimizing Paddle Speed in the Forced Feeder of a High-Speed Rotary Tablet Press. *Drug Dev. Ind. Pharm.* **2015**, *41*, 530–539. [CrossRef]
7. Schwedes, J. Review on Testers for Measuring Flow Properties of Bulk Solids. *Granul. Matter* **2003**, *5*, 1–43. [CrossRef]
8. Zhou, Y.C.; Xu, B.H.; Yu, A.B.; Zulli, P. An Experimental and Numerical Study of the Angle of Repose of Coarse Spheres. *Powder Technol.* **2002**, *125*, 45–54. [CrossRef]
9. Ileleji, K.E.; Zhou, B. The Angle of Repose of Bulk Corn Stover Particles. *Powder Technol.* **2008**, *187*, 110–118. [CrossRef]
10. Saker, A.; Cares-Pacheco, M.G.; Marchal, P.; Falk, V. Powders Flowability Assessment in Granular Compaction: What about the Consistency of Hausner Ratio? *Powder Technol.* **2019**, *354*, 52–63. [CrossRef]
11. Saw, H.Y.; Davies, C.E.; Paterson, A.H.J.; Jones, J.R. Correlation between Powder Flow Properties Measured by Shear Testing and Hausner Ratio. *Procedia Eng.* **2015**, *102*, 218–225. [CrossRef]
12. Zhou, X.; Nauka, E.; Narang, A.; Mao, C. Flow Function of Pharmaceutical Powders at Low-Stress Conditions Can Be Inferred Using a Simple Flow-Through-Orifice Device. *J. Pharm. Sci.* **2020**, *109*, 2009–2017. [CrossRef]
13. Nalluri, V.R.; Kuentz, M. Flowability Characterisation of Drug-Excipient Blends Using a Novel Powder Avalanching Method. *Eur. J. Pharm. Biopharm.* **2010**, *74*, 388–396. [CrossRef]
14. Pirard, S.L.; Lumay, G.; Vandewalle, N.; Pirard, J.P. Motion of Carbon Nanotubes in a Rotating Drum: The Dynamic Angle of Repose and a Bed Behavior Diagram. *Chem. Eng. J.* **2009**, *146*, 143–147. [CrossRef]
15. Freeman, R. Measuring the Flow Properties of Consolidated, Conditioned and Aerated Powders-A Comparative Study Using a Powder Rheometer and a Rotational Shear Cell. *Powder Technol.* **2007**, *174*, 25–33. [CrossRef]
16. Koynov, S.; Glasser, B.; Muzzio, F. Comparison of Three Rotational Shear Cell Testers: Powder Flowability and Bulk Density. *Powder Technol.* **2015**, *283*, 103–112. [CrossRef]
17. Shah, R.B.; Tawakkul, M.A.; Khan, M.A. Comparative Evaluation of Flow for Pharmaceutical Powders and Granules. *AAPS PharmSciTech* **2008**, *9*, 250–258. [CrossRef]
18. Stavrou, A.G.; Hare, C.; Hassanpour, A.; Wu, C.Y. Investigation of Powder Flowability at Low Stresses: Influence of Particle Size and Size Distribution. *Powder Technol.* **2020**, *364*, 98–114. [CrossRef]
19. Wu, C.Y.; Dihoru, L.; Cocks, A.C.F. The Flow of Powder into Simple and Stepped Dies. *Powder Technol.* **2003**, *134*, 24–39. [CrossRef]
20. Mendez, R.; Muzzio, F.; Velazquez, C. Study of the Effects of Feed Frames on Powder Blend Properties during the Filling of Tablet Press Dies. *Powder Technol.* **2010**, *200*, 105–116. [CrossRef]
21. Mehrotra, A.; Chaudhuri, B.; Faqih, A.; Tomassone, M.S.; Muzzio, F.J. A Modeling Approach for Understanding Effects of Powder Flow Properties on Tablet Weight Variability. *Powder Technol.* **2009**, *188*, 295–300. [CrossRef]
22. Lee, W.B.; Widjaja, E.; Heng, P.W.S.; Chan, L.W. The Effect of Rotation Speed and Particle Size Distribution Variability on Mixability: An Avalanche Rheological and Multivariate Image Analytical Approach. *Int. J. Pharm.* **2020**, *579*. [CrossRef]
23. Lumay, G.; Boschini, F.; Traina, K.; Bontempi, S.; Remy, J.C.; Cloots, R.; Vandewalle, N. Measuring the Flowing Properties of Powders and Grains. *Powder Technol.* **2012**, *224*, 19–27. [CrossRef]
24. Boschini, F.; Delaval, V.; Traina, K.; Vandewalle, N.; Lumay, G. Linking Flowability and Granulometry of Lactose Powders. *Int. J. Pharm.* **2015**, *494*, 312–320. [CrossRef] [PubMed]
25. Geldart, D. Types of Gas Fhidization. *Powder Technol.* **1973**, *7*, 285–292. [CrossRef]
26. Aulton, M.E.; Taylor, K.M.G. Powder flow. In *Aulton's Pharmaceutics E-Book: The Design and Manufacture of Medicines*; Churchill Livingstone (Elsevier): Edinburgh, Scotland, 2013; pp. 189–200. ISBN 978-0702042904.
27. Goh, H.P.; Heng, P.W.S.; Liew, C.V. Comparative Evaluation of Powder Flow Parameters with Reference to Particle Size and Shape. *Int. J. Pharm.* **2018**, *547*, 133–141. [CrossRef]
28. Fu, X.; Huck, D.; Makein, L.; Armstrong, B.; Willen, U.; Freeman, T. Effect of Particle Shape and Size on Flow Properties of Lactose Powders. *Particuology* **2012**, *10*, 203–208. [CrossRef]
29. Rajchenbach, J. Flow in Powders: From Discrete Avalanches to Continuous Regime. *Phys. Rev. Lett.* **1990**, *65*, 2221–2224. [CrossRef]

30. Castellanos, A.; Valverde, J.M.; Pérez, A.T.; Ramos, A.; Watson, P.K. Flow Regimes in Fine Cohesive Powders. *Phys. Rev. Lett.* **1999**, *82*, 1156–1159. [CrossRef]
31. Alexander, A.W.; Chaudhuri, B.; Faqih, A.M.; Muzzio, F.J.; Davies, C.; Tomassone, M.S. Avalanching Flow of Cohesive Powders. *Powder Technol.* **2006**, *164*, 13–21. [CrossRef]
32. Neveu, A.; Janssen, P.; Lumay, G. The Flowability of Lactose Powders to Optimise Tableting Processes. *ONdrugDelivery* **2020**, *2020*, 58–62.
33. Francia, V.; Yahia, L.A.A.; Ocone, R.; Ozel, A. From Quasi-Static to Intermediate Regimes in Shear Cell Devices: Theory and Characterisation. *KONA Powder Part. J.* **2021**, *38*, 3–25. [CrossRef]
34. Tardos, G.I.; McNamara, S.; Talu, I. Slow and Intermediate Flow of a Frictional Bulk Powder in the Couette Geometry. *Powder Technol.* **2003**, *131*, 23–39. [CrossRef]
35. Amidon, G.E.; Secreast, P.J.; Mudie, D. Particle, Powder, and Compact Characterization. *Dev. Solid Oral Dos. Forms* **2009**, 163–186. [CrossRef]
36. Shi, H.; Lumay, G.; Luding, S. Stretching the Limits of Dynamic and Quasi-Static Flow Testing on Cohesive Limestone Powders. *Powder Technol.* **2020**, *367*, 183–191. [CrossRef]
37. Allenspach, C.; Timmins, P.; Lumay, G.; Holman, J.; Minko, T. Loss-in-Weight Feeding, Powder Flow and Electrostatic Evaluation for Direct Compression Hydroxypropyl Methylcellulose (HPMC) to Support Continuous Manufacturing. *Int. J. Pharm.* **2021**, *596*, 120259. [CrossRef] [PubMed]
38. Hickey, A.J.; Mansour, H.M.; Telko, M.J.; Xu, Z.; Smyth, H.D.C.; Mulder, T.; McLean, R.; Langridge, J.; Papadopoulos, D. Physical Characterization of Component Particles Included in Dry Powder Inhalers. II. Dynamic Characteristics. *J. Pharm. Sci.* **2007**, *96*, 1302–1319. [CrossRef]
39. Molerus, O.; Nywlt, M. The Influence of the Fine Particle Content of the Flow Behaviour of Bulk Materials. *Powder Technol.* **1984**, *37*, 145–154. [CrossRef]
40. Pillitteri, S.; Opsomer, E.; Lumay, G.; Vandewalle, N. How Size Ratio and Segregation Affect the Packing of Binary Granular Mixtures. *Soft Matter* **2020**, *16*, 9094–9100. [CrossRef]
41. Kojima, T.; Elliott, J.A. Incipient Flow Properties of Two-Component Fine Powder Systems and Their Relationships with Bulk Density and Particle Contacts. *Powder Technol.* **2012**, *228*, 359–370. [CrossRef]

pharmaceutics

MDPI

Article

Bulk Flow Optimisation of Amorphous Solid Dispersion Excipient Powders through Surface Modification

Danni Suhaidi [1], Yao-Da Dong [2], Paul Wynne [3], Karen P. Hapgood [4] and David A. V. Morton [1,*]

1 School of Engineering, Deakin University, Waurn Ponds, VIC 3216, Australia; danni.suhaidi@deakin.edu.au
2 Drug Delivery, Disposition and Dynamics, Monash Institute of Pharmaceutical Sciences, Monash University, Parkville, VIC 3052, Australia
3 Medicines Manufacturing Innovation Centre, Monash University, Clayton, VIC 3168, Australia
4 School of Engineering, Swinburne University, Hawthorn, VIC 3122, Australia
* Correspondence: david.morton@deakin.edu.au

Abstract: Particulate amorphous solid dispersions (ASDs) have been recognised for their potential to enhance the performance of various solid dose forms, especially oral bioavailability and macromolecule stability. However, the inherent nature of spray-dried ASDs leads to their surface cohesion/adhesion, including hygroscopicity, which hinders their bulk flow and affects their utility and viability in terms of powder production, processing, and function. This study explores the effectiveness of L-leucine (L-leu) coprocessing in modifying the particle surface of ASD-forming materials. Various contrasting prototype coprocessed ASD excipients from both the food and pharmaceutical industries were examined for their effective coformulation with L-leu. The model/prototype materials included maltodextrin, polyvinylpyrrolidone (PVP K10 and K90), trehalose, gum arabic, and hydroxypropyl methylcellulose (HPMC E5LV and K100M). The spray-drying conditions were set such that the particle size difference was minimised, so that it did not play a substantial role in influencing powder cohesion. Scanning electron microscopy was used to evaluate the morphology of each formulation. A combination of previously reported morphological progression typical of L-leu surface modification and previously unreported physical characteristics was observed. The bulk characteristics of these powders were assessed using a powder rheometer to evaluate their flowability under confined and unconfined stresses, flow rate sensitivities, and compactability. The data showed a general improvement in maltodextrin, PVP K10, trehalose and gum arabic flowability measures as L-leu concentrations increased. In contrast, PVP K90 and HPMC formulations experienced unique challenges that provided insight into the mechanistic behaviour of L-leu. Therefore, this study recommends further investigations into the interplay between L-leu and the physico-chemical properties of coformulated excipients in future amorphous powder design. This also revealed the need to enhance bulk characterisation tools to unpack the multifactorial impact of L-leu surface modification.

Keywords: amorphous solid dispersions; spray drying; L-leucine; bulk powder characterization; particle engineering; flowability enhancement; surface modification

Citation: Suhaidi, D.; Dong, Y.-D.; Wynne, P.; Hapgood, K.P.; Morton, D.A.V. Bulk Flow Optimisation of Amorphous Solid Dispersion Excipient Powders through Surface Modification. *Pharmaceutics* **2023**, *15*, 1447. https://doi.org/10.3390/pharmaceutics15051447

Academic Editor: Colin Hare

Received: 31 March 2023
Revised: 4 May 2023
Accepted: 5 May 2023
Published: 9 May 2023

1. Introduction

In recent years, there has been substantial interest in the pharmaceutical particle engineering research community for optimising spray-dried formulation powders and conditions to synthesise smaller-sized ($\leq$10 µm) amorphous solid dispersions (ASDs) [1,2]. Typically, ASD formulations are used to improve the stability of biomacromolecules [3] and dissolution characteristics of poorly water-soluble active pharmaceutical ingredients (APIs) [4,5]. Because of their small particle size and amorphous physical nature, these spray-dried ASD powders have both a high surface-area-to-volume ratio and high surface energy, which increases the impact of cohesive interparticular forces, such as van der Waals interactions, on bulk behaviours [6]. Most industrially relevant ASD powders are postprocessed in downstream operations consisting of feeding, blending, filling, and tableting [7,8].

This poses a challenge, as formulated highly cohesive fine ASD powders tend to agglomerate and suffer from low bulk powder flowability. These ramifications include transport issues such as bridging or arching, which results in decreased operational efficiency and reduced yield [9]. Furthermore, benefits in the increased surface area for enhanced dissolution can be lost due to stronger agglomeration. Relatively few publications have explored potential solutions to this issue [10].

One promising avenue is the use of a coprocessing approach to modify the surface characteristics of the ASD powder materials. Zhou et al. demonstrated that the surface coating of fine lactose powders with magnesium stearate using mechanofusion successfully improved powder bulk flowability [11]. However, magnesium stearate has been reported to reduce powder dissolution rates, rendering it a less appealing option [12]. An alternative is the introduction of coexcipient materials during the spray-drying particle formulation process to achieve surface modification. Previous studies have explored the utilisation of amino acids, including arginine, aspartic acid, L-leucine (L-leu), L-isoleucine, and phenylalanine, to enhance pulmonary performance [13]. Among them, L-leu has been consistently found to have a substantial effect on improving the aerosolisation potential, attributed to its surface activity [14].

L-leu is an aliphatic α-amino acid with a hydrophobic isobutyl side chain frequently used in inhalation powder formulation research to improve powder dispersibility for pulmonary delivery [15]. In research studies in this area, it was reported and proposed that during the early stages of drying, the relative L-leu droplet surface concentration increased until it reached a supersaturation point, where it underwent a self-assembly process to form the observed partially ordered crystalline layers [16,17]. This shell interrupts the mass transfer of water vapour to the external environment, consequently creating a distinct corrugated morphology [18]. The effectiveness of L-leu was theorised to be related to its observed surface crystallinity, and its initial formulation concentration would be a deciding factor in influencing bulk flowability resulting from surface impact [16].

Published state-of-the-art research reports coformulating L-leu with APIs to leverage its properties as a potential dispersion, stability, and dissolution enhancer [4,5,15,19–21]. Mangal et al. highlighted the potential of developing a universal multifunction tablet excipient platform when their research showed the dramatic impact of different concentrations of L-leu on the flowability of polyvinylpyrrolidone K10 (PVP K10)-based formulations [22]. As a follow-up, this study aimed to explore the potential powder flow enhancement of a coformulation of L-leu with standard amorphous-forming excipients ubiquitously used in the pharmaceutical industry. Therefore, this study addresses this research gap by studying and comparing candidate excipients with differing physico-chemical properties to determine bulk powder outcomes from cospray drying with or without L-leu. The intention was to develop knowledge to underpin potential multifunctional particle-engineered delivery platforms suitable for the ASD powder formulation needs of future APIs.

Mannitol is one of the most used sugar-based excipients in commercial products; consequently, there has been a glut of scientific research exploring its physico-chemical properties and compatibility with L-leu [23,24]. However, mannitol in this form rapidly recrystallises from its initial amorphous nature, making it less attractive for ASD powder formation. Thus, this study will prototype various alternative coprocessed prototypes using emerging excipients from both the food and pharmaceutical industries with greater potential to remain amorphous and then examine the resulting bulk powder behaviours on coformulation with L-leu.

Maltodextrin was selected as a polysaccharide consisting of D-glucose subunits typically used as a binding agent in tableting and carrier during spray drying [25]. Maltodextrin has been successfully formulated as an amorphous carrier to improve the bioavailability of various nutraceutical products [26,27]. Trehalose is a disaccharide nonreducing sugar with a relatively high glass transition temperature and has been proposed as a strong candidate to remain in a metastable amorphous form. Currently, it is coformulated in several therapeutics such as: Herceptin®, Avastin®, Lucentis®, and Advate® as a protein-stabilising agent [28]. Recently, it has gained academic interest because of its potential as a spray-drying carrier for the effective pulmonary delivery of next-generation APIs [29–31]. Gum arabic is a heterogeneous collection of polysaccharides and glycoproteins with applications as an emulsifying agent, stabilising agent, and tablet binder [32]. Additional interest stems from its reported physico-chemical benefits as an antioxidant and its role in lipid metabolism [33].

PVP, a water-soluble polymer composed of N-vinylpyrrolidone subunits, is a promising amorphous solid excipient. PVP is a nontoxic and stable excipient formulated into a large range of novel controlled delivery systems [34]. PVPs are considered desirable excipients in solid dispersions because of their mucoadhesive properties that prolong their retention time in mucosal membranes [35,36]. Published studies have already shown the effectiveness of L-leu surface modification in improving PVP K10 bulk flowability. To extend our understanding, this study investigated the use of PVP K10 (10,000 Da) and the larger molecular weight PVP K90 (1,300,000 Da), which differ in their viscosity forming characteristics. These PVPs serve as benchmarks from past work to evaluate the flowability changes of other excipient powders.

The last excipients explored were two grades of soluble methylcellulose ether hydroxypropyl methylcellulose (HPMC), which is commonly used as a tablet binder or in amorphous matrices for extended release [37,38]. The two HPMC grades investigated in this study were E5LV and K100M. They differ in their chemical structures because of the different functional groups substituted in their structures. A prominent difference between the two HPMCs is their viscosity. HPMC E5LV has a lower viscosity of 6 mPa·s than HPMC K100M at 100,000 mPa·s. This difference in viscosity can be attributed to the difference in their substitution levels as well as the size and weight of their respective molecules [39].

Currently, conventional approaches for characterising potential ASDs involve understanding their behaviour at the molecular level using thermal and spectroscopic techniques and evaluating their pharmacokinetic behaviour through in vitro and in vivo studies [40,41]. These may provide some insight into the mechanism underlying the coprocessing behaviour with, for example, L-leu, but these techniques do not inform research on the holistic bulk behaviours of these powders [17,22]. Therefore, this study evaluates each powder composition formulated with or without L-leu using a range of standard and modified bulk powder tests from the Freeman FT4 powder rheometer, which were selected to gain an understanding of the effect of L-leu on the resulting bulk powder cohesion properties. In addition, scanning electron microscopy (SEM) and particle sizing were used to further differentiate the formulations for a better perception of structure–performance properties from coprocessing, and powder X-ray diffraction (XRD) was employed to examine the crystalline-amorphous nature.

2. Materials and Methods

2.1. Materials

Maltodextrin DE18 and gum arabic were acquired from the Melbourne Food Depot (Melbourne, Australia). L-leu was purchased from Sigma-Aldrich (Castle Hill, Australia), and trehalose Powder was sourced from Swanson (Melbourne, Australia). Polyvinylpyrrolidone K10 (Average molecular weight 10,000 Da) and K90 (Average molecular weight 90,000 Da) were sourced from Sigma-Aldrich (Castle Hill, Australia). Hydroxypropyl methylcellulose E5LV (Methocel E5 Premium LV) was purchased from DuPont chemicals (Mississauga, ON, Canada), and hydroxypropyl methylcellulose K100M was purchased from Sigma-Aldrich (Castle Hill, Australia).

2.2. Formulation and Spray Drying

Table 1 lists the formulation composition of each excipient. These were dissolved in aqueous solutions in accordance with the desired L-leu concentrations (0, 2.5, 5, 7.5, 10, 15, 20, 25, and 30 wt% dry basis). The mass of each component was weighed using a mass balance to within ±5% of the ideal target mass, and subsequently mixed with 400 mL of demineralized water (≤10 ppm total dissolved solids) using a magnetic stirrer (800 rpm, 35 °C). Spray drying was conducted using a Büchi B290 mini benchtop spray dryer (Büchi Laboratory Equipment, Flawil, Switzerland) with a standard 0.5 mm two-fluid nozzle. The standard operating conditions were as follows: T_{inlet}, 125 °C; aspirator rate, 35 m^3/h; feedstock flow rate, 7.5 mL/min; and T_{outlet}, 76 ± 2 °C. After spray drying, the powders were quickly collected in a sealed container and stored away from direct sunlight to minimise environmental exposure.

Table 1. List of formulation compositions for spray drying.

Excipient	Abbreviation	Formulation (Excipient/L-Leu Mixture)	L-leucine Content Percentage (L-Leu wt% Dry Basis)
Maltodextrin	MD	MD/L-leu	0, 2.5, 5, 7.5, 10, 15, 20, 30%
Polyvinylpyrrolidone K10	PVP K10	PVP K10/L-leu	0, 2.5, 5, 7.5, 10, 15, 20, 30%
Polyvinylpyrrolidone K90	PVP K90	PVP K90/L-leu	0, 2.5, 5, 7.5, 10, 15, 20, 30%
Trehalose	Trh	Trh/L-leu	0, 2.5, 5, 7.5, 10, 15, 20, 30%
Gum Arabic	GA	GA/L-leu	0, 2.5, 5, 7.5, 10, 15, 20, 30%
Hydroxypropyl methylcellulose E5LV	HPMC E5LV	HPMC E5LV/L-leu	0, 2.5, 5, 7.5, 10, 15, 20, 30%
Hydroxypropyl methylcellulose K100M	HPMC K100M	HPMC K100M/L-leu	0, 2.5, 5, 7.5, 10, 15, 20, 30%

2.3. Scanning Electron Microscopy

The surface morphologies of the formulations were imaged using a JSM-IT300 microscope (JEOL, Tokyo, Japan). A small amount of powder was fixed onto carbon tape on an aluminium stub, and the excess powder was removed using compressed air. It was subsequently given a 5 nm platinum coating using an ACE600 sputter coater (Leica, Wetzlar, Germany). The platinum-coated stubs were loaded into the SEM and imaged using an electron voltage of 10 kV at a working distance of 12 mm.

2.4. Particle Size Analysis

The particle size distribution data were acquired using a Camsizer X2 with an X-Dry Module and X-Jet cartridge (Retch Technology GmbH, Haan, Germany). Principally, Camsizer X2 functions based on ISO 1332202, where two cameras (a zoom for smaller particles and a base for larger particles) worked in conjunction with digital image analysis to derive particle size and shape distributions. Briefly, the powder was fed into the system using a hopper mechanism, where it encountered a compressed air stream that dispersed the powder into individual particles flying perpendicular to the sightlines of both cameras. These particles were illuminated using a light source, and the generated shadows were imaged using two cameras. Spray-dried powders generally adopt a spherical morphology; therefore, image analysis was conducted using the Camsizer X2's X_{area} protocol. The X_{area} protocol determined the particle diameter by calculating the diameter of an equivalent circle with the volume of a sphere of diameter X_{area} (Equation (1)).

$$X_{area} = \sqrt{\frac{4A}{\pi}} \tag{1}$$

Each formulation also had velocity adaptation performed to eliminate the risk of overlapping particle measurements. Three measurements were taken for each formulation and the average value was calculated to obtain the desired volume-based D_{10}, D_{50}, D_{90}, and span.

2.5. Powder X-ray Diffraction

Powder X-ray diffraction (P-XRD) analysis was performed using a Shimadzu 7000 L X-Ray Powder Diffractometer (Shimadzu, Kyoto, Japan). Scanning was performed from 5° to 35° at 2θ. The crystalline status of several formulations was evaluated using diffraction patterns.

2.6. Powder Bulk Flowability Characterization

Powder bulk flowability was characterised using a Freeman FT4 powder rheometer (Freeman Technology, Worcestershire, UK). This well-established instrument quantifies powder flowability through accessories, such as blades and shear heads, which are axially inserted into a powder bed while simultaneously rotating them [42]. For analytical purposes, four separate test protocols available for FT4 were used to quantify each spray-dried formulation: shear cell testing, stability and variable flow rate testing, modified stability testing, and permeability testing.

2.6.1. Shear Cell Testing

The shear cell test is based on the principle of applying both vertical and rotational stresses using a shear head on a sample powder bed [43]. Initially, an appropriate volume of each formulation was loaded into a standard 25 × 10 mL split-ring vessel, where a vented piston was used to precondition the powder bed to 9 kPa. Subsequently, the shear head was lowered into the powder bed until the predesignated normal stresses of 3, 4, 5, 6, and 7 kPa were reached. Slow rotation was then applied to the powder bed at a set rate until the powder bed experienced incipient flow. The maximum shear stress observed prior to the incipient flow was then recorded for each normal stress examined, and the data were plotted to generate a yield locus (Equation (2)).

$$\tau = \sigma \tan(\eta) + C \quad (2)$$

where τ denotes the shear stress, σ is the normal stress, η is the angle of internal friction (AIF), and C is the cohesion force. For this analysis, the flow function coefficient (ffc), AIF and Cohesion parameters were used to compare each formulation. The cohesion parameter was derived by extrapolating the yield loci to identify the y-axis shear-stress intercept, which represents the strength of a powder under zero confining stress [6]. The ffc represents the ratio between the major principal stress (MPS) and unconfined yield strength (UYS). The AIF was defined as the angle of the line drawn from the origin of the shear-stress vs. normal-stress graph towards the preshear data point.

2.6.2. Stability and Variable Flow Rate Testing

The stability test was based on the principle of axially inserting a blade into a powder bed while simultaneously rotating it at a set rate and angle. Each formulation was loaded into a 25 × 25 mL split-ring vessel and underwent two conditioning cycles prior to testing. The conditioning was intended to return the powder bed to a similar initial compaction stage to minimise errors between runs of the same sample. The protocol dictated that each test comprised 7 identical consecutive test cycles, in which the blade operated at 100 mm/s. Subsequent cycles were referred to as the variable flow rate test, where the blade speed was varied as 100 mm/s, 70 mm/s, 40 mm/s, and 10 mm/s.

Several data metrics were derived from this protocol: the basic flow energy (BFE), specific energy (SE), and flow rate index (FRI). The BFE represents the total aggregated energy required for the blade to progress through the powder bed during its downward motion. This was calculated using the seventh downwards test cycle of the protocol. SE measures the total energy needed to move the blade in an upward direction out of the powder bed; compared to BFE, it does not account for the force exerted to compress the powder bed during downwards motion. SE was calculated using the average value of the total energy required during the upward traverse of conditioning cycles six and seven then divided by mass. Finally, the FRI is a measure of powder bed sensitivity to changes in shear rate, defined as the ratio between the total energy recorded at a blade speed of 10 mm/s vs. 100 mm/s.

2.6.3. Modified Stability Testing

FRI data from the previously mentioned flow rate test highlighted the poor differentiating behaviour between the formulations. A previous review indicated that different formulations experience different shear sensitivities. Therefore, the standard FT4 stability protocol was modified in consultation with Freeman Technology staff to explore this phenomenon. The standard test was altered from using 100 mm/s to 60 mm/s. It should be noted that procedures for data collection, such as BFE and SE, were not altered to ensure that comparison between datasets was possible.

2.6.4. Permeability Testing

The permeability testing comprised three initial conditioning steps to ensure identical initial compaction states in the powder beds. Subsequently, a vented piston was used to bring the powder bed to specific normal stresses (1, 2, 4, 6, 8, 10, 12, and 15 kPa), and the pressure drop incurred from a specific air velocity was measured. The formulations were loaded into a 25 × 10 mL split-ring vessel for testing. The air supply module was cleaned and calibrated between every formulation to minimise the impact of powder accumulation. Experiments were conducted in triplicate for each formulation, and the reported results are the averaged values.

2.6.5. Statistical Analysis

Statistical analyses were performed using Microsoft Excel (Microsoft Corporation, Redmond, WA, USA). Changes in bulk flowability of the different excipient powder co-formulations with and without L-leu were evaluated using a one-way ANOVA test with $p < 0.05$, regarded as significant.

3. Results

3.1. Scanning Electron Microscopy

Selected representative electron microscopy images of the formulations are shown in Figures 1–6. Morphologically, the maltodextrin/L-leu formulations displayed the most visually observable changes with increasing leucine concentrations (Figure 1). At 0 wt% L-leu, a substantial number of the spray dried powders appeared as typical collapsed spheres which could be described as 'blood cells'. The powders physically presented as more agglomerated than those containing L-leu.

At L-leu concentrations of 5 wt%, some unusual structures appeared, notably a unique 'cupcake' morphology not seen in other formulations (Figure 1B). We propose that this resulted from L-leu influencing the mechanical properties of the surface film of the drying droplet. As the internal vapour pressure increased, a localised puncture appears to have occurred, which likely led to an irregular structure. Starting from a 10 wt% concentration, the powders began to exhibit a more heavily indented and wrinkled morphology, typically associated with L-leu surface-modified powders. However, above 20 wt%, the wrinkled morphology was accompanied by more spherical structures with apparent surface flaking.

PVP K10/L-leu formulations displayed degrees of morphological changes with increasing L-leu, as previously reported (Figure 2). At 0 wt%, the powder exhibited a smooth-dimpled morphology, which progressed to a more corrugated surface at 5 wt%. At concentrations of ≥10 wt%, 'collapsed' morphologies were observed. As previously theorised, above a certain concentration threshold, L-leu was able to form an outer shell with sufficient mechanical and transport resistance to entrap the escaping water vapour. This results in an increased internal vapour pressure, leading to expansion and eventual structural failure, resulting in the observed collapsed morphology [22].

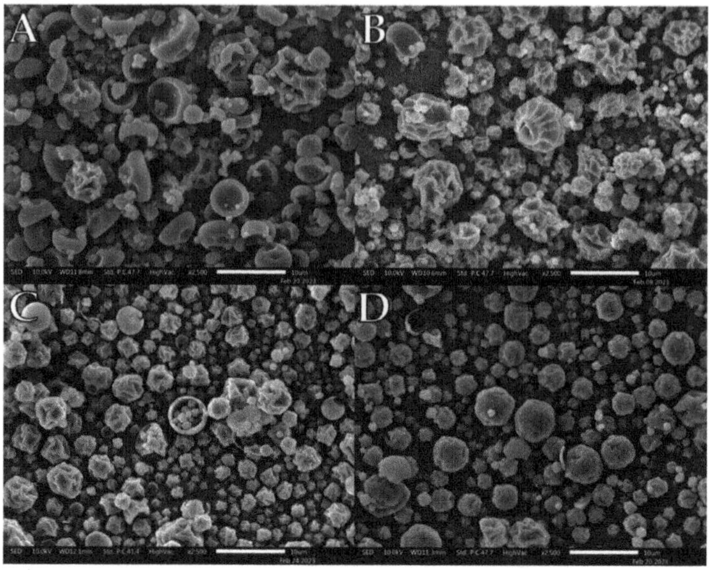

Figure 1. Representative SEM images of spray-dried maltodextrin/L-leu formulations: (**A**) MD/L-leu (0%), (**B**) MD/L-leu (5%), (**C**) MD/L-leu (10%), (**D**) MD/L-leu (20%).

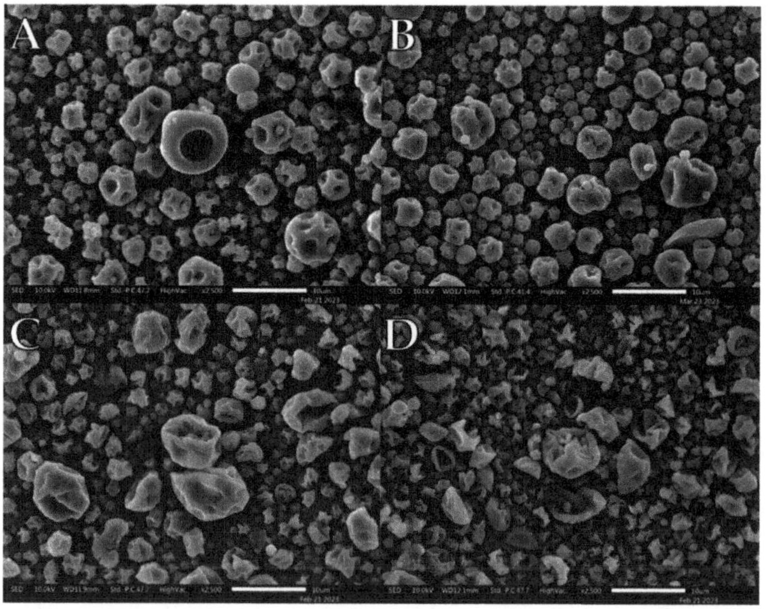

Figure 2. Representative SEM images of spray dried PVP K10/L-leu formulations: (**A**) PVP K10/L-leu (0%), (**B**) PVP K10/L-leu (5%), (**C**) PVP K10/L-leu (10%), (**D**) PVP K10/L-leu (20%).

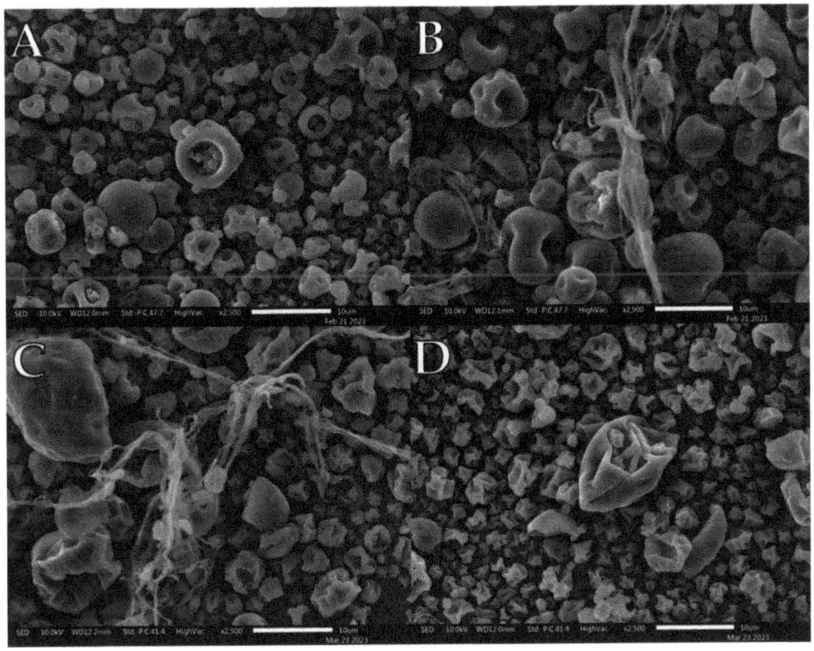

Figure 3. Representative SEM images of spray-dried PVP K90/L-leu formulations: (**A**) PVP K90/L-leu (0%), (**B**) PVP K90/L-leu (5%), (**C**) PVP K90/L-leu (10%), (**D**) PVP K90/L-leu (20%).

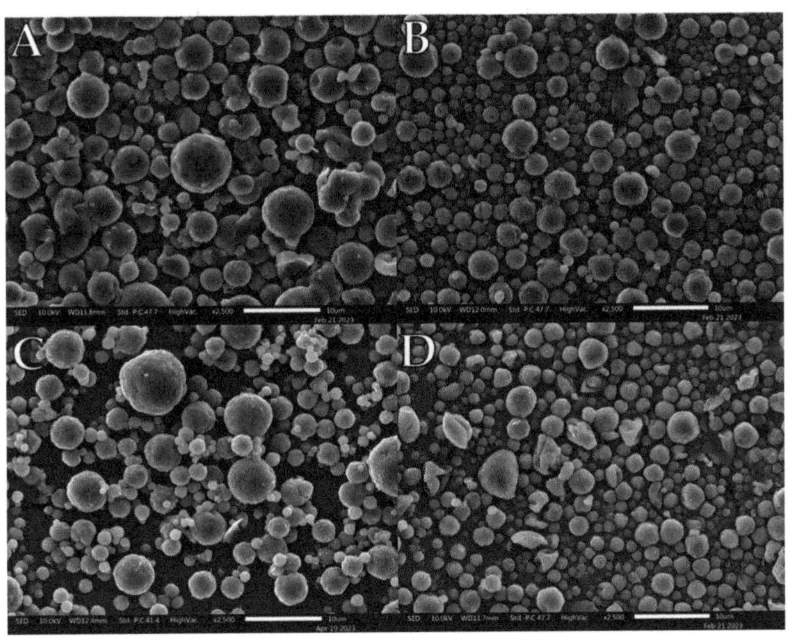

Figure 4. Representative SEM images of spray-dried trehalose/L-leu formulations: (**A**) Trh/L-leu (0%), (**B**) Trh/L-leu (5%), (**C**) Trh /L-leu (10%), (**D**) Trh /L-leu (20%).

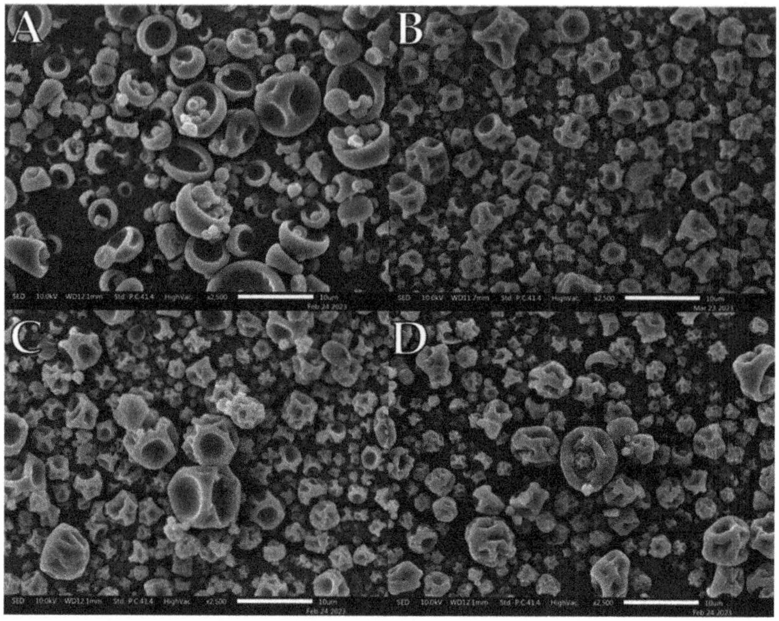

Figure 5. Representative SEM images of spray-dried gum arabic/L-leu formulations: (**A**) GA/L-leu (0%), (**B**) GA/L-leu (5%), (**C**) GA/L-leu (10%), (**D**) GA/L-leu (20%).

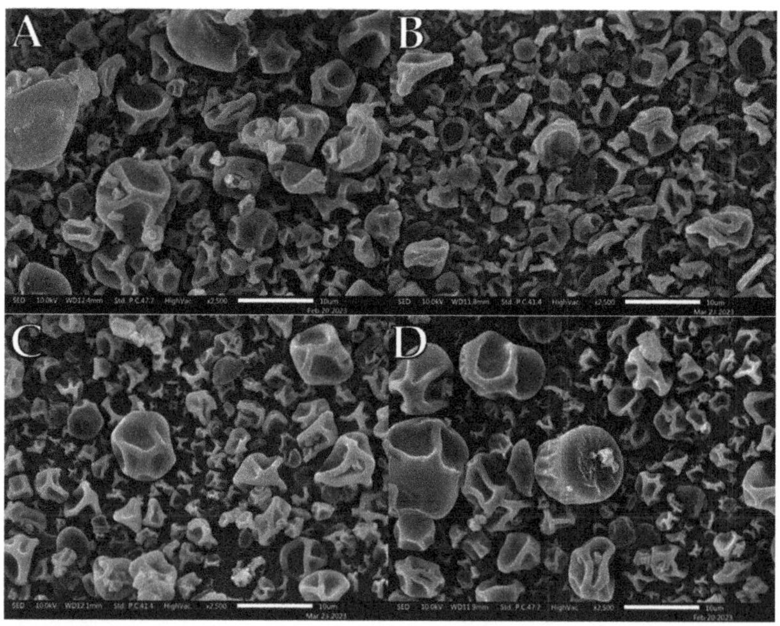

Figure 6. Representative SEM images of spray dried HPMC E5LV/L-leu formulations: (**A**) HPMC E5LV/L-leu (0%), (**B**) HPMC E5LV/L-leu (5%), (**C**) HPMC E5LV/L-leu (10%), (**D**) HPMC E5LV /L-leu (20%).

Unlike PVP K10, PVP K90 demonstrated previously unreported complications with L-leu during spray drying (Figure 3). At concentrations of 2.5–10 wt%, spray drying generated macroscale fibrous byproducts at the spray nozzle which severely affected the yield (Figure 3B,C). In each case, long fibrous spindles of material reminiscent of silk fibres formed within the drying chamber, preventing the recovery of most free powder. Consequently, no powder was recovered at 2.5 wt% and 7.5 wt% L-leu concentrations, preventing any bulk powder characterisation of these concentrations. However, no fibrous byproducts were generated at concentrations of ≥15 wt%, with the collected powders sharing corrugated morphologies similar to those of the PVP K10 formulations. Notably, these fibrous formations did not occur during the initial feedstock formulation or in the pure PVP K90 spray-drying runs.

Trehalose-based powders displayed a different morphological progression from the other powders (Figure 4). SEM of individual particles did not display an apparent transition from smooth dimpled spheres to corrugated structures. Instead, the powders maintained a mostly lightly dimpled spherical shape from 0 wt% to 30 wt%. Gum arabic is a heterogeneous material that is composed of polysaccharides and glycoproteins. SEM images showed that gum arabic exhibited a gradual morphological change as L-leu concentration increased (Figure 5). At 0 wt%, it exhibited similar 'blood-cell' like particles, like pure maltodextrin powder. Increasing the L-leu concentration resulted in the same wrinkled corrugated morphologies associated with L-leu. However, unlike homogeneous excipients, gum arabic required a higher L-leu concentration to confer the same morphological shifts.

Finally, electron microscopy showed that there were no discernible morphological differences between spray-dried pure HPMC and those coprocessed with L-leu. At all L-leu concentrations, both the HPMC E5LV and K100M (K100M in Appendix A as Figure A1) powders presented themselves as smooth-dimpled powders (Figure 6).

3.2. Particle Size and Size Distribution

The dispersion pressure was optimised through a pressure-titration process, which showed that 100 kPa provided sufficient force to deagglomerate each formulation. All formulations displayed a narrow monomodal particle size distribution; the particle size data are summarised in Table 2. Most spray-dried powders had a relatively small particle size distribution (D_{50}: ~4–6 μm), whereas known viscosity-modifying excipients, such as PVP K90, HPMC E5LV, and K100M, had a relatively larger particle size distribution (D_{50}: ~8–10 μm). Formulating these excipients in the same volume of solvent would result in higher feedstock viscosity. This is believed to impact the atomisation efficiency in the spray nozzle, leading to larger droplet sizes, which increase the mean particle size of the dried powders [44]. There are conflicting reports on the effect on the particle size distribution. Previous reports by Mangal et al. showed that higher L-leu concentrations resulted in larger particle size distributions [22]. However, Ferdynand et al. observed that L-leu decreased the particle size distribution owing to its surfactant properties [45]. This was observed in both HPMC E5LV and K100M, where the particle sizes decreased slightly with more coprocessed L-leu. Furthermore, the particle size is known to play an integral role in powder cohesion. However, the particle size data for nonviscosity influencing excipients showed no substantial difference; therefore, size should play no substantial role in influencing interparticular interactions. Hence, this study can directly attribute any bulk flowability improvement to L-leu surface modification.

Table 2. Particle size data of the spray-dried formulations. Data represent the mean ± SD (n = 3).

Formulation Code	D10 (μm)	D50 (μm)	D90 (μm)	Span	Formulation Code	D10 (μm)	D50 (μm)	D90 (μm)	Span
MD/L-leu (0%)	3.17 ± 0.01	4.69 ± 0.03	8.79 ± 0.42	1.2 ± 0.15	GA/L-leu (0%)	3.78 ± 0.02	6.55 ± 0.04	11.61 ± 0.21	1.2 ± 0.09
MD/L-leu (2.5%)	3.55 ± 0.01	5.94 ± 0.01	11.07 ± 0.25	1.27 ± 0.09	GA/L-leu (2.5%)	3.74 ± 0.01	6.52 ± 0.02	11.54 ± 0.09	1.2 ± 0.04
MD/L-leu (5%)	3.14 ± 0.02	4.72 ± 0.01	8.72 ± 0.11	1.18 ± 0.04	GA/L-leu (5%)	3.67 ± 0.03	6.17 ± 0.07	10.76 ± 0.15	1.15 ± 0.08
MD/L-leu (7.5%)	3.49 ± 0.01	5.62 ± 0.07	10.2 ± 0.11	1.2 ± 0.06	GA/L-leu (7.5%)	3.78 ± 0.02	6.57 ± 0.06	11.27 ± 0.14	1.14 ± 0.08
MD/L-leu (10%)	3.52 ± 0.01	5.83 ± 0.05	10.46 ± 0.21	1.19 ± 0.09	GA/L-leu (10%)	3.48 ± 0.03	5.38 ± 0.06	9.55 ± 0.08	1.13 ± 0.05
MD/L-leu (15%)	2.92 ± 0	4.49 ± 0.01	7.96 ± 0.13	1.12 ± 0.05	GA/L-leu (15%)	3.49 ± 0.01	5.45 ± 0.01	9.66 ± 0.06	1.13 ± 0.03
MD/L-leu (20%)	2.95 ± 0.01	4.54 ± 0.02	8.47 ± 0.26	1.21 ± 0.09	GA/L-leu (20%)	3.58 ± 0.01	6.28 ± 0.03	11.23 ± 0.13	1.22 ± 0.05
MD/L-leu (25%)	3.08 ± 0.04	4.71 ± 0.01	9.23 ± 0.04	1.31 ± 0.03	GA/L-leu (25%)	3.42 ± 0.01	5.4 ± 0.02	10.23 ± 0.08	1.26 ± 0.04
MD/L-leu (30%)	3.42 ± 0.02	5.78 ± 0.11	10.87 ± 0.22	1.29 ± 0.12	GA/L-leu (30%)	3.66 ± 0.02	6.63 ± 0.02	16.38 ± 0.79	1.92 ± 0.27
PVP K10/L-leu (0%)	3.69 ± 0.02	6.12 ± 0.06	10.38 ± 0.1	1.09 ± 0.06	HPMC E5LV/L-leu (0%)	4.05 ± 0.03	8.1 ± 0.14	17.77 ± 0.69	1.69 ± 0.29
PVP K10/L-leu (2.5%)	3.49 ± 0.02	5.48 ± 0.04	9.84 ± 0.05	1.16 ± 0.03	HPMC E5LV/L-leu (2.5%)	4 ± 0.02	7.95 ± 0.06	17.01 ± 0.54	1.63 ± 0.21
PVP K10/L-leu (5%)	3.19 ± 0	4.75 ± 0.02	8.88 ± 0.07	1.2 ± 0.03	HPMC E5LV/L-leu (5%)	4.18 ± 0.01	9.13 ± 0.1	19.74 ± 0.71	1.7 ± 0.27
PVP K10/L-leu (7.5%)	3.67 ± 0.03	6.13 ± 0.02	10.4 ± 0.06	1.1 ± 0.04	HPMC E5LV/L-leu (7.5%)	4.16 ± 0.06	9.09 ± 0.13	22.11 ± 0.94	1.98 ± 0.38
PVP K10/L-leu (10%)	3.41 ± 0.02	5.02 ± 0.05	8.98 ± 0.08	1.11 ± 0.05	HPMC E5LV/L-leu (10%)	3.93 ± 0.02	7.77 ± 0.07	17.64 ± 0.59	1.76 ± 0.23
PVP K10/L-leu (15%)	3.37 ± 0.01	5.01 ± 0.04	9.01 ± 0.06	1.12 ± 0.04	HPMC E5LV/L-leu (15%)	4.08 ± 0.01	8.33 ± 0.07	18.82 ± 0.76	1.77 ± 0.28
PVP K10/L-leu (20%)	3.42 ± 0.01	5.02 ± 0.04	9.04 ± 0.14	1.12 ± 0.06	HPMC E5LV/L-leu (20%)	3.95 ± 0.02	7.91 ± 0.1	17.43 ± 0.42	1.7 ± 0.18
PVP K10/L-leu (25%)	3.29 ± 0.01	4.75 ± 0.01	8.47 ± 0.12	1.09 ± 0.05	HPMC E5LV/L-leu (25%)	3.97 ± 0.01	7.84 ± 0.01	17.93 ± 0.29	1.78 ± 0.1
PVP K10/L-leu (30%)	3.39 ± 0.02	4.88 ± 0.02	8.75 ± 0.08	1.1 ± 0.04	HPMC E5LV/L-leu (30%)	3.89 ± 0.03	7.57 ± 0.12	17.66 ± 1.35	1.82 ± 0.5
Trh/L-leu (0%)	3.77 ± 0.09	6.21 ± 0.15	12.64 ± 0.54	1.43 ± 0.26	HPMC K100M/L-leu (0%)	4.29 ± 0.01	11.43 ± 0.15	26.4 ± 0.49	1.93 ± 0.22
Trh/L-leu (2.5%)	3.57 ± 0.04	6.06 ± 0.07	11.93 ± 0.48	1.38 ± 0.19	HPMC K100M/L-leu (2.5%)	4.39 ± 0.02	11.01 ± 0.17	24.97 ± 0.48	1.87 ± 0.22
Trh/L-leu (5%)	3.15 ± 0.03	4.72 ± 0.04	9.71 ± 0.53	1.39 ± 0.2	HPMC K100M/L-leu (5%)	3.94 ± 0.01	8.71 ± 0.11	20.79 ± 0.33	1.93 ± 0.15
Trh/L-leu (7.5%)	3.13 ± 0.06	4.68 ± 0.04	8.99 ± 0.14	1.25 ± 0.08	HPMC K100M/L-leu (7.5%)	4.29 ± 0.02	10.75 ± 0.3	25.28 ± 1.34	1.95 ± 0.55
Trh/L-leu (10%)	3.14 ± 0.02	4.72 ± 0.01	9.82 ± 0.22	1.42 ± 0.08	HPMC K100M/L-leu (10%)	3.76 ± 0.02	8.38 ± 0.19	20.99 ± 0.95	2.06 ± 0.39
Trh/L-leu (15%)	3.07 ± 0.11	4.88 ± 0.36	13.33 ± 4.13	2.1 ± 1.53	HPMC K100M/L-leu (15%)	4.1 ± 0.02	9.82 ± 0.22	24.12 ± 0.81	2.04 ± 0.35
Trh/L-leu (20%)	3.36 ± 0.03	5.28 ± 0.03	11.11 ± 0.18	1.47 ± 0.08	HPMC K100M/L-leu (20%)	4.07 ± 0.03	8.99 ± 0.11	20.62 ± 1.3	1.84 ± 0.48
Trh/L-leu (25%)	3.64 ± 0.03	6.37 ± 0.06	12.49 ± 0.24	1.39 ± 0.11	HPMC K100M/L-leu (25%)	4.14 ± 0.02	9.49 ± 0.09	21.62 ± 0.47	1.84 ± 0.19
Trh/L-leu (30%)	3.76 ± 0.03	6.37 ± 0.05	11.07 ± 0.06	1.15 ± 0.04	HPMC K100M/L-leu (30%)	4.09 ± 0.02	9.08 ± 0.13	20.11 ± 0.22	1.76 ± 0.12
PVP K90/L-leu (0%)	3.96 ± 0.03	7.99 ± 0.07	16.3 ± 0.12	1.54 ± 0.07					
PVP K90/L-leu (5%)	4.32 ± 0.04	8.93 ± 0.04	18.13 ± 0.31	1.55 ± 0.13					
PVP K90/L-leu (10%)	4.06 ± 0.66	8.96 ± 0.62	18.41 ± 0.91	1.6 ± 0.73					
PVP K90/L-leu (15%)	3.98 ± 0.01	8.19 ± 0.02	16.18 ± 0.15	1.49 ± 0.06					
PVP K90/L-leu (20%)	3.98 ± 0.02	8 ± 0.04	15.72 ± 0.23	1.47 ± 0.09					
PVP K90/L-leu (25%)	4.16 ± 0.02	8.89 ± 0.11	18.12 ± 0.53	1.57 ± 0.22					
PVP K90/L-leu (30%)	4.09 ± 0.02	8.27 ± 0.05	16.41 ± 0.53	1.49 ± 0.2					

3.3. Powder X-ray Diffraction

The P-XRD diffractogram analysis of several formulations was conducted to evaluate the solid-state characteristics of L-leu (Figure 7). Figure 7I shows the diffractogram of pure unspray-dried L-leu powder, which was apparent based on the existence of several distinctive peaks of its crystalline nature. The spray-dried powders that were not coprocessed with L-leu did not contain any distinct diffraction peaks, indicating that the powder existed in an amorphous form (Figure 7A,C,E,G). A diffuse peak was noted at 10° for all excipients. We theorise that this diffuse peak likely represents the aggregate chain length of polymeric excipients. This broadening effect was attributed to the random unordered conformations adopted by each individual polymeric chain [46]. Conversely, at a L-leu concentration of 20 wt%, all formulations displayed distinct 6° and 19° 2θ diffraction peaks along with a diffuse background (Figure 7B,D,F,H). Compared to the profile in Figure 7I, this is indicative of an amorphous core surrounded by a partially ordered L-leu crystalline shell. Sou et al. theorised that L-leu arranges itself into a strong two-dimensional lamellar form, but it failed to fully achieve a three-dimensional order [17]. Unlike other excipients, the HPMCs exhibited a degree of peak broadening in their diffractograms (Figure 7H). We theorise that this was a consequence of the physico-chemical properties of HPMC interfering with L-leu's ability to self-assemble into a coherent crystalline structure.

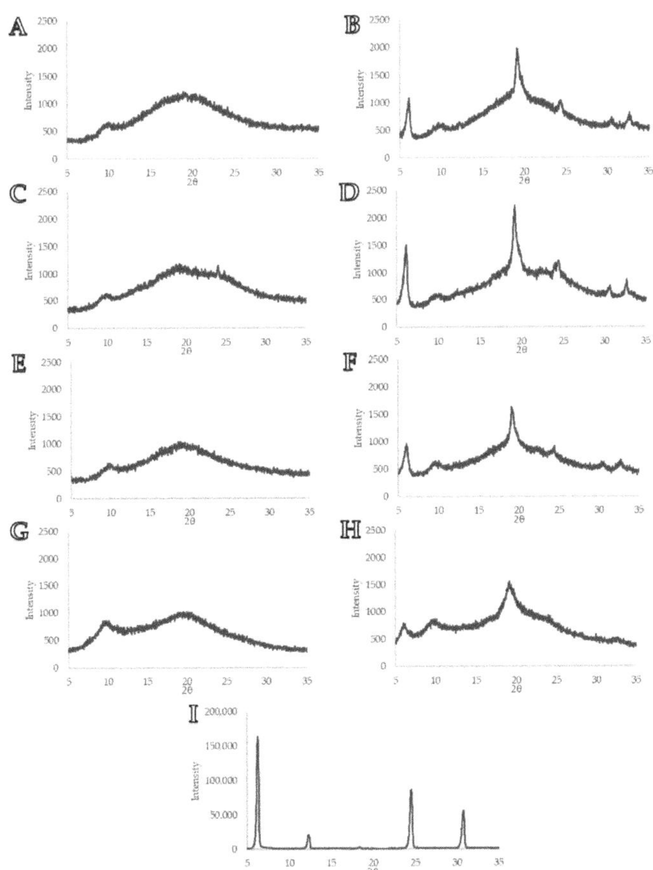

Figure 7. P-XRD diffractograms of various spray dried formulations: (**A**) MD/L-leu (0%), (**B**) MD/L-leu (20%) (**C**) Trh/L-leu (0%), (**D**) Trh/L-leu (20%), (**E**) GA/L-leu (0%), (**F**) GA/L-leu (20%), (**G**) HPMC E5LV/L-leu (0%), (**H**) HPMC E5LV/L-leu (20%), (**I**) Pure unspray-dried L-leu.

3.4. Shear Cell

The bulk powder flowability of each formulation under confined stress conditions was evaluated using an FT4 powder rheometer shear cell test, and the results are summarised in Figure 8. The examined data points were cohesion, flow function coefficient (ffc), and AIF.

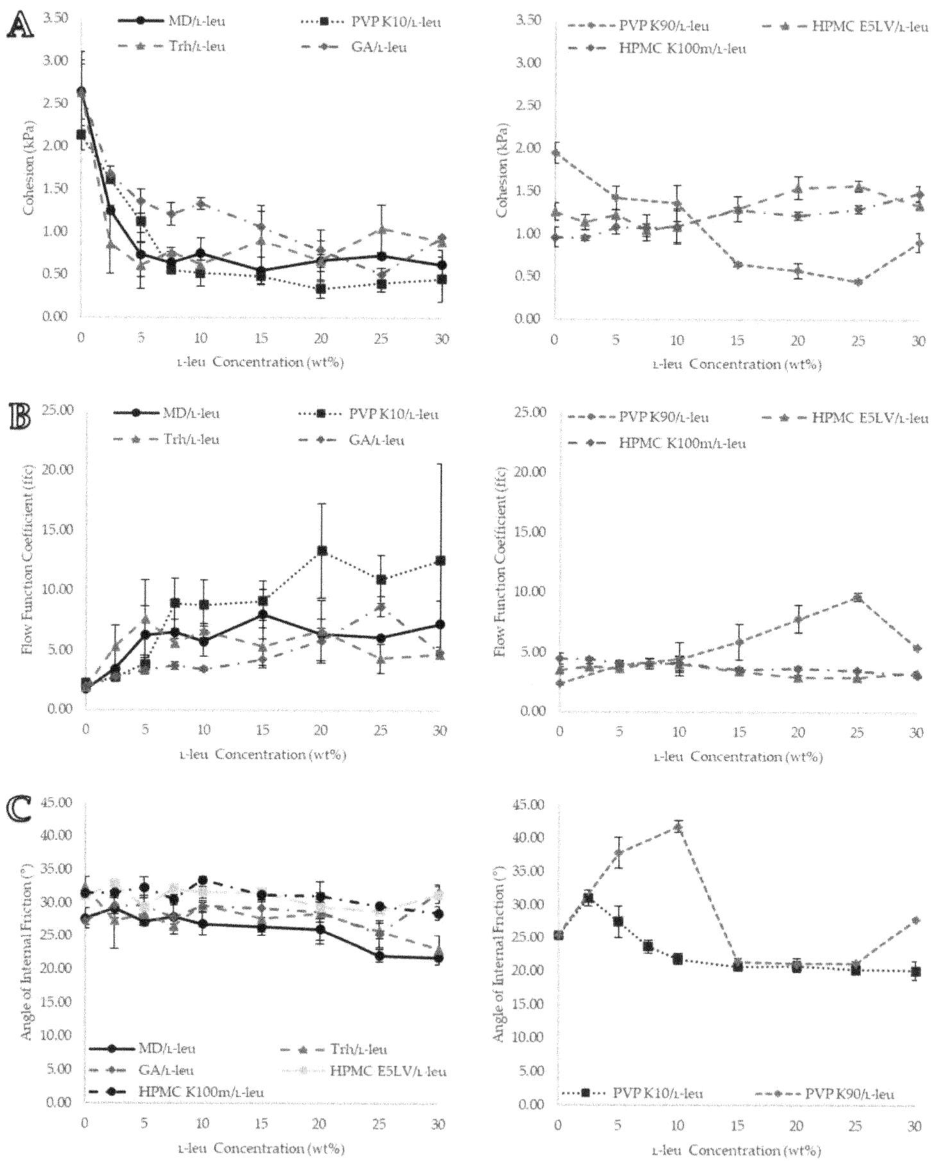

Figure 8. Visual representation of formulation shear cell data from the Freeman FT4 powder rheometer: (**A**) cohesion data, (**B**) flow function coefficient (ffc) data, and (**C**) angle of internal friction (AIF) data. All data points are displayed as mean ± SD (n = 3).

3.4.1. Cohesion Data

Cohesion data displayed a differentiation of each powder as L-leu content increased, with two distinct behaviour patterns observed (Figure 8A). Most pure spray-dried excipients displayed a higher degree of cohesiveness, which was expected for micron-sized fine powders (cohesion values ≥2 kPa). The exceptions were the HPMC formulations (E5LV and K100M), which exhibited a relatively low degree of cohesion (≤1.5 kPa). For both HPMCs, increasing L-leu concentration did not result in improved flowability. Instead, it slightly increased cohesion values at concentrations ≥15 wt%; for all other excipient powders, cohesion values decreased more substantially, indicating flowability improvements directly tied to increases in L-leu concentration.

Both maltodextrin and trehalose formulations achieved the most substantial flowability improvements at L-leu concentrations as low as 2.5 wt% ($p < 0.05$). These cohesion improvements plateaued at ≥5 wt% ($p < 0.05$) with maltodextrin (0.74 ± 0.26 kPa and trehalose (0.61 ± 0.27 kPa respectively). Further increases in concentration did not confer additional benefits, and in the case of trehalose, concentrations ≥10 wt% appeared to gradually decrease its flowability performance from 0.62 ± 0.053 kPa at 10 wt% L-leu to 0.89 ± 0.04 kPa at 30 wt% L-leu ($p < 0.05$). Additionally, increasing L-leu concentrations led to flowability improvements in PVP K10 formulations, reaching a maximum reduction of cohesion at 7.5 wt% and reaching a cohesion of 0.56 ± 0.025 kPa ($p < 0.05$). Consistent with the conclusions reported by Mangal et al., increasing the concentration above 10 wt% did not result in any further substantial improvement in bulk powder flowability [22]. Although the required concentration of L-leu was relatively higher, the maximum cohesion reduction for PVP K10 formulations marginally exceeded that of maltodextrin and trehalose powders ($p < 0.05$). Formulations of both PVP K90 and gum arabic required a greater level of L-leu to achieve similar cohesion value improvements. At L-leu concentrations of ≥20 wt%, flowability improvements were comparable to those of other excipients such as maltodextrin, trehalose, and PVP K10 ($p < 0.05$).

3.4.2. Flow Function Coefficient (ffc)

ffc has been considered a benchmark measure of powder flowability in the materials industry [47], and the most common interpretation of this value is shown in Table 3. In this study, only the preshear consolidation stress of 9 kPa was investigated. This largely followed the same bulk property trend as the cohesion data; in general, an increase in L-leu concentrations resulted in an increase in flowability (Figure 8B).

Table 3. Guidelines correlating ffc values with powder flow characteristics.

Flow Function Coefficient	
Ratio	**Flowability**
<2	Very Cohesive
2–4	Cohesive
4–10	Easy-flowing
>10	Free-flowing

The relationship between cohesion and ffc values was previously reported by Wang et al., who showed that these parameters derived from low-cohesion powders with the same preconsolidation stress had a statistically substantial inverse correlation with low-cohesion powders [48]. The results of this study further support this relationship.

HPMC formulations could be classified under the 'cohesive' category, as the majority of their ffc values remained below four. For maltodextrin, trehalose, and PVP K10, a L-leu concentration of ≥5 wt% could be classified as easily flowing. Gum arabic only became 'easy-flowing' after a concentration of 15 wt%. No powders could be definitively identified as 'free-flowing" owing to the high degree of variability in the recorded data between the different runs.

3.4.3. Angle of Internal Friction (AIF)

AIF was derived to represent the bulk friction of a powder bed during incipient flow. It was determined by drawing a linear line of the best fit from the origin of a shear-stress vs. normal-stress graph towards the preshear data point. Contrary to both cohesion and ffc, AIF did not clearly differentiate between formulations (Figure 8C). Increasing the L-leu content resulted in only a slight reduction in the AIF of maltodextrin and trehalose at ≥20 wt%. No substantial changes were noted for gum arabic- and HPMC-based formulations. Instead, L-leu had a pronounced impact on the AIF of the PVP K10 formulations. Initially, the AIF increased from 0 wt% to 2.5 wt% (25.34° ± 0.18, 0 wt% vs. 31.08° ± 1.21, 2.5%) and then decreased to a value substantially below its starting point (25.34° ± 0.18, 0 wt% vs. 20.11° ± 1.37, 30 wt%) ($p < 0.05$). In the case of PVP K90, at L-leu concentrations of 5 wt% and 10 wt%, the fibrous byproduct appeared to have contributed substantially to the frictional resistance of the powder bed, greatly increasing the AIF (25.44° ± 0.39, 0 wt% vs. 41.79° ± 0.89, 10 wt%) ($p < 0.05$). This was followed by a significant reduction at concentrations of ≥15 wt%, reaching an AIF of ~21° ($p < 0.05$).

3.5. Stability and Variable Flow Rate

3.5.1. Basic Flow Energy (BFE)

The BFE data derived for each formulation using the FT4 standard stability test protocol are shown in Figure 9A. The results showed that increasing the L-leu content led to a general decrease in the BFE. The BFE itself was defined as the amount of total energy required by the blade to move through the powder bed in a downward anticlockwise motion. Therefore, the BFE data were effectively the inverse of the bulk flowability. Thus, when the BFE values decreased, the flowability of the powder increased. Unlike the previous shear cell data, the maltodextrin-based formulations experienced a more gradual decrease in BFE. A L-leu concentration of ≥10 wt% was used to achieve the maximum BFE reduction (≤20 mJ, $p < 0.05$). Conversely, PVP K10 required a comparatively lower concentration of L-leu (≥5 wt%) to achieve the same BFE reduction (≤20 mJ, $p < 0.05$). Trehalose powder exhibited the most substantial reduction in BFE, reaching a maximum reduction (≤20 mJ, $p < 0.05$) at concentrations as low as ≥2.5 wt%. Gum arabic formulations required a high amount of coprocessed L-leu (≥20 wt%) to achieve a BFE of ≤20 mJ ($p < 0.05$). Interestingly, the BFE was able to differentiate the fibrous byproduct found in PVP K90 powders at L-leu concentrations of 2.5–10 wt%, where BFE values increased to 134 ± 60 mJ at 5 wt% from 53 ± 4 mJ of pure spray-dried PVP K90. Initially, it was presumed that the BFE could have provided insight into the formulation behaviour at a lower-stress regime. However, it became apparent that the BFE provided insufficient differentiation between each excipient once sufficiently surface-modified with L-leu. Regardless of excipient or additional L-leu coprocessing, BFE was not able to detect any further improvements below the 20 mJ threshold.

3.5.2. Specific Energy (SE)

In contrast to the BFE, the SE measured the upward clockwise motion of the blade as it left the powder bed. In effect, it provided an evaluation of the flow characteristics in unconfined and uncompacted states, unlike the shear test and downwards motion-based BFE value. Therefore, it eliminated the energy component required to compact the bed and represented a different powder characteristic. In addition, it was also normalised for the mass of the powder bed, minimising the impact of differences in the material bulk density on the total energy measured. As a general rule, a low SE value is associated with low material cohesion, with an SE value <5 classified as low cohesion.

As shown in Figure 9C, contrary to cohesion data from the shear cell test, under unconfined flow, maltodextrin-based formulations required a noticeably higher concentration of L-leu (≥10 wt% vs. ≥5 wt%) to achieve low cohesion (3.53 ± 0.45 mJ/g, $p < 0.05$). In contrast, PVP K10 formulations reached low cohesion (3.79 ± 0.16 mJ/g) at a lower L-leu concentration compared to unconfined flow (≥5 wt% vs. ≥ 10 wt%). The threshold for maximum flowability improvement through L-leu coprocessing remained the same for trehalose and gum arabic powders, requiring concentrations of ≥5 wt% and ≥20 wt%, respectively. These results highlight the differences in the behaviour of these surface-modified powders under confined and unconfined flow conditions.

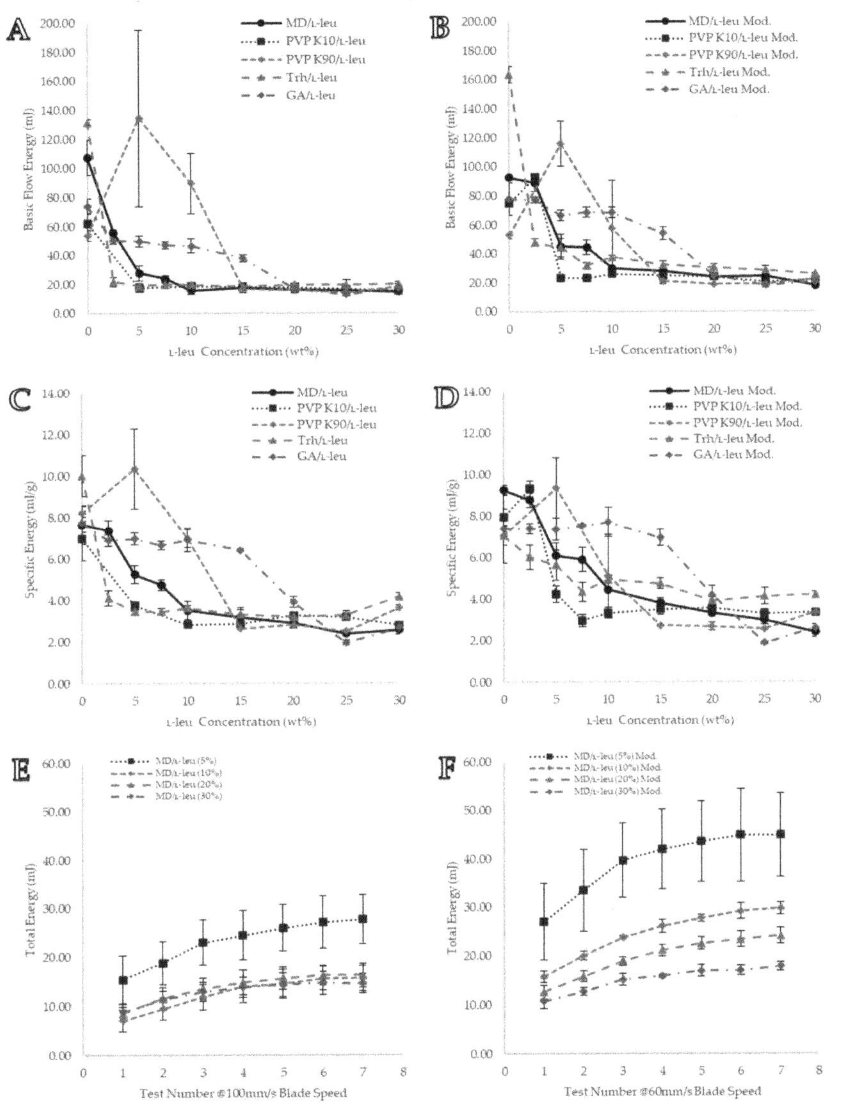

Figure 9. Visual representation of formulation stability data from the Freeman FT4 powder rheometer: (**A**) BFE value derived from the standard protocol; (**B**) BFE value derived from the modified protocol; (**C**) SE value derived from the standard protocol; (**D**) SE value derived from the modified protocol; (**E**) MD/L-leu stability data from the standard protocol; (**F**) MD/L-leu stability data from the modified protocol. All data points are displayed as mean ± SD (n = 3).

3.5.3. Flow Rate Index (FRI)

FRI is a flow property used to characterise bulk powder sensitivity to variations in the flow rate. Generally, cohesive powders are more sensitive to flow rate changes, with an FRI value of >3 associated with high cohesion. Figure 10A shows that coprocessing maltodextrin and trehalose powders with low concentrations of L-leu dramatically increased their FRI values (FRI≥3 at 5 wt% and 2.5 wt%, respectively). However, the FRI values gradually decreased as the L-leu concentration increased, eventually dipping back to below 3 at 30 wt%. This pattern is apparent in Figure 10B,C, where increasing L-leu in maltodextrin and trehalose powders reduced the gradient of these graphs, signifying a decreasing sensitivity towards flow rate changes.

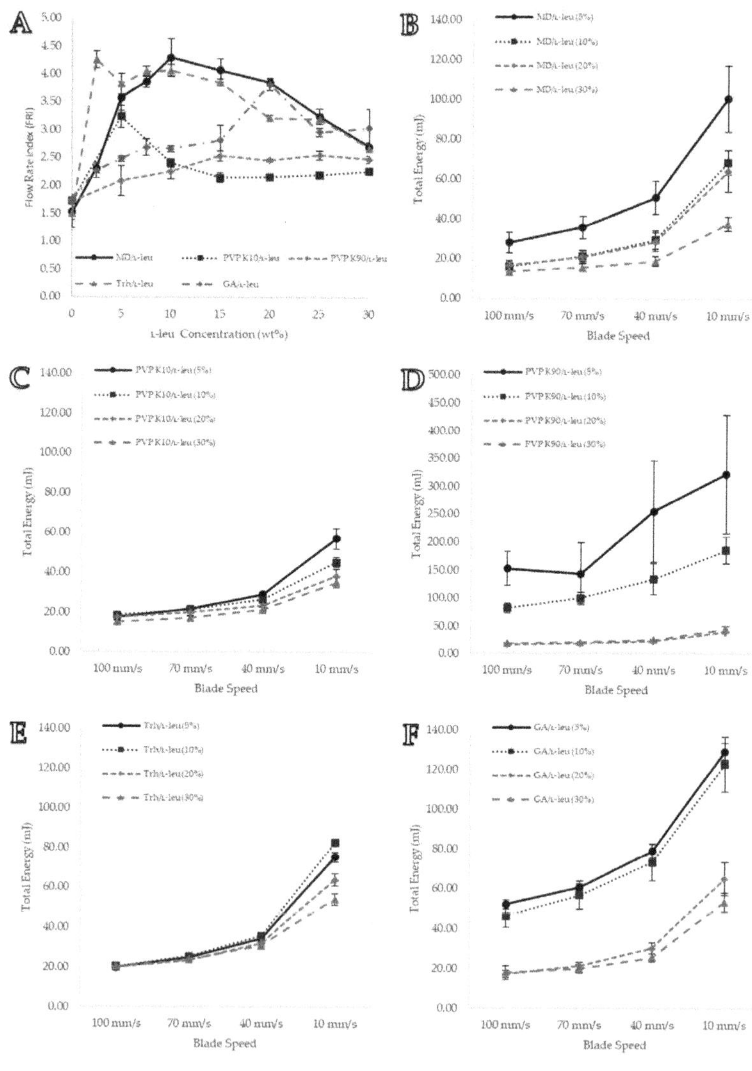

Figure 10. Visual representation of formulation variable flow rate data from the Freeman FT4 powder rheometer: (**A**) FRI values of each formulation tested; (**B**) MD/L-leu powder FRI data; (**C**) PVP K10/L-leu powder FRI data; (**D**) PVP K90/L-leu powder FRI data; (**E**) Trh/L-leu powder FRI data; and (**F**) GA/L-leu powder FRI data. All data points are displayed as mean ± SD (n = 3).

L-leu coprocessing mostly did not adversely affect the flowrate sensitivity of the PVP K10 and PVP K90 powders. Figure 10A shows that, at 5 wt%, the PVP K10 powders experienced a substantial increase in the FRI values; however, this value plateaued at its original level at ≥10 wt%. Furthermore, Figure 10C highlights PVP K10s with a generally low flowrate sensitivity at most L-leu concentrations. A pattern repeated with PVP K90 powders, regardless of L-leu concentration, flowrate sensitivity, and FRI values, remained consistent (Figure 10F). Conversely, gum arabic powder became increasingly more sensitive to the flowrate as the L-leu concentration increased, reaching a high of 3.82 ± 0.27 at 20 wt% (Figure 10A,F).

It was clear that the flowrate sensitivity was a bulk powder characteristic affected by L-leu coprocessing. Further investigations are required to better understand the mechanistic underpinnings of this phenomenon. This was coupled with another noteworthy behaviour: during the stability test, formulations coprocessed with L-leu required several test cycles before they reached a steady state (Figure 9E). We theorise that this was a consequence of the powders needing to reach a steady state of compaction when exposed to downward motion of the blade. Because the FRI value only took one total energy measurement at every blade speed, it was considered that these powders did not reach a steady state. Therefore, in response to these two observed powder behaviour issues, a modified stability test was conducted to determine whether it could provide a better powder characterisation.

3.6. Modified Stability

Here, the standard stability test was modified to run at a lower blade speed; instead of seven cycles operating at 100 mm/s, they would operate at 60 mm/s. Maltodextrin powder was used to evaluate the effectiveness of the modified approach. A comparison of Figure 9E,F shows that this blade speed provides better powder differentiation. Figure 9E shows that the original higher flowrate did not distinguish between maltodextrin coprocessed with 10, 20, and 30 wt% L-leu. The same test, operating at lower speeds, was able to distinguish between these concentrations (Figure 9F). In addition, it was confirmed that these formulations still required several test cycles before they provided a steady-state total energy measurement. Altering the flowrates also led to increased differentiation in the extracted BFE and SE values. Trehalose exhibited the most substantial reduction in BFE, reaching a maximum reduction (≤20 mJ) at concentrations as low as ≥2.5 wt%. However, at a slower flow rate, the trehalose powders had higher BFEs than the maltodextrin and PVP K10 formulations (Figure 9A,B). A similar result was observed for SE data. The specific energy was derived from the upward motion of the blade during conditioning cycles six and seven, there were no alterations in its flow rate between the original and modified tests. A comparison of Figure 9C,D shows the differences between the measured values. We theorised that this was because the different flowrates during the test cycles led to different consolidation states of the powder, thereby indirectly influencing the specific energy measurement.

3.7. Permeability

Permeability testing combines both compaction and air entrapment to characterise each formulation. It provided a relative assessment of the packing efficiency of each powder under different normal-stress regimes, as packing controls the bed porosity and thus the resistance to airflow. Figure 11 summarises the permeability test results for several excipients of interest. It shows two types of bulk behaviour profiles. The first, where the packing efficiency improved from L-leu coprocessing, was more noticeable under higher normal-stress regimes. PVP K10, trehalose, and HPMC were classified under this category. Except for 0 wt% and 2.5 wt%, PVP K10 formulations experienced a narrow range of pressure drop across the bed (8.6 ± 1.7 mBar, 25 wt% to 12.3 ± 2.1 mBar, 10 wt%). However, as the normal pressure increased, substantial differentiation was observed (Figure 11B). A similar pattern was also observed with the trehalose and HPMC formulations; at low normal pressures (1 kPa), the powders had a narrow range of pressure drops between the different L-leu concentrations. At higher normal pressures (15 kPa), the recorded pressure drops of each trehalose formulation were distinct from one another (Figure 11C,E). Although relatively modest in its improvement of packing efficiency, data from Figure 11E suggest that L-leu coprocessing had a positive impact on the performance of HPMC powders.

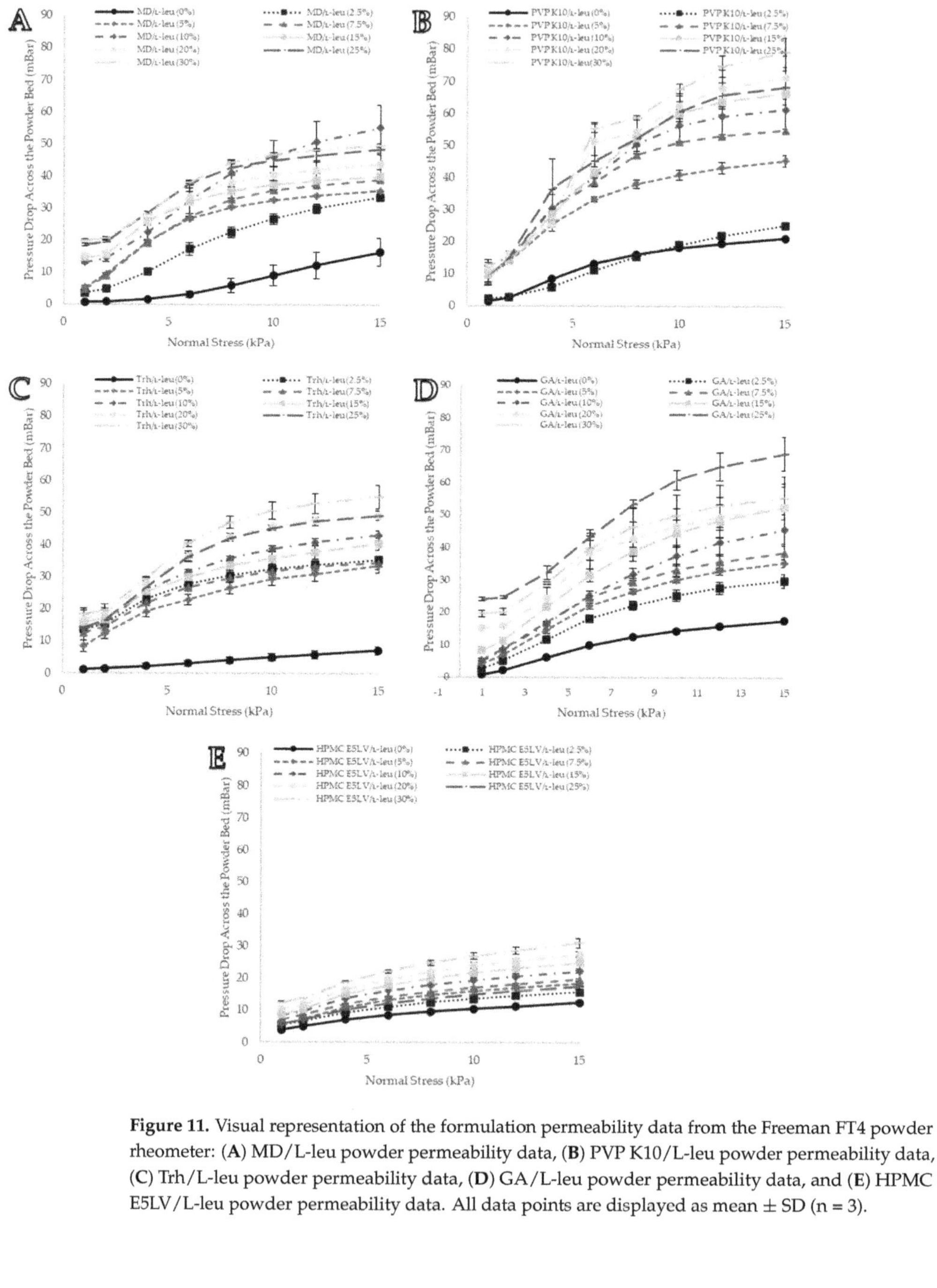

Figure 11. Visual representation of the formulation permeability data from the Freeman FT4 powder rheometer: (**A**) MD/L-leu powder permeability data, (**B**) PVP K10/L-leu powder permeability data, (**C**) Trh/L-leu powder permeability data, (**D**) GA/L-leu powder permeability data, and (**E**) HPMC E5LV/L-leu powder permeability data. All data points are displayed as mean ± SD (n = 3).

The second category of powders had great differentiation between each L-leu concentration, regardless of the applied normal stress. Both maltodextrin and gum arabic powders displayed this behaviour (Figure 11A,D). At both low and high normal pressures, increasing the L-leu concentration generally resulted in a higher pressure drop across the powder bed. From these data, it could be inferred that the larger the increase in pressure drop as the L-leu concentration increased, the greater the improvements conferred towards packing efficiency. Therefore, PVP K10 and gum arabic benefitted the most from L-leu coprocessing.

4. Discussion

The recent focus of academic research on L-leu in the coformulation of powders has been on understanding the mechanistic reasons behind its surface deposition and the ongoing self-assembly and crystallisation processes. However, studies have not evaluated its comparative effects when coprocessed into powders with different amorphous-forming excipient groups, especially its effect on powder flow. Therefore, the primary goal of this study was to investigate the effect of L-leu on the bulk flowability of excipients coprocessed at varying levels.

Qualitative observation of the SEM images compiled during this study showed that L-leu surface modification resulted in a wide range of different morphologies based on the level of the coformulated excipient and excipient nature. Many morphologies have been previously observed, while some of our observed morphologies relating to L-leu levels have not been previously reported. The deleterious impact of a low concentration (2.5–10 wt%) L-leu on PVP K90, forming fibrous spindle byproducts, has not been reported. In addition, the formation of flaky spherical structures in maltodextrin does not conform to the previously reported patterns of a more corrugated or collapsed morphology.

Another example of morphological differences is the comparison between PVP K10 and the trehalose-based formulations. The PVP K10 particles exhibited increased corrugation up to $\geq$10 wt% L-leu, where the collapsed morphologies became prominent. In contrast, the trehalose powder retained a corrugated dimpled spherical morphology, regardless of L-leu content. The morphological progression of both excipients had been examined in past publications [22,49]. Vehring et al. theorised that spray-dried powder morphology could be predicted using the Péclet number [18]. The Péclet number is defined as the ratio between the droplet surface evaporation rate and the diffusivity of the excipient material within the droplet. Given that spray-drying conditions remained constant for all formulations, the only differences were the excipient physico-chemical properties and L-leu concentrations. PVP K10 has a larger molecular structure than trehalose and is a known film-forming material [50]. Because molecular size affects the diffusivity of a material, PVP-based formulations would have a higher Péclet number than trehalose. Therefore, we theorise that a combination of PVP K10 surface accumulation and a coherent L-leu crystalline shell that forms at $\geq$10 wt% concentration offers stronger resistance, which restricts the ability of water vapour to escape from the droplet core, leading to collapsed morphologies. In contrast, trehalose is not classified as a film-forming agent. Hence, trehalose should diffuse more readily within the drying droplet, resulting in the formation of more spherical particles.

Increasing the L-leu content substantially improved the bulk flowability performance of the excipients tested, with the notable exception of HPMCs. Two mechanisms of action have historically been used to account for flowability improvements: (a) surface corrugation, which results in decreased particle contact areas, and (b) a coherent L-leu shell, which decreases the surface energy of spray-dried formulations. These mechanisms are not mutually exclusive and can work in complement, but our results presented here indicated that improvements across the different excipients were not simply explained by observed surface corrugation, and that a coherent L-leu shell is a more appropriate explanation for flowability improvements. For example, consider the case of trehalose and HPMC. Increasing the concentration of L-leu did not result in observed morphological changes. However, bulk characterisation data showed that coprocessing improved trehalose flow characteristics. In contrast, HPMC powders did not exhibit any substantial bulk flowability with increasing L-leu concentration. Therefore, this study highlighted that the relationship between flowability enhancement and morphological changes cannot be consistently predicted or generalised.

Bulk characterisation data have previously shown that above a certain L-leu concentration, flowability improvements reached an optimum and plateaued [22,47], which is supported by our results. However, our results showed that the threshold concentration differed between the excipients. Two flowability tests showed this progression: shear cell and stability tests. Shear cell testing showed the behaviour of these powders in a confined high-stress environment, reminiscent of what could be observed in a hopper feed system [47]. PVP K10, maltodextrin, and trehalose exhibited this plateau behaviour in both the cohesion and ffc data. Finally, the AIF data did not provide substantial insights into the differences between formulations.

The stability test represented the behaviour of these powders in the case of an unconfined powder flow. The BFE obtained from the standard stability test was insufficient to differentiate the formulations. Unlike the shear cell, BFE did not differentiate any formulation once L-leu had sufficiently minimised the interparticular cohesion ($\leq$20 mJ). Therefore, we conclude that SE provides a more representative measure of the flow behaviour of a formulated powder under unconfined stress. As this test measured the blade leaving the powder bed, it did not consider the energy required to compact the bed. The results were normalised by sample mass, eliminating bulk density differences and giving a better account of powder cohesion.

A comparison of both confined and unconfined data shows how these surface-modified powders perform differently under different consolidation conditions. For example, shear cell data showed that coprocessed maltodextrin achieved maximum flowability improvement at a lower L-leu concentration than PVP K10 ($\geq$5 wt% vs. $\geq$10 wt%). This trend was reversed in the SE data, where PVP K10 performed better than maltodextrin, achieving an improvement plateau at $\geq$5 wt% compared to $\geq$10 wt%.

In addition, the FRI data showed how altering L-leu level coprocessing altered the flowrate sensitivity of the powder. This is attributed to the differences in packing behaviour and efficiency of each formulation. At a lower speed, greater sensitivity is shown; therefore, we propose that the slower application of force allowed more time and opportunity for the materials to consolidate and form a bulk resistance towards the blade, increasing the total energy measured. The FRI data showed that PVP K10 was not greatly affected by alterations in blade speed, whereas maltodextrin and trehalose exhibited substantial flowrate sensitivity as a function of blade speed.

To further explore this observation, a modified stability test was used, where the blade speed was reduced from 100 mm/s to 60 mm/s. It showed better differentiation for each formulation at a lower speed. For example, at a faster speed, trehalose displayed the lowest BFE among all the excipients. However, in the lower speed regime, trehalose formulations plateaued at a BFE higher than most other excipients. In addition, the stability test was unable to differentiate maltodextrin formulations in the 10–30 wt% L-leu range, whereas this modified test could differentiate the powders. We believe that this provides an improved approach to enhance the characterisation and differentiation of these powders based on their flow-rate sensitivities.

The permeability test provided further insights into the compactability and air entrapment efficiency of leucine coprocessed powders. The powders were evaluated under a range of normal-stress regimes. The greatest difference between L-leu modifications occurred in the higher-stress regime (15 kPa). In general, a higher L-leu concentration resulted in an increase in the pressure drop across the powder bed. We attributed this to the increased air resistance and entrapment to their surface state and morphologies, which allowed them to pack more efficiently when subjected to an external force. All of these bulk tests showed that there were multiple exceptions and variations in the generalised preconception that L-leu improves bulk powder flowability.

The excipients maltodextrin, PVP K10, and trehalose were homogeneous in nature and exhibited substantial surface modification from L-leu coformulation. Gum arabic coprocessed with L-leu was the only excipient investigated with a heterogeneous nature, containing a mixture of polysaccharides and glycoproteins [51]. It displayed morphological progression, such as maltodextrin, blood cell-like structures when purely spray dried, and rugose morphologies with substantial (≥20 wt%) L-leu. There was also a gradual increase in flowability at relatively higher L-leu concentrations. It is understood that other amino acids have surface activity similar to that of L-leu; however, they do not display the same degree of flowability improvements [17]. Since gum arabic contains a substantial proportion of amino acids within its composition, it was speculated that this may lead to a degree of interruption in the formation of the historically reported L-leu shell. This example of gum arabic and L-leu formulations highlights the potential challenges and uncertainties in outcomes when coprocessed with heterogeneous formulations, with one or more APIs and other components, which may be expected in pharmaceutical and nutraceutical products.

Larger macromolecule excipients, such as PVP K90 and HPMCs, displayed contrasting results when coprocessed. PVP K90 displayed severe incompatibility at low L-leu concentrations (2.5–10 wt%) with the formation of fibrous byproducts, which necessitates future investigation. However, observing the morphologies of the free powders showed that they exhibited the same morphologies as the PVP K10 powders. Furthermore, at ≥15 wt% L-leu, PVP K90 formulations had bulk characteristics similar to those of PVP K10. This indicated that the large difference in molecular size did not greatly impact the ability of L-leu to modify the powders formed from these excipients.

The HPMC E5LV and K100M formulations displayed contrasting characteristics compared to the other excipients. Morphologically, most bulk flowability tests showed that L-leu had little to no effect on the particles formed. However, XRD analysis showed the presence of the characteristic reported 6° and 19° 2θ peak, associated with a crystalline L-leu surface shell. These peaks in the HPMC XRDs displayed a degree of broadening compared with the other formulations. This may be consistent with the differences in the L-leu formation and the alternative form on the surface. HPMC is a known viscosity-modifying material [44]. Feng et al. reported that viscosity may play a role in hindering the ability of L-leu to form a crystalline shell [16]. The HPMCs used were also the largest molecules examined here and therefore would have a higher Péclet number relative to L-leu. HPMC was also reported to have surface activity, where it was able to improve the flowability of various ASDs in common with L-leu [52]. Therefore, a combination of such factors may result in HPMC having a physico-chemical interaction with L-leu, preventing it from adequately forming a well-ordered crystalline structure. Data from this study showed the contextual interaction between L-leu's mechanism of action and the different physico-chemical characteristics of the coformulated excipients. Furthermore, it highlighted how the different FT4 powder rheometer tests characterised the different physical properties of each formulation.

This study demonstrated the impact of L-leu surface modification on the bulk flowability performance of excipients in spray-dried formulations, but it also raised questions regarding the underlying mechanisms of L-leu surface crystallisation. In future work, to gain a deeper understanding of these mechanisms, we propose that various spectroscopic techniques, such as Fourier-transform infrared spectroscopy, X-ray photoelectron spectroscopy, and time-of-flight secondary ion mass spectrometry, can be used to investigate the surface characteristics of the modified materials. Additional techniques, such as inverse gas chromatography, can also provide insight into the surface characteristics of these surface modified powders. Future studies may focus on gum arabic and HPMC formulations, as their interaction with L-leu may offer insights into their mechanism of action through their disruption. Future work should also incorporate model materials with additional studies to examine aspects such as chemical stability and dissolution to further explore the impact of L-leu surface modification on the behaviour of spray-dried formulations.

5. Conclusions

Overall, this study showed the effects of L-leu coprocessing via spray drying on the bulk powder characteristics of different excipient formulations. Effects, such as improved flow and altered compaction, are likely to lead to better control of postprocessability. The addition of L-leu substantially modified the performance of excipients, such as PVP K10, maltodextrin, and trehalose, which may help in the design of potential ASDs. However, we observed novel behaviours of other common excipients, such as PVP K90, HPMCs, and gum arabic, which exhibited differing behaviours. This highlights that L-leu behaviour in particle formation differs with the differing physico-chemical properties of the coformulated excipients and will not be directly predictable for all materials. This work highlights the need for a better understanding of its mechanistic behaviour in multicomponent systems, especially with increasing complexity of the composition. This also demonstrates the variability of results from different standard bulk characteristics of such formulated powders and the importance of suitable test design and control in identifying the nature of bulk character changes due to surface modification with a coexcipient, such as L-leu.

Author Contributions: Conceptualization, D.S., D.A.V.M. and K.P.H.; methodology, D.S., Y.-D.D., P.W. and D.A.V.M.; validation, D.S. and D.A.V.M.; investigation, D.S. and Y.-D.D.; resources, D.S., D.A.V.M., P.W. and Y.-D.D.; data curation, D.S.; writing—original draft preparation, D.S.; writing—review and editing, D.S. and D.A.V.M.; visualization, D.S.; supervision, D.A.V.M. and K.P.H.; project administration, D.S., D.A.V.M. and K.P.H. All authors have read and agreed to the published version of the manuscript.

Funding: The authors would like to acknowledge the Australian Government for their support through the Australian Government Research Training Program Scholarship.

Institutional Review Board Statement: Not applicable.

Informed Consent Statement: Not applicable.

Data Availability Statement: Not applicable.

Acknowledgments: Special mention to Reza Parvizi who was extraordinary in ensuring the smooth safe operation in the laboratories. We would also like to thank Guy Stimpson and Amalia Thomas from Freeman Technologies for their technical insight during methodology development. Additionally, we would like to thank John Long for his insights into X-ray Diffraction. Finally, the author would like to personally thank Wai Mun Wong for providing support in automating the data entry process, saving a substantial amount of effort and sanity in the process.

Conflicts of Interest: The authors declare no conflict of interest.

Appendix A

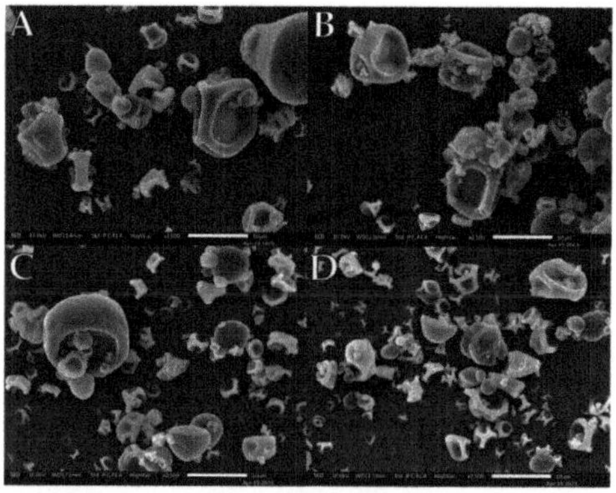

Figure A1. Representative SEM images of spray dried HPMC K100M/L-leu formulations: (**A**) HPMC K100M/L-leu (0%), (**B**) HPMC K100M/L-leu (5%), (**C**) HPMC K100M/L-leu (10%), (**D**) HPMC K100M/L-leu (20%).

References

1. Pourshahab, P.S.; Gilani, K.; Moazeni, E.; Eslahi, H.; Fazeli, M.R.; Jamalifar, H. Preparation and Characterization of Spray Dried Inhalable Powders Containing Chitosan Nanoparticles for Pulmonary Delivery of Isoniazid. *J. Microencapsul.* **2011**, *28*, 605–613. [CrossRef] [PubMed]
2. Mehanna, M.M.; Mohyeldin, S.M.; Elgindy, N.A. Rifampicin-Carbohydrate Spray-Dried Nanocomposite: A Futuristic Multiparticulate Platform for Pulmonary Delivery. *Int. J. Nanomed.* **2019**, *14*, 9089–9112. [CrossRef] [PubMed]
3. Sou, T.; Forbes, R.T.; Gray, J.; Prankerd, R.J.; Kaminskas, L.M.; McIntosh, M.P.; Morton, D.A.V. Designing a Multi-Component Spray-Dried Formulation Platform for Pulmonary Delivery of Biopharmaceuticals: The Use of Polyol, Disaccharide, Polysaccharide and Synthetic Polymer to Modify Solid-State Properties for Glassy Stabilisation. *Powder Technol.* **2016**, *287*, 248–255. [CrossRef]
4. Mangal, S.; Nie, H.; Xu, R.; Guo, R.; Cavallaro, A.; Zemlyanov, D.; Zhou, Q. (Tony) Physico-Chemical Properties, Aerosolization and Dissolution of Co-Spray Dried Azithromycin Particles with L-Leucine for Inhalation. *Pharm. Res.* **2018**, *35*, 28. [CrossRef]
5. Shetty, N.; Park, H.; Zemlyanov, D.; Mangal, S.; Bhujbal, S.; Zhou, Q. (Tony) Influence of Excipients on Physical and Aerosolization Stability of Spray Dried High-Dose Powder Formulations for Inhalation. *Int. J. Pharm.* **2018**, *544*, 222–234. [CrossRef]
6. Shi, H.; Mohanty, R.; Chakravarty, S.; Cabiscol, R.; Morgeneyer, M.; Zetzener, H.; Ooi, J.Y.; Kwade, A.; Luding, S.; Magnanimo, V. Effect of Particle Size and Cohesion on Powder Yielding and Flow. *KONA Powder Part. J.* **2018**, *2018*, 226–250. [CrossRef]
7. Démuth, B.; Nagy, Z.K.; Balogh, A.; Vigh, T.; Marosi, G.; Verreck, G.; Van Assche, I.; Brewster, M.E. Downstream Processing of Polymer-Based Amorphous Solid Dispersions to Generate Tablet Formulations. *Int. J. Pharm.* **2015**, *486*, 268–286. [CrossRef]
8. Tabor, D. *Surface Forces and Surface Interactions*; Academic Press, Inc.: New York, NY, USA, 1977.
9. Jain, S. Mechanical Properties of Powders. *Bulk Solids Handl.* **1988**, *8*, 615–624.
10. Singh, A.; Van den Mooter, G. Spray Drying Formulation of Amorphous Solid Dispersions. *Adv. Drug Deliv. Rev.* **2016**, *100*, 27–50. [CrossRef]
11. Qu, L.; Zhou, Q.; Denman, J.A.; Stewart, P.J.; Hapgood, K.P.; Morton, D.A.V. Influence of Coating Material on the Flowability and Dissolution of Dry-Coated Fine Ibuprofen Powders. *Eur. J. Pharm. Sci.* **2015**, *78*, 264–272. [CrossRef]
12. Ariyasu, A.; Hattori, Y.; Otsuka, M. Delay Effect of Magnesium Stearate on Tablet Dissolution in Acidic Medium. *Int. J. Pharm.* **2016**, *511*, 757–764. [CrossRef] [PubMed]
13. Seville, P.C.; Learoyd, T.P.; Li, H.Y.; Williamson, I.J.; Birchall, J.C. Amino Acid-Modified Spray-Dried Powders with Enhanced Aerosolisation Properties for Pulmonary Drug Delivery. *Powder Technol.* **2007**, *178*, 40–50. [CrossRef]
14. Mah, P.T.; O'Connell, P.; Focaroli, S.; Lundy, R.; O'Mahony, T.F.; Hastedt, J.E.; Gitlin, I.; Oscarson, S.; Fahy, J.V.; Healy, A.M. The Use of Hydrophobic Amino Acids in Protecting Spray Dried Trehalose Formulations against Moisture-Induced Changes. *Eur. J. Pharm. Biopharm.* **2019**, *144*, 139–153. [CrossRef]
15. Alhajj, N.; O'Reilly, N.J.; Cathcart, H. Leucine as an Excipient in Spray Dried Powder for Inhalation. *Drug Discov. Today* **2021**, *26*, 2384–2396. [CrossRef] [PubMed]
16. Feng, A.L.; Boraey, M.A.; Gwin, M.A.; Finlay, P.R.; Kuehl, P.J.; Vehring, R. Mechanistic Models Facilitate Efficient Development of Leucine Containing Microparticles for Pulmonary Drug Delivery. *Int. J. Pharm.* **2011**, *409*, 156–163. [CrossRef] [PubMed]
17. Sou, T.; Kaminskas, L.M.; Nguyen, T.H.; Carlberg, R.; McIntosh, M.P.; Morton, D.A.V. The Effect of Amino Acid Excipients on Morphology and Solid-State Properties of Multi-Component Spray-Dried Formulations for Pulmonary Delivery of Biomacromolecules. *Eur. J. Pharm. Biopharm.* **2013**, *83*, 234–243. [CrossRef]
18. Vehring, R. Pharmaceutical Particle Engineering via Spray Drying. *Pharm. Res.* **2008**, *25*, 999–1022. [CrossRef]
19. Lu, W.; Rades, T.; Rantanen, J.; Chan, H.K.; Yang, M. Amino Acids as Stabilizers for Spray-Dried Simvastatin Powder for Inhalation. *Int. J. Pharm.* **2019**, *572*, 118724. [CrossRef]
20. Chang, R.Y.K.; Chan, H.K. Advancements in Particle Engineering for Inhalation Delivery of Small Molecules and Biotherapeutics. *Pharm. Res.* **2022**, *39*, 3047–3061. [CrossRef]
21. Ordoubadi, M.; Shepard, K.B.; Wang, H.; Wang, Z.; Pluntze, A.M.; Churchman, J.P.; Vehring, R. On the Physical Stability of Leucine-Containing Spray-Dried Powders for Respiratory Drug Delivery. *Pharmaceutics* **2023**, *15*, 435. [CrossRef]
22. Mangal, S.; Meiser, F.; Tan, G.; Gengenbach, T.; Denman, J.; Rowles, M.R.; Larson, I.; Morton, D.A.V. Relationship between Surface Concentration of L-Leucine and Bulk Powder Properties in Spray Dried Formulations. *Eur. J. Pharm. Biopharm.* **2015**, *94*, 160–169. [CrossRef] [PubMed]
23. Malamatari, M.; Somavarapu, S.; Kachrimanis, K.; Buckton, G.; Taylor, K.M.G. Preparation of Respirable Nanoparticle Agglomerates of the Low Melting and Ductile Drug Ibuprofen: Impact of Formulation Parameters. *Powder Technol.* **2017**, *308*, 123–134. [CrossRef]
24. Sou, T.; Orlando, L.; McIntosh, M.P.; Kaminskas, L.M.; Morton, D.A.V. Investigating the Interactions of Amino Acid Components on a Mannitol-Based Spray-Dried Powder Formulation for Pulmonary Delivery: A Design of Experiment Approach. *Int. J. Pharm.* **2011**, *421*, 220–229. [CrossRef]
25. Hofman, D.L.; van Buul, V.J.; Brouns, F.J.P.H. Nutrition, Health, and Regulatory Aspects of Digestible Maltodextrins. *Crit. Rev. Food Sci. Nutr.* **2016**, *56*, 2091–2100. [CrossRef] [PubMed]
26. Sansone, F.; Mencherini, T.; Picerno, P.; D'Amore, M.; Aquino, R.P.; Lauro, M.R. Maltodextrin/Pectin Microparticles by Spray Drying as Carrier for Nutraceutical Extracts. *J. Food Eng.* **2011**, *105*, 468–476. [CrossRef]

27. Caliskan, G.; Nur Dirim, S. The Effects of the Different Drying Conditions and the Amounts of Maltodextrin Addition during Spray Drying of Sumac Extract. *Food Bioprod. Process.* **2013**, *91*, 539–548. [CrossRef]
28. Ohtake, S.; Wang, Y.J. Trehalose: Current Use and Future Applications. *J. Pharm. Sci.* **2011**, *100*, 2020–2053. [CrossRef]
29. Maury, M.; Murphy, K.; Kumar, S.; Shi, L.; Lee, G. Effects of Process Variables on the Powder Yield of Spray-Dried Trehalose on a Laboratory Spray-Dryer. *Eur. J. Pharm. Biopharm.* **2005**, *59*, 565–573. [CrossRef]
30. Focaroli, S.; Mah, P.T.; Hastedt, J.E.; Gitlin, I.; Oscarson, S.; Fahy, J.V.; Healy, A.M. A Design of Experiment (DoE) Approach to Optimise Spray Drying Process Conditions for the Production of Trehalose/Leucine Formulations with Application in Pulmonary Delivery. *Int. J. Pharm.* **2019**, *562*, 228–240. [CrossRef]
31. Sou, T.; Morton, D.A.V.; Williamson, M.; Meeusen, E.N.; Kaminskas, L.M.; McIntosh, M.P. Spray-Dried Influenza Antigen with Trehalose and Leucine Produces an Aerosolizable Powder Vaccine Formulation That Induces Strong Systemic and Mucosal Immunity after Pulmonary Administration. *J. Aerosol Med. Pulm. Drug Deliv.* **2015**, *28*, 361–371. [CrossRef]
32. Osman, M.E.; Williams, P.A.; Menzies, A.R.; Phillips, G. Characterization of Commercial Samples of Gum Arabic. *J. Agric. Food Chem.* **1993**, *41*, 71–77. [CrossRef]
33. Ali, B.H.; Ziada, A.; Blunden, G. Biological Effects of Gum Arabic: A Review of Some Recent Research. *Food Chem. Toxicol.* **2009**, *47*, 1–8. [CrossRef] [PubMed]
34. Kurakula, M.; Rao, G.S.N.K. Pharmaceutical Assessment of Polyvinylpyrrolidone (PVP): As Excipient from Conventional to Controlled Delivery Systems with a Spotlight on COVID-19 Inhibition. *J. Drug Deliv. Sci. Technol.* **2020**, *60*, 102046. [CrossRef] [PubMed]
35. De Caro, V.; Murgia, D.; Seidita, F.; Bologna, E.; Alotta, G.; Zingales, M.; Campisi, G. Enhanced in Situ Availability of Aphanizomenon Flos-Aquae Constituents Entrapped in Buccal Films for the Treatment of Oxidative Stress-Related Oral Diseases: Biomechanical Characterization and In Vitro/Ex Vivo Evaluation. *Pharmaceutics* **2019**, *11*, 35. [CrossRef] [PubMed]
36. Malik, S.; Kumar, A.; Ahuja, M. Synthesis of Gum Kondagogu-g-Poly(N-Vinyl-2-Pyrrolidone) and Its Evaluation as a Mucoadhesive Polymer. *Int. J. Biol. Macromol.* **2012**, *51*, 756–762. [CrossRef] [PubMed]
37. Deshmukh, K.; Basheer Ahamed, M.; Deshmukh, R.R.; Khadheer Pasha, S.K.; Bhagat, P.R.; Chidambaram, K. *Biopolymer Composites with High Dielectric Performance: Interface Engineering*; Elsevier Inc.: Amsterdam, The Netherlands, 2017; ISBN 9780081009741.
38. Chowhan, Z.T. Role of Binders in Moisture-induced Hardness Increase in Compressed Tablets and Its Effect on In Vitro Disintegration and Dissolution. *J. Pharm. Sci.* **1980**, *69*, 1–4. [CrossRef] [PubMed]
39. Oh, C.M.; Heng, P.W.S.; Chan, L.W. A Study on the Impact of Hydroxypropyl Methylcellulose on the Viscosity of PEG Melt Suspensions Using Surface Plots and Principal Component Analysis. *AAPS PharmSciTech* **2015**, *16*, 466–477. [CrossRef]
40. Ma, X.; Williams, R.O. Characterization of Amorphous Solid Dispersions: An Update. *J. Drug Deliv. Sci. Technol.* **2019**, *50*, 113–124. [CrossRef]
41. Pandi, P.; Bulusu, R.; Kommineni, N.; Khan, W.; Singh, M. Amorphous Solid Dispersions: An Update for Preparation, Characterization, Mechanism on Bioavailability, Stability, Regulatory Considerations and Marketed Products. *Int. J. Pharm.* **2020**, *586*, 119560. [CrossRef]
42. Freeman, R. Measuring the Flow Properties of Consolidated, Conditioned and Aerated Powders—A Comparative Study Using a Powder Rheometer and a Rotational Shear Cell. *Powder Technol.* **2007**, *174*, 25–33. [CrossRef]
43. Freeman, R.E.; Cooke, J.R.; Schneider, L.C.R. Measuring Shear Properties and Normal Stresses Generated within a Rotational Shear Cell for Consolidated and Non-Consolidated Powders. *Powder Technol.* **2009**, *190*, 65–69. [CrossRef]
44. Jin, C.; Wu, F.; Hong, Y.; Shen, L.; Lin, X.; Zhao, L.; Feng, Y. Updates on Applications of Low-Viscosity Grade Hydroxypropyl Methylcellulose in Coprocessing for Improvement of Physical Properties of Pharmaceutical Powders. *Carbohydr. Polym.* **2023**, *311*, 120731. [CrossRef] [PubMed]
45. Ferdynand, M.S.; Nokhodchi, A. Co-Spraying of Carriers (Mannitol-Lactose) as a Method to Improve Aerosolization Performance of Salbutamol Sulfate Dry Powder Inhaler. *Drug Deliv. Transl. Res.* **2020**, *10*, 1418–1427. [CrossRef] [PubMed]
46. Jiang, Z.; Bai, X. Effects of Polysaccharide Concentrations on the Formation and Physical Properties of Emulsion-Templated Oleogels. *Molecules* **2022**, *27*, 5391. [CrossRef]
47. Schulze, D. *Powders and Bulk Solids*; Springer: Berlin/Heidelberg, Germany, 2007; Volume 13, ISBN 978-3-540-73767-4.
48. Wang, Y.; Koynov, S.; Glasser, B.J.; Muzzio, F.J. A Method to Analyze Shear Cell Data of Powders Measured under Different Initial Consolidation Stresses. *Powder Technol.* **2016**, *294*, 105–112. [CrossRef]
49. Ordoubadi, M.; Gregson, F.K.A.; Wang, H.; Nicholas, M.; Gracin, S.; Lechuga-Ballesteros, D.; Reid, J.P.; Finlay, W.H.; Vehring, R. On the Particle Formation of Leucine in Spray Drying of Inhalable Microparticles. *Int. J. Pharm.* **2021**, *592*, 120102. [CrossRef]
50. Franco, P.; De Marco, I. The Use of Poly(N-Vinyl Pyrrolidone) in the Delivery of Drugs: A Review. *Polymers* **2020**, *12*, 1114. [CrossRef]

51. Williams, P.A.; Phillips, G.O. Gum Arabic. In *Handbook of Hydrocolloids*; Elsevier: Amsterdam, The Netherlands, 2021; pp. 627–652.
52. Lin, X.; Chyi, C.W.; Ruan, K.F.; Feng, Y.; Heng, P.W.S. Development of Potential Novel Cushioning Agents for the Compaction of Coated Multi-Particulates by Co-Processing Micronized Lactose with Polymers. *Eur. J. Pharm. Biopharm.* **2011**, *79*, 406–415. [CrossRef]

pharmaceutics

Article

Effect of Mixer Type on Particle Coating by Magnesium Stearate for Friction and Adhesion Modification

Wei Pin Goh, Ana Montoya Sanavia and Mojtaba Ghadiri *

Faculty of Engineering and Physical Sciences, University of Leeds, Leeds LS2 9JT, UK; w.p.goh@leeds.ac.uk (W.P.G.); pm19ams@leeds.ac.uk (A.M.S.)
* Correspondence: m.ghadiri@leeds.ac.uk

Abstract: Glidants and lubricants are often used to modify interparticle friction and adhesion in order to improve powder characteristics, such as flowability and compactability. Magnesium stearate (MgSt) powder is widely used as a lubricant. Shear straining causes MgSt particles to break, delaminate, and adhere to the surfaces of the host particles. In this work, a comparison is made of the effect of three mixer types on the lubricating role of MgSt particles. The flow behaviour of α-lactose monohydrate, coated with MgSt at different mass percentages of 0.2, 0.5, 1, and 5 is characterised. The mixing and coating process is carried out by dry blending using Turbula, ProCepT, and Mechanofusion. Measures have been taken to operate under equivalent mixing conditions, as reported in the literature. The flow resistance of the coated samples is measured using the FT4 rheometer. The results indicate that the flow characteristics of the processed powders are remarkably similar in the cases of samples treated by Turbula and Mechanofusion, despite extreme conditions of shear strain rate. The least flow resistance of samples is observed in the case of samples treated by the ProCepT mixer. High-velocity collisions of particles round off the sharp corners and edges, making them less resistant to flow. The optimal percentage of magnesium stearate is found to be approximately 1% by weight for all mixer types, as the addition of higher amounts of lubricant does not further improve the flowability of the material.

Keywords: mixing; coating; magnesium stearate; lubricant; Turbula; ProCepT; Mechanofusion; flowability; rheometry; flow energy

Citation: Goh, W.P.; Montoya Sanavia, A.; Ghadiri, M. Effect of Mixer Type on Particle Coating by Magnesium Stearate for Friction and Adhesion Modification. *Pharmaceutics* **2021**, *13*, 1211. https://doi.org/10.3390/10.3390/pharmaceutics13081211

Academic Editor: Axel Zeitler

Received: 17 July 2021
Accepted: 2 August 2021
Published: 5 August 2021

1. Introduction

Lubricants and glidants are commonly used in powder processing to reduce bulk powder friction in order to promote flowability and impart other desirable attributes, such as tabletability and compactability [1–3]. Good examples are magnesium stearate (MgSt) acting as a lubricant [4,5], and nano-particles of silica acting as a glidant [6,7]. Their use is highly desirable for preparing appropriate formulations for further processing, such as compaction and tabletting in the pharmaceutical industry [8]. The coating process is carried out by mixing, for which a wide variety of mixers are available [9]. By adding a small quantity of lubricant/glidant to a powder system, the energetic sites on the host particle surfaces get separated by a thin layer of the lubricant, thereby reducing the van der Waals interactions and sliding friction significantly [10], thus, the term 'flow aid' is used to describe their functionality. Due to the huge size difference between the host and guest particles, the strong intermolecular forces between them cause the guest particles to adhere strongly to the surfaces of the host particles. Moreover, shear straining in the presence of lubricants and glidants modifies the surface characteristics of the host powder, and produces new functionalities and features [11,12]. The quantity of flow aids that is necessary to achieve an optimal flow performance remains unknown to date. Conesa et al. [13] found that at 0.3 wt%, silica-coated polyester-based particles have optimal flowability. Castellanos [14] later converted this gravimetric value into surface area coverage (SAC) and found that it was approximately 20%. The work of Fulchini et al. [15] led to the same conclusion as that of Castellanos [14]. They used the image analysis technique to quantify the SAC from SEM micrographs of zeolite-coated silanised glass beads, and found that the optimal SAC for maximum flow improvement to be approximately 20%. Beyond this optimal value, the flowabilility decreased, as the system was then dominated by guest–guest rather than host–guest particle interactions. However, the work of Jallo et al. [16], on the coating of APIs with nano-silica particles, found an optimal SAC value of 293% instead, implying a multilayer coating gives rise to better flow performance. Sato et al. [17] coated sugar granules with MgSt powder in a Cyclomix mixer and reported that the guest particles were initially covered discretely, and then formed a film over the surfaces of the host particles. In addition, attrition was induced by excessive operating conditions, which resulted in the worsening of dry coating performance.

Through coating α-lactose monohydrate (α-LM) crystals with MgSt using the Mechanofusion device, Zhou et al. [18] found that the optimum SAC to be 64.5%, estimated from X-ray photoelectron spectroscopy and time-of-flight of the secondary ion mass spectrometry (Tof-SIMS). However, MgSt delaminates upon shearing [19–21] and the optimal amount would depend on the mixer type and shearing conditions. Moreover, Hussain et al. [19] propose that a Langmuir-type coating is first achieved with MgSt, followed by a patchy coverage in powder form. Therefore, the characterisation of SAC of MgSt on lactose crystals has yet to be satisfactorily established.

The preparation of a powder formulation is usually carried out on a small scale in a laboratory mixer, and extension to a large-scale operation is very challenging. Therefore, a 'recipe' for mixing equivalency is highly desirable. The attention here is on batch mixing, for which the literature is vast, addressing numerous aspects. These include the evaluation of the performance of various mixer types, assessment of mixing time and operational scale, operating conditions, the optimal amount of flow aids, and the development of models [5,12,22,23]. Horibe et al. [24] recently assessed the effect of mixing time and the concentration of flow aids on the flow properties of the end product. They found that the mixing time has a higher correlation to the flow properties than the concentration of flow aids, signifying the importance of the process and mixing conditions in improving the flow behaviour of a particulate system. As quoted from Bridgwater's review paper [9], process design and operation are largely based on judgement rather than science. Finding the optimal mixing conditions for new materials is often a time-consuming and challenging task. The quality of a mixture depends on the mixer selected. Asachi et al. [25] critically evaluated current techniques for the evaluation of powder mixing. Of special interest here is the work of Barling et al. [26] who carried out an experimental evaluation of various mixer types, and presented a methodology for establishing an 'equivalent extent' of batch mixing amongst various mixer types at different scales. They used fine iron oxide powder as a tracer, and evaluated its spreading over the surfaces of α-lactose monohydrate in several batch mixing devices by monitoring the changes in the colour intensity and hue through a colour measurement methodology. The powder blend changed colour as a result of mixing and coating. By quantifying changes in the intensity and hue of the colour of the mixture, the extent of mixing and coating of iron oxide was determined, based on which a mixing 'equivalency' was proposed for different mixer types. It was found that under specific operating conditions, a mixing equivalency could be established among different coating devices, namely Turbula T2F (figure-eight tumbler type), Key International KG5 (high shear mixer, two sizes, 1 L and 5 L), Hosokawa AMS-MINI Mechanofusion (extreme shear rate), T.K. Fielder high shear mixers (TRV25 and PMA65), and Diosna P100 (high shear mixer type).

In this work, the above equivalency criterion is followed and applied to the coating of α-LM by MgSt powder. In particular, samples treated by different mixers, but with equivalent mixing conditions, are prepared and analysed for the effect of MgSt on the flow behaviour of α-LM. Three types of mixers, Turbula, Mechanofusion, and ProCepT high shear mixer are used. The last one is close in geometry and operation to the T.K. Fielder TRV25, 1 L high shear mixer. The equivalent mixing conditions, estimated based on the work of Barling et al. [26], are given in Table 1. The flowability of the coated samples, prepared at different MgSt compositions and mixer types, is then evaluated based on the mechanical work expended to penetrate a rotating impeller into a confined powder bed using the Freeman Technology FT4 rheometer. This will also enable the sensitivity to the shear strain rate to be quantified. The outcome provides a guideline on the amount of MgSt to be used for its effect on the flowability.

Table 1. Mixing speed and time for achieving mixing equivalency adapted from Barling et al. [26], Elsevier, 2015.

Coating Device	Speed (RPM)	Mixing Time (Minutes)
Turbula	72	45
TRV25	235	20
Mechanofusion	600	1

2. Materials and Methods

2.1. α-Lactose Monohydrate (α-LM) Crystals

α-lactose monohydrate ($C_{12}H_{22}C_{11}\cdot H_2O$) is the hydrate crystal form of lactose sugar, widely used in the pharmaceutical industry as an excipient for tablet formulation as well as a host for the delivery of the active pharmaceutical ingredient (API) in the dry powder inhaler. The α-LM crystals used in this work were 'Pharmatose 50M' donated by DFE Pharma. They were first classified by sieving into the size range from 200 to 400 μm using BS410 sieves. Large particles were used deliberately so that the fluid drag of the medium does not affect the flowability testing. The surfaces of α-LM are very energetic with heterogeneous energy distribution, referred to as hot spots [27]. They attract a considerable amount of very fine particles of the same material, due to attrition during manufacturing, herein referred to as debris [28]. In fact, for dry powder inhalation, additional fine α-LM powder is often added to blind these hot spots, so that the adhesion and detachment of the active pharmaceutical ingredients are more controllable [29]. The crystals were therefore washed and rinsed in a fine sieve to separate the adhered lactose debris with isopropyl alcohol (propa-2-ol, supplied by VWR, Leighton Buzzard, UK). Crystals of α-LM are not soluble in this liquid carrier, and the process removes fines that are adhered to the crystal surfaces [30]. Following this stage, the crystals were left in a fume hood to air dry at room temperature prior to further analysis.

2.2. Magnesium Stearate (MgSt)

Magnesium stearate, a solid soap with a chemical composition of Mg $(C_{18}H_{35}O_2)_2$, is a white cohesive powder, commonly used as a lubricant in tablet and capsule formulations to reduce friction, and to a lesser extent as a glidant and anti-adherent. In the latter case, it is also used in dry powder inhaler formulations. Pure MgSt is often not available commercially, as it is very cohesive and does not possess the best lubricating properties [20]. Commercial MgSt powder is usually a blend of several different fatty acids (for example stearic acid and palmitic acid), which performs better as a lubricant compared to its unblended counterpart. The MgSt powder (MF-2-V Premium grade, vegetable source) used in this work was donated by Peter Greven (Bad Münstereifel, Germany). It has stearic acid and palmitic acid contents of between 63% and 27%, respectively. It is in crystalline form and has plate-shaped morphology. It is easily deformed and smeared over surfaces by frictional traction, so its size changes during the mixing.

2.3. Coating α-Lactose Monohydrate Crystals with Magnesium Stearate Powder

The washed and dried α-LM crystals were divided into a series of smaller samples, each having a fixed mass of 30 g. These samples were subsequently subjected to blending with MgSt powder of different weight percentages (0.2, 0.5, 1, 2, and 5% w/w) through mechanical means using a Turbula T2F mixer (GlenMills, Clifton, NJ, USA), high shear granulator (ProCepT, Zelzate, Belgium), and Mechanofusion mixer (Hosokawa Micron Ltd., Cheshire, UK). The operating conditions used for these mixing and coating devices (Table 1) were selected according to the 'mixing equivalency' procedure proposed by Barling et al. [26], assuming that the ProCepT design and operation are similar to those of the TRV25 mixer.

Blending of the α-LM and MgSt powder mixtures in the Turbula mixer was performed through a three-dimensional harmonic interaction of rotation, translation, and inversion of a 1 L glass vessel, in which they were tumbled around and mixed. The second mixer was the ProCepT high shear granulator. It has a three-bladed impeller in a 250 mL glass vessel. In the Mechanofusion device, the powder is pressed against the wall by centrifugal action and sheared against a press head, by which MgSt is 'smeared' over the surfaces of α-LM.

2.4. Flowability Measurement Using FT4 Rheometer

The influence of MgSt coating on the flowability of α-LM particles was assessed using an FT4 Rheometer (Freeman Technology, Tewkesbury, UK). This is a dynamic powder tester, which measures the mechanical work (commonly referred to as the 'flow energy') expended by a two-bladed impeller, as it rotates and penetrates through a bed of powder of known volume in a glass vessel. In the downward motion, the impeller rotates anti-clockwise and applies compressive and shear stresses on the powder bed. Expressing the work per unit mass of the powder, which has undergone shear straining, is termed the specific downward flow energy (SDFE). It is indicative of the ease with which bulk powder flows under compressive and shear stresses. Moreover, as the impeller is retracted from the base to the free surface, whilst cutting and lifting the bed, the associated work per unit mass of the sheared bed is commonly referred to as the specific upward flow energy (SUFE). Under this configuration, the powder bed is relatively unconfined and is subjected to tensile stresses, hence the energy expenditure is highly dependent on the bulk cohesion. Bulk friction and cohesion play a role here, but the former is the more dominant contributor in the downward test, as it will be shown in the difference of the results between SDFE and SUFE later below. Samples of α-LM crystals were treated with five different mass percentages of MgSt (0.2, 0.5, 1.0, and 5.0%) using the three mixing technologies described above. Triplicate measurements of both specific downward and upward flow energies were carried out and the results are reported below.

2.5. Powder Stability and Sensitivity to Shear Strain Rate

Powder beds may undergo segregation and attrition under shear straining, depending on the mechanical strength and particle size distribution. By repeating the measurement of powder flow energy in a series of consecutive tests a number of times, and comparing the energy recorded for each repetition, the stability of a powder can be inferred. The measurement can also be extended to assess the shear strain rate sensitivity of a powder by varying the impeller speed. A "stable" and "strain rate insensitive" powder should result in little to no change in the flow energy regardless of the number of times the measurement is repeated or different impeller speeds used. The standard procedure, as recommended by Freeman Technology, is followed here [31]. It involves repeating the flow energy measurement of the same sample at an impeller speed of 100 RPM eight times (referred to Test 1 to 8), followed by another three measurements (Test 8 to 11) at decreasing impeller speeds of 70, 40, and 10 RPM, as shown in Figure 1. A conditioning cycle is run between two consecutive measurements to remove any residual strains and stresses in the powder from the previous measurement.

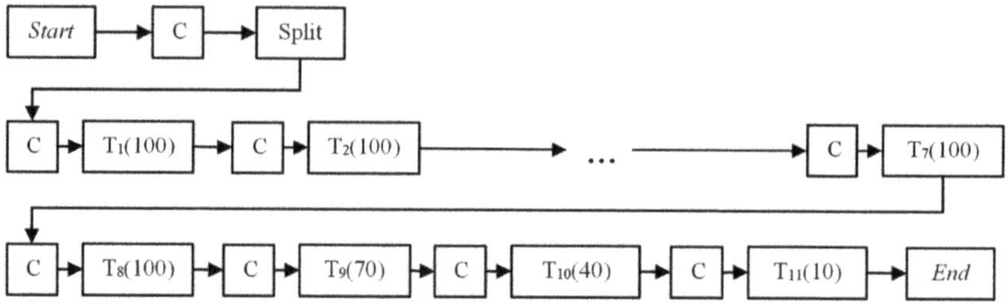

Figure 1. Flow chart of 11 standard FT4 tests (C: Conditioning; T_N(RPM)): Test number N at specific RPM).

2.6. Quantification of Particle Breakage through Sieving

Depending on the stresses experienced, the relative motion of particles, with respect to each other, could cause surface wear, attrition, or even fragmentation if the particles are weak. As the α-LM crystals are subjected to different levels of mechanical stresses during the blending process, a certain extent of damage is inflicted. Size classification by sieving of crystals, treated with the three mixing technologies discussed earlier, indicates their breakage of the crystals. In addition to the sieve corresponding to the lower limit of the sieve cuts used to initially classify the crystals (200 μm), two smaller sieve sizes (180 and 150 μm) were also used to analyse the debris formed. The size analysis was repeated three times for each of the mixing technologies.

3. Experimental Results and Discussions

3.1. Effect of Washing on the Flow Behaviour of α-LM

The presence of fines in a granular assembly changes its flow properties (see for example [32]). The consequences of having fines in a particulate system could be double-edged. In some cases, fines could provide a gliding effect by three-body rolling [15]. Alternatively, the presence of fines could result in a more compact and cohesive packing of the powder bed, leading to poor flow [33,34]. A comparison of the specific downward flow energy (SDFE) measured for twelve samples is shown in Figure 2; eight samples as supplied, i.e., unwashed, and four washed samples, showing test to test variation. For each sample, eleven consecutive tests were carried out to test their stability and strain rate sensitivity, with the first eight tests conducted at 100 RPM and the rest, i.e., nine to eleven were run at decreasing impeller speeds of 70, 40, and 10 RPM. The SDFE recorded for the unwashed α-LM samples (filled symbols) varies widely. The data recorded in Test 1 (*cf.* Figure 1) for the fresh samples are taken as the most representative of all the tests. The powder bed carries no previous measurement history in this case and has not experienced attrition. Despite this, the SDFE of the eight unwashed samples measured in Test 1 still ranges from approximately 9 mJ/g to ~120 mJ/g, an approximately 12-fold difference. In addition, the observed trend also varies greatly when the test is repeated (*cf.* Sample 5). This shows that the presence of fines in α-LM samples has an adverse effect on the stability of the powder flow in terms of temporal repeatability of the test results. This could be attributed to the detachment of the debris and its segregation. The strain rate sensitivity tests (Tests 8 to 11) show a fairly consistent upward trend of the SDFE as the impeller speed is reduced from 100 to 10 RPM. Compared to the unwashed samples, washed α-LM samples (empty symbols) exhibit a more repeatable pattern. Using the same sample, the resistance of the powder bed to flow is found to only increase marginally from Tests 1 to8. Decreasing the impeller speed (Tests 9 to 11), the SDFE increases, similar to the trend of the unwashed samples. In general, the SDFE recorded for the washed samples is much lower than that of the unwashed ones, indicating better flowability.

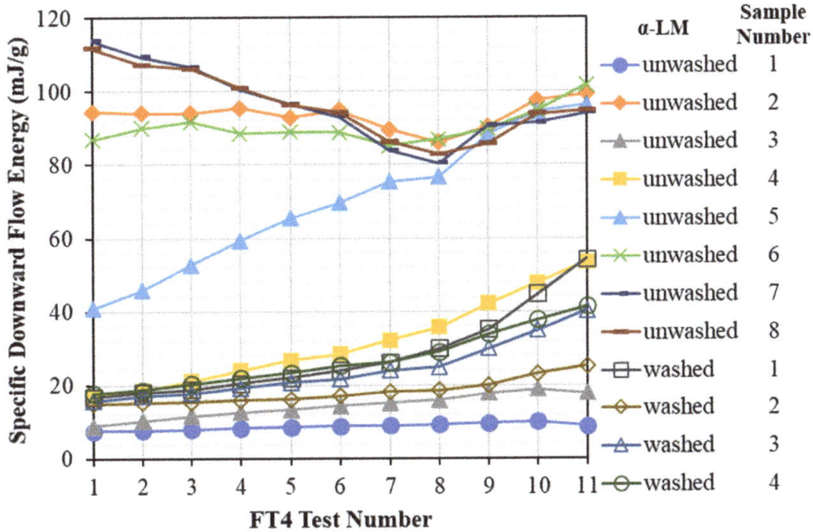

Figure 2. Specific downward flow energy of α-lactose monohydrate (α-LM) for twelve samples, eight samples as received and unwashed, and four samples washed with isopropyl alcohol, showing sample to sample variation. Data are for eleven consecutively repeated tests for each sample, showing test stability and strain rate sensitivity, the first eight at 100 RPM and the last three at 70, 40, and 10 RPM.

3.2. Effect of MgSt on α-LM Bulk Flow

Coating of MgSt onto α-LM crystals changes the flow behaviour of the powder bed. However, the way coating is carried out is influential, as the interactions are no longer limited to particles of the same kind, but rather a more complex interaction of binary mixture of MgSt and α-LM. The effect of mixing MgSt with α-LM using the three mixers on the flow behaviour is shown in Figure 3. Note that each data point shown in the plot is a mean value of three measurements taken from Test 1 for every MgSt composition tested, and the associated error bar represents the standard deviation of the three repetitions.

The diamond data point legends ◆ and ◇ in Figure 3a,b represent the specific downward and upward flow energies, respectively, recorded for unwashed and untreated α-LM samples (mixing done without the addition of α-LM). With the addition of MgSt, the ProCepT-treated samples exhibit the least resistance to flow. Samples treated by Turbula and Mechanofusion exhibit remarkably similar specific flow energy for all MgSt concentrations, despite having undergone extreme shear strain rates. For all mixer types, increasing the composition of MgSt to 1 wt% improves the flow performance. Above 1 wt%, the specific upward and downward flow energies both converge to asymptotic levels, suggesting that the flow behaviour of the α-LM samples has now stabilised and is dominated by MgSt. Notably, the concentration of 1 wt% MgSt appears to be the threshold value, beyond which there is no significant improvement in the flow performance of α-LM, regardless of the type of mixer used.

An interesting phenomenon is observed in the strain rate sensitivity of the α-LM samples when MgSt is added. For α-LM samples alone, reducing the impeller speed leads to an increase in the specific flow energies (Tests 8–11 in Figure 2). The opposite trend prevails for the samples coated with MgSt for all three cases of mixer type, as shown in Supplementary Material, Figures S1–S3.

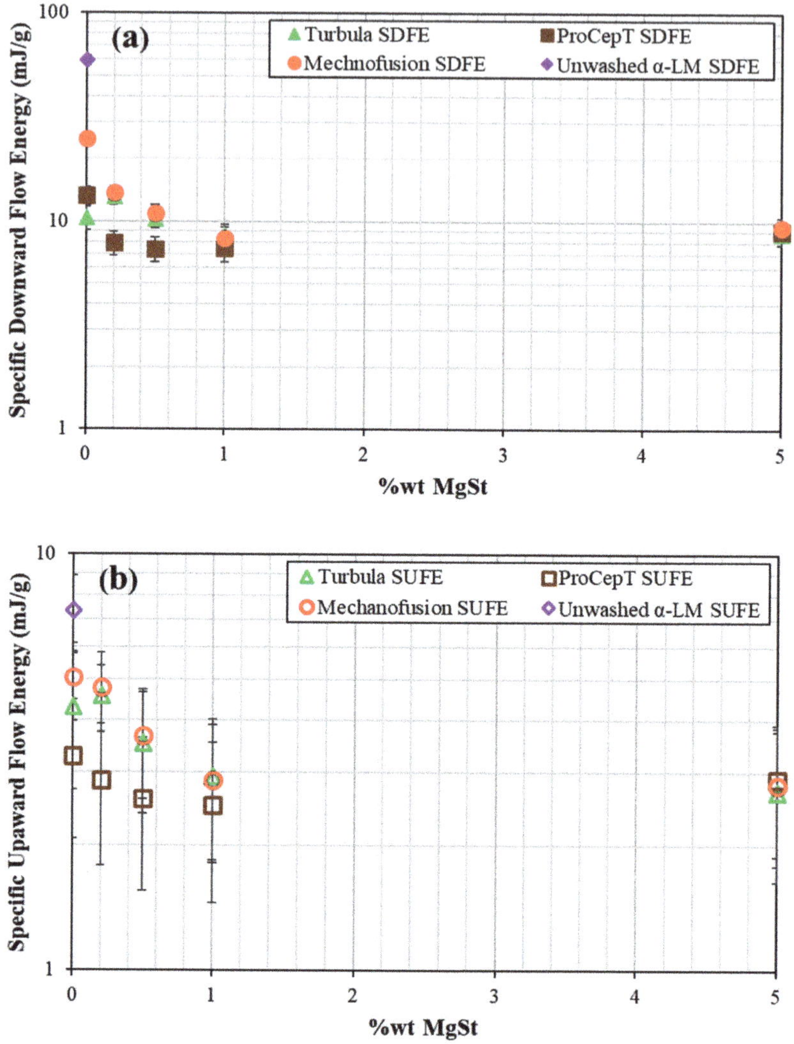

Figure 3. Specific flow energies of α-lactose monohydrate treated with different percentages of magnesium stearate in the three mixers. (**a**) Downward test; (**b**) Upward test. Each data point represents the mean specific flow energy of the triplicate FT4 measurements taken from Test 1. The error bar represents the standard deviation. Note: Error bars that cannot be seen are smaller than the symbols.

3.3. Effect of Mixer Type on Powder Flow

The operating procedure used for mixing α-LM and MgSt follows the work of Barling et al. [26]. According to their work, a mixing equivalency is achieved amongst the three mixers when the mixing is done under the operating conditions specified in Table 1. A comparison of the SDFE and SUFE recorded for the three mixers at different mass percentages of MgSt is shown in Figure 3a,b, respectively. There exists a large difference in the SDFE among the three mixers at 0% MgSt. The α-LM samples treated with Mechanofusion have the highest SDFE (~25 mJ/g) and those treated with Turbula have the lowest SDFE (~10 mJ/g). The difference among the three mixers diminishes with the MgSt wt% and converges at 1 wt% MgSt. Apart from 0% MgSt, samples treated with Turbula and Mechanofusion show almost identical behaviour, suggesting a mixing equivalency is achieved using the proposed operating procedure. Unfortunately, the same remark could not be made for ProCepT-treated samples. In this case, the difference in SDFE between the initial two consecutive cases (0 and 0.1 wt% MgSt) diminishes quickly, and then reaches a plateau. The results seem to suggest that ProCepT is a more efficient coating device compared to Turbula and Mechanofusion since a lower amount of MgSt is needed to improve the flow behaviour of the α-LM bulk solids. However, the 'improved' flowability seen in ProCepT at low MgSt wt% is likely due to the rounding of the corners and edges of the crystals, rather than the efficiency of the mixer in spreading MgSt over the surfaces of α-LM crystals. The latter is most effective when a dense particle bed is strongly sheared. This facilitates the 'smearing' of MgSt particles over the surfaces of the host particles. In the case of the ProCepT mixer, the impeller action aerates the bed, thereby reducing the bulk friction and shear viscosity of the particle bed. The scanning electron microscope images, taken from a Hitachi Benchtop SEM TM3030 Plus, as shown in Figure 4, indicate that chips and fragments are present in ProCepT-treated samples, and, also, to a lesser extent in the case of Mechanofusion, as it can be implicitly inferred from the extent of breakage reported below. The magnification of the images is shown as a line bar with a 1 mm scale.

Figure 4. Scanning electron microscope images of isopropyl-alcohol-washed α-lactose monohydrate crystals treated with Turbula, Mechanofusion, and ProCepT with 0.2 wt% MgSt, showing notable breakage in the case of ProCepT mixer.

In terms of SUFE, the trend of the results of samples treated with MgSt by Turbula and Mechanofusion almost overlap. ProCepT-treated samples generally have much better flow performance compared to Turbula- and Mechanofusion-treated samples at the same MgSt wt%.

3.4. Extent of Breakage

α-LM crystals experience different levels of mechanical stresses due to agitation in the three mixers. Vigorous relative motion between the particles during the mixing process damages the particles, and a reduction in particle size is expected. Following the mixing process, a sieving analysis is performed to quantify and compare the extent of size reduction of the α-LM particles in the three mixers. The size distributions of the α-LM samples treated with Turbula, ProCepT, and Mechanofusion with 0.2 wt% MgSt are shown in Table 2. The size of the α-LM particles is reduced as a result of mixing, but only to a small extent. The following criterion is adopted to quantify the extent of breakage: the mass of particles collected by a sieve that is two sizes (according to BS410 sieves) below the feed sieve size is used as a breakage indicator. The extent of particle breakage is found to be in the range of ~0.1%. This suggests that the breakage mode of the α-LM particles in the three mixers of interest is dominated by attrition and surface abrasion. The amount of debris collected in Turbula-treated samples is the least amongst the three mixers. This is expected as the mixing action is the gentlest amongst the three mixers and involves no moving mechanical part that comes into direct contact with the particles in the mixing chamber.

Table 2. Particle size distribution of α-lactose monohydrate particles, treated by Turbula, ProCepT, and Mechanofusion with 0.2 wt% MgSt under equivalent mixing conditions according to Barling et al. [26].

Sieve Size, μm	% Mass					
	Mechanofusion		ProCepT		Turbula	
	Mean	SD	Mean	SD	Mean	SD
<150	0.13	0.03	0.19	0.10	0.08	0.05
150–180	0.18	0.18	0.15	0.00	0.19	0.11
180–200	0.62	0.33	1.11	0.24	0.41	0.15
>200	99.06	0.29	98.55	0.29	99.32	0.23

4. Conclusions

The influence of mixer type on the coating of α-LM crystals with a small quantity of MgSt powder has been assessed. Three different types of mixers, i.e., Turbula, ProCepT, and Mechanofusion have been used. The assessment method is based on the effect of MgSt coating on flowability, as measured by the mechanical work expended by penetrating a rotating impeller into a powder bed by using an FT4 rheometer. The flowability of α-LM crystals is affected by the presence of debris, i.e., fine particles of the same material, adhered to their surfaces, causing a wide sample to sample variation. Washing the α-LM crystals using a liquid carrier, which does not dissolve them, reduces the variation as well as the magnitude of the expended work significantly, implying improved flowability. The addition of MgSt to α-LM particles further improves the flowability for powders coated in all of the three mixers under equivalent mixing conditions, described in the work of Barling et al. [26]. There exists a threshold at ~1 wt% MgSt above which the further addition of MgSt does not improve the bulk flow performance of α-LM. Samples coated by both Turbula and Mechanofusion mixers show remarkably similar flow performance, despite extreme shear strain rate. At the same MgSt composition, the ProCepT-treated α-LM samples consistently perform better in terms of flow behaviour. This is attributed to collisional impacts in this mixer, rounding off sharp corners and edges. Mixing equivalency is best achieved between the Turbula and Mechanofusion mixers.

A breakage analysis due to mixing is also performed, and the results suggest that α-LM samples coated using Turbula experience the least damage. The particle size distribution of the treated samples reveals that attrition and surface abrasion create the dominant breakage mode, particularly in the case of Turbula and Mechanofusion mixers, with their breakage extent being found to be in the range of approximately 0.1%.

Supplementary Materials: The following are available online at https://www.mdpi.com/article/10.3390/pharmaceutics13081211/s1, Figure S1. Specific downward flow energy of α-lactose monohydrate coated with different mass percentages of magnesium stearate, prepared by Turbula, for 11 consecutively-repeated tests for each sample. Figure S2. Specific downward flow energy of α-lactose monohydrate coated with different mass percentages of magnesium stearate, prepared by ProCepT, for 11 consecutively-repeated tests for each sample. Figure S3. Specific downward flow energy of α-lactose monohydrate coated with different mass percentages of magnesium stearate, prepared by Mechanofusion, for 11 consecutively-repeated tests for each sample.

Author Contributions: Conceptualization, M.G.; methodology, W.P.G.; validation, A.M.S. and W.P.G.; formal analysis, A.M.S.; investigation, A.M.S.; data curation, A.M.S.; writing—original draft preparation, W.P.G.; writing—review and editing, W.P.G. and M.G.; visualization, A.M.S. and W.P.G.; supervision, W.P.G. and M.G.; funding acquisition, M.G. All authors have read and agreed to the published version of the manuscript.

Funding: This research was funded by Engineering and Physical Sciences Research Council, grant number EP/001766/1.

Institutional Review Board Statement: Not applicable.

Informed Consent Statement: Not applicable.

Acknowledgments: The financial support of the Engineering and Physical Sciences Research Council for the project (grant number EP/001766/1), as a part of 'Friction: The Tribology Enigma' Programme Grant (www.friction.org.uk, date accessed: 20 July 2021) is gratefully acknowledged. The authors are grateful to Hosokawa Micron Ltd., Runcorn, UK, for their support in providing access to the Mechanofusion device and are thankful to Kathryn Hipkins and Erin Holman for their help. The test material, crystals of α-lactose monohydrate were kindly donated by DFE Pharma, Goch, Germany, for which we wish to thank Bastiaan Dickhoff for facilitating it. Our thanks are also extended to Peter Greven (Bad Münstereifel, Germany) for kindly donating the magnesium stearate powder.

Conflicts of Interest: The authors declare no conflict of interest. The funders had no role in the design of the study, in the collection, analyses, or interpretation of data, in the writing of the manuscript, or in the decision to publish the results.

References

1. Chen, Y.; Yang, J.; Dave, R.N.; Pfeffer, R. Fluidization of coated group C powders. *AIChE J.* **2008**, *54*, 104–121. [CrossRef]
2. Quintanilla, M.A.S.; Castellanos, A.; Valverde, J.M. Correlation between bulk stresses and interparticle contact forces in fine powders. *Phys. Rev. E* **2001**, *64*, 031301. [CrossRef] [PubMed]
3. Espin, M.J.; Ebri, J.M.P.; Valverde, J.M. Tensile strength and compressibility of fine CaCO3 powders. Effect of nanosilica addition. *Chem. Eng. J.* **2019**, *378*, 122166. [CrossRef]
4. Faqih, A.M.N.; Mehrotra, A.; Hammond, S.V.; Muzzio, F.J. Effect of moisture and magnesium stearate concentration on flow properties of cohesive granular materials. *Int. J. Pharm.* **2007**, *336*, 338–345. [CrossRef]
5. Ouabbas, Y.; Chamayou, A.; Galet, L.; Baron, M.; Thomas, G.; Grosseau, P.; Guilhot, B. Surface modification of silica particles by dry coating: Characterization and powder ageing. *Powder Technol.* **2009**, *190*, 200–209. [CrossRef]
6. Chattoraj, S.; Shi, L.; Sun, C.C. Profoundly improving flow properties of a cohesive cellulose powder by surface coating with nano-silica through comilling. *J. Pharm. Sci.* **2011**, *100*, 4943–4952. [CrossRef] [PubMed]
7. Kunnath, K.; Huang, Z.; Chen, L.; Zheng, K.; Davé, R. Improved properties of fine active pharmaceutical ingredient powder blends and tablets at high drug loading via dry particle coating. *Int. J. Pharm.* **2018**, *543*, 288–299. [CrossRef]
8. Muzzio, F.J.; Shinbrot, T.; Glasser, B.J. Powder technology in the pharmaceutical industry: The need to catch up fast. *Powder Technol.* **2002**, *124*, 1–7. [CrossRef]
9. Bridgwater, J. Mixing of powders and granular materials by mechanical means—A perspective. *Particuology* **2012**, *10*, 397–427. [CrossRef]
10. Mullarney, M.P.; Beach, L.E.; Dave, R.N.; Langdon, B.A.; Polizzi, M.; Blackwood, D.O. Applying dry powder coatings to pharmaceutical powders using a comil for improving powder flow and bulk density. *Powder Technol.* **2011**, *212*, 397–402. [CrossRef]
11. Pingali, K.; Mendez, R.; Lewis, D.; Michniak-Kohn, B.; Cuitino, A.; Muzzio, F. Evaluation of strain-induced hydrophobicity of pharmaceutical blends and its effect on drug release rate under multiple compression conditions. *Drug Dev. Ind. Pharm.* **2011**, *37*, 428–435. [CrossRef] [PubMed]
12. Sato, A.; Serris, E.; Grosseau, P.; Thomas, G.; Chamayou, A.; Galet, L.; Baron, M. Effect of operating conditions on dry particle coating in a high shear mixer. *Powder Technol.* **2012**, *229*, 97–103. [CrossRef]

13. Conesa, C.; Saleh, K.; Thomas, A.; Guigon, P.; Guillot, N. Characterization of flow properties of powder coatings used in the automotive industry. *KONA Powder Part. J.* **2004**, *22*, 94–106. [CrossRef]
14. Castellanos, A. The relationship between attractive interparticle forces and bulk behaviour in dry and uncharged fine powders. *Adv. Phys.* **2005**, *54*, 263–376. [CrossRef]
15. Fulchini, F.; Zafar, U.; Hare, C.; Ghadiri, M.; Tantawy, H.; Ahmadian, H.; Poletto, M. Relationship between surface area coverage of flow-aids and flowability of cohesive particles. *Powder Technol.* **2017**, *322*, 417–427. [CrossRef]
16. Jallo, L.J.; Ghoroi, C.; Gurumurthy, L.; Patel, U.; Dave, R.N. Improvement of flow and bulk density of pharmaceutical powders using surface modification. *Int. J. Pharm.* **2012**, *423*, 213–225. [CrossRef] [PubMed]
17. Sato, A.; Serris, E.; Grosseau, P.; Thomas, G.; Galet, L.; Chamayou, A.; Baron, M. Experiment and simulation of dry particle coating. *Chem. Eng. Sci.* **2013**, *86*, 164–172. [CrossRef]
18. Zhou, Q.; Qu, L.; Gengenbach, T.; Denman, J.A.; Larson, I.; Stewart, P.J.; Morton, D.A. Investigation of the extent of surface coating via mechanofusion with varying additive levels and the influences on bulk powder flow properties. *Int. J. Pharm.* **2011**, *413*, 36–43. [CrossRef]
19. Hussain, M.S.H.; York, P.; Timmins, P. A study of the formation of magnesium stearate film on sodium chloride using energy-dispersive X-ray analysis. *Int. J. Pharm.* **1988**, *42*, 89–95. [CrossRef]
20. Marwaha, S.B.; Rubinstein, M.H. Structure-lubricity evaluation of magnesium stearate. *Int. J. Pharm.* **1988**, *43*, 249–255. [CrossRef]
21. Shah, A.C.; Mlodozeniec, A.R. Mechanism of surface lubrication: Influence of duration of lubricant-excipient mixing on processing characteristics of powders properties of compressed tablets. *J. Pharm. Sci.* **1977**, *66*, 1377–1382. [CrossRef]
22. Portillo, P.M.; Muzzio, F.J.; Ierapetritou, M.G. Characterizing powder mixing processes utilizing compartment models. *Int. J. Pharm.* **2006**, *320*, 14–22. [CrossRef] [PubMed]
23. Siraj, M.S.; Radl, S.; Glasser, B.J.; Khinast, J.G. Effect of blade angle and particle size on powder mixing performance in a rectangular box. *Powder Technol.* **2011**, *211*, 100–113. [CrossRef]
24. Horibe, M.; Sonoda, R.; Watano, S. Scale-Up of lubricant mixing process by using V-Type blender based on discrete element method. *Chem. Pharm. Bull.* **2018**, *66*, 548–553. [CrossRef]
25. Asachi, M.; Nourafkan, E.; Hassanpour, A. A review of current techniques for the evaluation of powder mixing. *Adv. Powder Technol.* **2018**, *29*, 1525–1549. [CrossRef]
26. Barling, D.; Morton, D.A.V.; Hapgood, K. Pharmaceutical dry powder blending and scale-up: Maintaining equivalent mixing conditions using a coloured tracer powder. *Powder Technol.* **2015**, *270*, 461–469. [CrossRef]
27. Ho, R.; Muresan, A.S.; Hebbink, G.A.; Heng, J.Y.Y. Influence of fines on the surface energy heterogeneity of lactose for pulmonary drug delivery. *Int. J. Pharm.* **2010**, *388*, 88–94. [CrossRef] [PubMed]
28. Bonakdar, T.; Ghadiri, M. Analysis of pin milling of pharmaceutical materials. *Int. J. Pharm.* **2018**, *552*, 394–400. [CrossRef]
29. Jones, M.D.; Price, R. The Influence of fine excipient particles on the performance of carrier-based dry powder inhalation formulations. *Pharm. Res.* **2006**, *23*, 1665–1674. [CrossRef]
30. Majd, F.; Nickerson, T.A. Effect of Alcohols on Lactose Solubility. *J. Dairy Sci.* **1976**, *59*, 1025–1032. [CrossRef]
31. Freeman Technology. Stability & Variable Flow Rate Method. In *FT4 Rheometer User Manual*; Freeman Technology: Gloucester, UK, 2007.
32. Madian, A.; Leturia, M.; Ablitzer, C.; Matheron, P.; Bernard-Granger, G.; Saleh, K. Impact of fine particles on the rheological properties of uranium dioxide powders. *Nucl. Eng. Technol.* **2020**, *52*, 1714–1723. [CrossRef]
33. Kurz, H.P.; Munz, G. The influence of particle size distribution on the flow properties of limestone powders. *Powder Technol.* **1975**, *11*, 37–40. [CrossRef]
34. Liu, Y.; Lu, H.; Guo, X.; Gong, X.; Sun, X.; Zhang, Z. The influence of fine particles on bulk and flow behavior of pulverized coal. *Powder Technol.* **2016**, *303*, 212–227. [CrossRef]

pharmaceutics

MDPI

Article

Discrete Element Method Evaluation of Triboelectric Charging Due to Powder Handling in the Capsule of a DPI

Francesca Orsola Alfano *, Alberto Di Renzo * and Francesco Paolo Di Maio

DIMES Department, University of Calabria, 87036 Rende, Italy; francesco.dimaio@unical.it
* Correspondence: francesca.alfano@unical.it (F.O.A.); alberto.direnzo@unical.it (A.D.R.)

Abstract: The generation and accumulation of an electrostatic charge from handling pharmaceutical powders is a well-known phenomenon, given the insulating nature of most APIs (Active Pharmaceutical Ingredients) and excipients. In capsule-based DPIs (Dry Powder Inhalers), the formulation is stored in a gelatine capsule placed in the inhaler just before inhalation. The action of capsule filling, as well as tumbling or vibration effects during the capsule life cycle, implies a consistent amount of particle–particle and particle–wall contacts. A significant contact-induced electrostatic charging can then take place, potentially affecting the inhaler's efficiency. DEM (Discrete Element Method) simulations were performed on a carrier-based DPI formulation (salbutamol–lactose) to evaluate such effects. After performing a comparison with the experimental data on a carrier-only system under similar conditions, a detailed analysis was conducted on two carrier–API configurations with different API loadings per carrier particle. The charge acquired by the two solid phases was tracked in both the initial particle settling and the capsule shaking process. Alternating positive–negative charging was observed. Particle charging was then investigated in relation to the collision statistics, tracking the particle–particle and particle–wall events for the carrier and API. Finally, an analysis of the relative importance of electrostatic, cohesive/adhesive, and inertial forces allowed the importance of each term in determining the trajectory of the powder particles to be estimated.

Keywords: triboelectric charging; electrostatics; DPI; DEM; inhalation; carrier-based formulation

Citation: Alfano, F.O.; Di Renzo, A.; Di Maio, F.P. Discrete Element Method Evaluation of Triboelectric Charging Due to Powder Handling in the Capsule of a DPI. *Pharmaceutics* **2023**, *15*, 1762. https://doi.org/10.3390/pharmaceutics15061762

Academic Editor: Colin Hare

Received: 18 April 2023
Revised: 15 June 2023
Accepted: 16 June 2023
Published: 18 June 2023

1. Introduction

Pharmaceutical powders are used in various stages of drug manufacturing and are an essential part of the pharmaceutical industry. As for most non-conductive materials, the handling and processing of these powders often leads to the transfer and accumulation of electrostatic charges. This, in turn, can cause issues in drug manufacturing, such as cross-contamination, intermittent powder flow, and even explosion hazards [1]. Poor powder flow can lead to difficulties in powder handling, filling, and packaging processes [2]. The agglomeration of particles can affect the powder's dissolution rate, which can impact the drug's bioavailability [3]. Segregation of the powder blend can result in non-uniformity of the drug content and, consequently, inconsistent dosing [4]. Therefore, understanding the mechanisms of electrostatic charge transfer and accumulation in pharmaceutical powders is crucial for ensuring the safety and quality of drug manufacturing processes.

Tribocharging is the process responsible for the charge transfer between contacting material surfaces, and it takes place upon detachment [5]. Carrier-based inhalation formulations used in Dry Powder Inhalers (DPIs) are composed of two solid components (i.e., the active pharmaceutical ingredient (API) and the carrier) with significantly different sizes and exhibit a complex tribocharging behavior. Effects such as the bipolar charging phenomena, i.e., particles the same material charged both positively and negatively, have been observed [6]. In some cases, experimental studies that focused on the tribocharging phenomenon [7–11] reported somewhat conflicting results, even in the polarity of the reported charge, calling for additional investigations at small scales. The charge polarity of lactose particles (a widely used carrier and excipient) depends on the chosen inhaler [8], on the material of the containing capsule [12], and even on their manufacturing process; for example, milled lactose tends to charge positively, while sieved lactose tends to charge negatively [8]. Salbutamol sulphate (a commonly used API in inhalation formulations) tends to acquire a positive charge if it is amorphous and a negative charge if it is in its crystalline form [11]. Peart [13] reports that salbutamol charges positively when in contact with PVC, while lactose charges negatively. However, when salbutamol particles detach from lactose, the opposite polarity is measured for the two materials (salbutamol's specific charge is about -3000 nC/g, and lactose's specific charge is about 100 nC/g).

When strong size polydispersity is present, the charge polarity also depends on particle size, with smaller particles that charge negatively and larger particles that charge positively [11,14–16]. This effect cannot be predicted using the triboelectric series, which is usually employed to rank the electronegativity of powder materials [1,2]. The concentration of the API plays a role as well, with the electrostatic charge of drug–carrier mixtures usually decreasing by increasing the concentration of the API [10,17,18].

Electrostatics also plays a major role in the design and use of hard capsules for capsule-based DPI. Chow et al. [19] found that mechanical vibration such as tapping induced significant static charge on lactose stored in a gelatine capsule. Hoe et al. [10,17] hypothesized that the surface charge on the capsule might be high enough to ionize the surrounding air. Understanding such a variety of observations requires a careful analysis of the links between microscopic charge transfer processes and the macroscopic manifestations. Particle-scale information is an important ingredient that is accessible only through simulation.

In recent years, Discrete Element Method (DEM) modeling has emerged as a powerful simulation tool for studying the behavior of powders in inhalation devices [20]. Coupled DEM-CFD (Computational Fluid Dynamics) studies showed the detailed motion of particles from the initial dispersion in air through to the mouthpiece [21–25]. In the last decade, model formulations have been introduced to extend the particle-scale contact tracking capability of DEM with surface–surface charge transfer and physical electrostatic interaction models (see, e.g., [26–28]), allowing triboelectric charging phenomena at the particle scale to be studied [27,29–33]. Naik et al. [34] studied the triboelectrification of binary mixtures of drug and excipient in a blender. They found that particle–particle interactions enhance the electrostatic interaction between the drug and excipient and decrease the overall charge transfer between particles and walls. The overall charging process that resulted was mitigated by this effect, leading to a lower total charge than that acquired by single components, and differences in excipient concentration, in some cases, caused charge polarity reversal. The importance of particle–particle contact charging was also highlighted by Chowdhury et al. [33]. Zhu et al. [35] studied the contact electrification effect of selected API agglomerates in the Turbuhaler®, finding a reduction in inhaler efficiency due to the triboelectrification of powders. Specific numerical studies on the motion of particles in the capsule of capsule-based DPIs are available in the literature [36–39], but the effects of charges generated upon contact on the release of powder from the capsule itself are not taken into account. Most of previous works in DPI applications focus on the influence of electrostatic interactions between previously charged particles, rather than the charge buildup process, and the charging of capsule walls is typically neglected.

The aim of the present study is to apply extended DEM simulations to evaluate triboelectric charging dynamics of a lactose–salbutamol binary mixture in the chargeable capsule of a DPI. The capsule is vibrated to simulate the routine handling operations to which a filled capsule is subjected during its life cycle, with the aim to assess the extent to which such movements give rise to electrostatic charge accumulation. The simulation tool offers the possibility to relate individual contact events and local charge transfer to the macroscopic influence that such a charge exerts on the material dynamics. The model formulation, simulation setup, and material parameter are presented first. After validation, the results of simulations of representative systems at low- and high-dosage conditions are discussed.

2. Materials and Methods

2.1. DEM Simulation Technique

DEM is a numerical method for simulating the behavior of a system of discrete particles by tracking their individual motions and interactions due to the forces and momenta acting on each of them. Spherical particles are considered in the present work.

The translational ($\vec{v}_i$) and rotational ($\vec{\omega}_i$) velocities of the i-th particle are calculated by integrating Newton's second law of motion:

$$m_i \frac{d\vec{v}_i}{dt} = \vec{F}_T = \sum_{j=1}^{Nc} \vec{F}^c_{i,j} + \sum_{j=1}^{Nc} \vec{F}^e_{i,j} + \vec{F}_{g,i} \quad (1)$$

$$I_i \frac{d\vec{\omega}_i}{dt} = \vec{T}_T = \sum_{j=1}^{Nc} \vec{T}^c_{i,j} + \vec{T}_r \quad (2)$$

where $\vec{F}_T$ is the sum of all the forces acting in the i-th particle—contact, electrostatic, and gravitational forces; and $\vec{T}_T$ is the sum of all the torques acting on the i-th particle—contact torque and rolling resistance, T_r.

Cohesive-contact forces are modeled following the JKR contact theory [40], which accounts for attractive forces due to van der Waals effects and is hysteretic, i.e., loading and unloading cases are different in that cohesive contact is initiated at zero distance, while detachment occurs at non-zero distance. A velocity-dependent dissipation is also introduced to be able to model (coefficient of) restitution after impacts. The magnitude of the normal contact force between two contacting particles is given by the following:

$$F^n_{c,ij} = 4\sqrt{\pi\gamma E_{eq}} a^{\frac{3}{2}} - \frac{4E_{eq}}{3R_{eq}} a^3 - \eta^H_n \delta_n^{\frac{1}{4}} v_n \quad (3)$$

where γ is the surface energy, a is the radius of the contact area, v_n is the normal velocity, δ_n is the normal overlap, η^H_n is the normal damping coefficient (related to the restitution coefficient; see, e.g., [41]), and E_{eq} and R_{eq} are the equivalent Young modulus and radius of the two contacting particles (i and j) [42].

The normal overlap, δ_n, is related to the radius of the contact area as follows:

$$\delta_n = \frac{a^2}{R_{eq}} - \sqrt{\frac{4\pi\gamma a}{E_{eq}}} \quad (4)$$

As mentioned above, according to the JKR theory, during the detachment phase, the contact remains active at negative overlaps between the spheres (as actual surfaces are elongated shapes that are still in contact) until a threshold overlap is reached. The maximum attractive force, usually referred to as the pull-off force, occurs at a negative overlap and is given by the following:

$$F_{pull-off} = 3\pi\gamma R_{eq} \quad (5)$$

The tangential contribution to the contact is considered following the no-slip solution of Mindlin and Deresiewicz [43] for the frictional–elastic part and a velocity-dependent dissipation term similar to the normal direction. The tangential contact force is calculated as follows:

$$F^t_{c,ij} = -\left(\mu_s F^n_{c,ij},\ 8G_{eq}\sqrt{R_{eq}}\delta_n^{\frac{1}{2}}\delta_t + \eta^H_t \delta_n^{\frac{1}{4}} v_t\right) \quad (6)$$

where μ_s is the static friction coefficient, G_{eq} is the equivalent shear modulus, δ_t is the tangential overlap, η^H_t is the tangential damping coefficient, and v_t is the tangential velocity.

In the rotational motion, the contact torque results from the action of the tangential contact force. The rolling friction torque is calculated according to the Constant Directional Torque model [44], introducing the rolling friction coefficient, μ_r, as material parameter.

All the models presented above are described in more detail in Alfano et al. [39,45].

The DEM simulations were carried out using an in-house customized version of the open-source code MFIX (NETL MFS, Department of Energy (Morgantown, WV, USA), version 18.1.5 [46]. Johnson-Kendall-Roberts (JKR) model for the cohesive force and constant directional torque (CDT) model for the rolling friction were implemented in the original version of the code (see [45] for more details). Moreover, a different approach for wall contacts [47] was preferred to the standard MFIX formulation. Special precautions have also been taken to prevent very fine particles from being unrealistically pushed out of the domain by the carrier particles, crossing the domain boundary. In terms of hardware resources, an own-managed cluster was extensively utilized by running parallel tests on up to 32 cores.

2.2. Triboelectric Charging Model and Electrostatic Forces

Consideration of the charge transfer, buildup, and corresponding modification to the powder dynamics requires two essential ingredients: a tribocharging model for the charge transfer between surfaces after contact and a model for the interaction force between charged bodies. Our selection for both models is introduced below.

The triboelectric charging model is based on the condenser model developed by Matsusaka et al. [48] in the formulation for DEM implementation proposed by Pei et al. [27]. It was extensively presented in a previous work [31] and is shortly summarized here.

Each particle and wall element is assigned a net scalar charge that is representative of its surface-charge distribution. This charge evolves as a result of particle–particle and particle–wall collisions and contacts (Figure 1a) upon surface detachment. The charge transferred after each particle–wall impact is calculated as follows:

$$\Delta q = kS_m \left(\frac{\Phi_i - \Phi_s}{e} + \xi \frac{z_s}{4\pi\varepsilon_0} \frac{q_i}{R_i^2} \right) \tag{7}$$

where Φ_i and Φ_s are the work functions of the particle and the wall surface, respectively; e is the elementary charge (1.602×10^{-19} C); R_i is the particle radius; z_s is the cutoff distance for particle–wall charge transfer (considered as 130 nm [27]); q_i is the charge on the particle before impact; ξ is the image correction factor [49], which is set to 2; ε_0 is the vacuum permittivity (8.854 pF/m); k is the charge efficiency, which is set to 10^{-4} C m^{-2} V^{-1} [27]; and S_m is the maximum contact area, which is calculated as follows:

$$S_m = \pi \delta_{n,max} R_{eq} \tag{8}$$

where $\delta_{n,max}$ is the maximum normal overlap and R_{eq} is the equivalent radius (harmonic mean) between the two contacting surfaces.

The first term of the sum in brackets in Equation (7) is the contact potential difference, i.e., the driving force for charge exchange, while the second term arises from the image effect. Note that charge transfer occurs, increasing one surface by some charge and decreasing the other one by the same amount, so that the total system charge remains constant.

The particle–particle charge transfer is calculated in a similar fashion [27]:

$$\Delta q = kS_m \left(\frac{\Phi_i - \Phi_j}{e} + \xi \frac{z_p}{4\pi\varepsilon_0} \left(\frac{q_i}{R_j^2} - \frac{q_j}{R_i^2} \right) \right) \tag{9}$$

where z_p is the cutoff distance for the particle–particle charge transfer, and it is set to 260 nm [27].

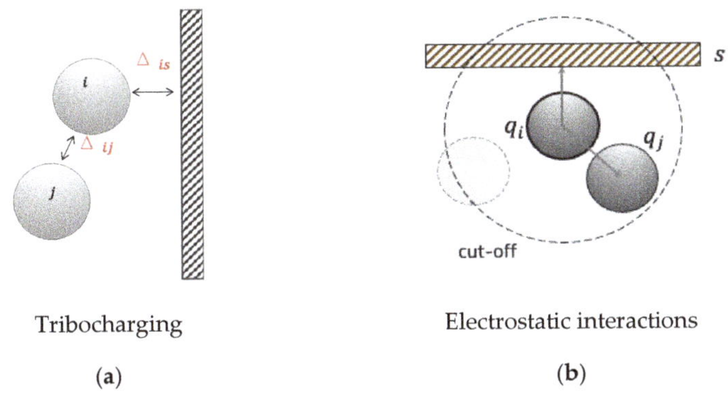

Figure 1. (**a**) Schematics of contact charging mechanism. (**b**) Graphical representation of particle–particle electrostatic forces evaluated within a cutoff distance.

Once each particle has its own charge, the second ingredient of the model, the electrostatic force between two charged particles (Figure 1b), is calculated according to Coulomb's law:

$$\vec{F}^{e}_{i,j} = \frac{1}{4\pi\varepsilon_0}\frac{q_i q_j}{r^2_{i,j}}\hat{n}_{i,j} \quad (10)$$

where $\hat{n}_{i,j}$ is the unit vector defined along the direction connecting the two particles' centers, and $r_{i,j}$ is the distance between these centers. The cutoff distance for the calculation of the electrostatic interactions is set to 1.2 times the sum of the two particles' radii.

Coulombic interactions between charged particles and the walls of the capsule were considered according to the method of mirror charges [50]:

$$\vec{F}^{e}_{i,s} = \frac{1}{4\pi\varepsilon_0}\frac{q_i^2}{(2r_{i,s})^2}\hat{n}_{i,s} \quad (11)$$

where $r_{i,s}$ is the distance between the wall surface and the center of the particle, and $\hat{n}_{i,s}$ is the unit vector perpendicular to the surface and passing through the particle center. It is useful to note that while particle–particle interactions may be attractive or repulsive depending on the charge polarity, particle–wall electrostatic interactions are always attractive.

2.3. Simulation Parameters

The reference materials for API and carrier particles are salbutamol and lactose, respectively. The diameter of the salbutamol particles is set to 5 μm, while the diameter of the lactose particles is 100 μm, considering the commercially available Inhalac® 230 as a reference.

The DEM simulation properties are reported in Table 1. The mechanical data are selected according to typical values found in the literature [51,52].

Table 1. Particles' mechanical and physical properties.

Property	Carrier	API	Capsule
Reference material	Lactose	Salbutamol	Gelatine
Diameter, d (μm)	100	5	-
Density, ρ (kg/m^3)	1500	1200	-
Sliding friction coefficient, μ_s (−)	0.5	0.5	0.5
Rolling friction coefficient, μ_r (−)	0.05	0.05	0.05
Restitution coefficient, e_n (−)	0.85	0.85	0.85
Young modulus, E (GPa)	0.2	0.2	0.2
Poisson's ratio, ν (−)	0.35	0.35	0.35
Work function, Φ (eV)	5.18	7.70	4.60

The work functions were calculated from molecular orbital calculation (MOPAC) by Naik et al. [9] for the lactose and by Zellnitz et al. [53] for salbutamol. The work function of the gelatine capsule was not found in the literature. Pinto et al. [12] observed that the charge acquired by the capsules in contact with stainless steel is about 60% of the charge acquired by the capsules in contact with PVC. Since the work function of PVC is 5.33 eV [34] and that of stainless steel is 5.05 eV [54], and since the charge is proportional to the difference between work functions (Equation (9)), the value $\Phi = 4.60$ eV was estimated.

By looking at the work functions, salbutamol is expected to become negatively charged, while the capsule is expected to become positively charged. The behavior of the carrier particles is less predictable: lactose will acquire a positive charge after contact with the API and a negative charge if in contact with the capsule.

The cohesion properties are reported in Table 2 for each material pair [52,55].

Table 2. Surface energy and reference pull-off force values.

	Pull-Off Force, $F_{pull\text{-}off}$ (nN)	Surface Energy, γ (mJ/m^2)
API–API	11	0.93
API–carrier	180	8.02
Carrier–carrier	150	0.64
Particle–wall	60	2.55

2.4. Geometry and Powder Configurations

The geometry used in the simulations is modeled using a size 3 hard-shell gelatine capsule [56] as a reference. The capsule is filled with a carrier–API blend. Two powder configurations are considered, as described in Table 3 and shown in Figure 2.

Table 3. Characteristics of the powder in the two simulated configurations.

	A	B
Sample total mass	25 mg	1 mg
Total no. of particles in sample	123,858	117,233
No. of API particles	92,883	115,825
No. of carrier particles	30,975	1408
API-to-carrier ratio (*w/w*)	1:3332	1:124
API loading (*w/w*)	0.03%	0.80%
Total surface area (cm^2)	9.80	0.52

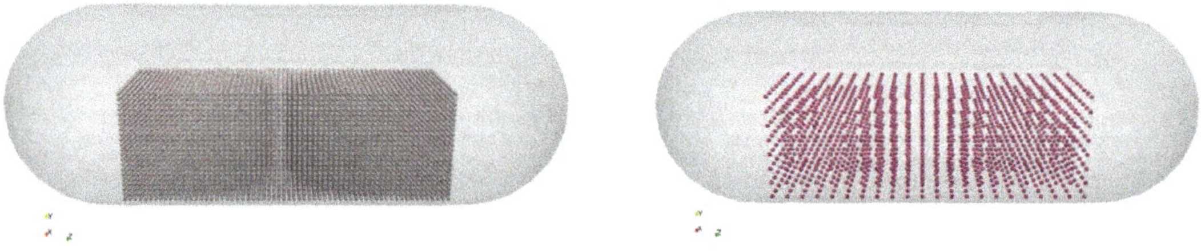

(a) (b)

Figure 2. Initial location of powder with the two powder loadings: (**a**) 25 mg and (**b**) 1 mg.

Configuration A (Figure 2a) has a total mass of 25 mg. It consists of about 31k carrier particles and 93k API particles. Configuration B (Figure 2b) has a total mass of 1 mg and is made of about 1500 carrier particles and 117k of API particles. The total number of particles is similar, but the mixing ratio between the two solid phases differs consistently: 1:3332 (0.03%, *w/w*) for Configuration A, and 1:124 (0.80%, *w/w*) for Configuration B.

The approach for the dry powder coating of the carrier with API is described in Alfano et al. [22]. The difference in the coating degree between Configuration A and B can be noticed in Figure 3, which shows two coated carrier particles.

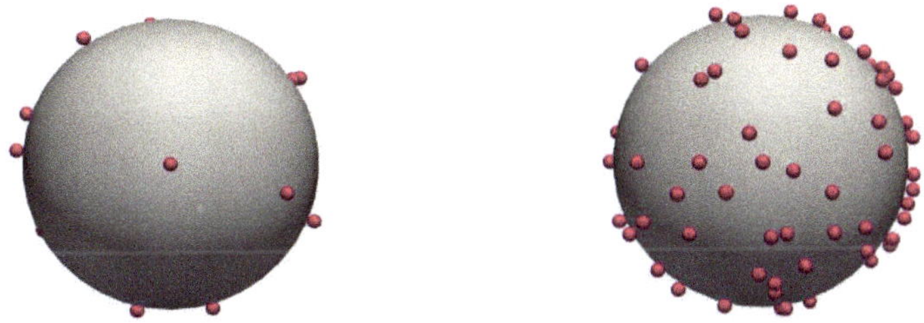

Figure 3. API-coated carrier particles in the two configurations considered: (**a**) 25 mg, 1:3332 (*w/w*); and (**b**) 1 mg, 1:124 (*w/w*).

The coated carrier particles are initially placed in a regular cubic arrangement, which was visualized in Figure 2. The initial random condition is obtained by letting them settle under gravity. Then, the capsule is subject to a periodic oscillatory translation to reproduce a tapping motion. The shaking frequency is 120 Hz, and the shaking amplitude is 0.8 mm. The shaking direction changes during the simulations, so that the capsule shakes alternatively in both horizontal and vertical directions (x and y, respectively). In implementation, rather than computing the actual motion of the capsule, the motion is tracked in the frame of reference of the capsule by adding the equivalent corresponding fictitious forces on the particles. More details of this implementation in MFIX can be found in Alfano et al. [39].

3. Results and Discussion

3.1. Model Validation

To verify the reliability of the tribocharging model with the selected parameters, a comparison was made with the data reported by Chow et al. [19], who studied the effect of powder handling by repeatedly tapping a gelatine capsule filled with 25 mg of lactose (InhaLac® 230, d_{50} = 97.2 µm). The specific charge of the sample was measured after 0, 10, 40, 100, and 200 taps.

To reproduce the tapping, a shaking motion with a 120 Hz frequency and 0.8 mm amplitude was considered. The simulation time was set to 1.67 s in order to have 200 full oscillations, corresponding to 200 taps. The capsule was filled with 25 mg of 100 µm lactose particles, initially settled at the bottom of the capsule. No API was included in this case. The initial charge of the particles was set according to a normal distribution with a mean of -0.025 pC (corresponding to the initial specific charge in the reference article, -0.779 nC/g, multiplied by the weight of a single particle) and standard deviation of 25% times the mean. The initial charge distribution is shown in Figure 4a.

Figure 4b shows the total charge of the sample as a function of the number of taps, which also corresponds to the time evolution. The solid line is the result of the simulation, and the dots are the reference experimental data [19]. The increase in negative total charge follows the trend exhibited by the experimental measurements, well within the experimental uncertainty, except for the second point. Indeed, the data by Chow et al. [19] show a sudden jump in the charge, followed by a somewhat stationary stage between 10 and 40 taps. The simulations show a more gradual increase in the acquired charge, which may be helped by the ability to track the charge at every instant during the simulation. Attempts have been made to better reproduce the charge corresponding to the second point by considering tri-disperse particle sizes (10%vol 60 µm, 80%vol 100 µm, and 10%vol 150 µm) by changing the capsule wall's work function and the charging efficiency coefficient. In no case, however, did the abrupt initial change and subsequent less pronounced increase turn out to be predictable. Overall, the agreement can be judged to be sufficiently good to proceed with the more complex carrier–API cases.

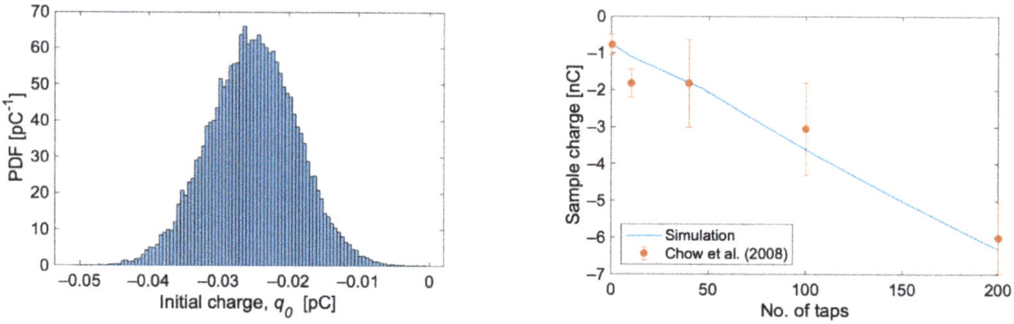

(**a**) (**b**)

Figure 4. (**a**) Initial charge distribution and (**b**) sample charge as a function of the number of taps. The line is the result of the simulations, and the points are experimental data reported by Chow et al. [19].

3.2. Gravity Settling in the Capsule

Starting from the ordered configurations of the powder shown in Figure 2, with both carrier and API particles, DEM gravity settling simulations have been performed to evaluate the charge buildup due to operations such as capsule filling and storage. Figure 5 shows the final configuration after the settling simulations. Figure 5a shows Configuration A (25 mg) after 40 ms, with the powder bed stable at the bottom of the capsule. A much longer simulation was required for Configuration B (1 mg), as after 40 ms, the particles were still consistently bouncing off the walls of the capsule. The simulation was extended up to 200 ms when most of the powder bed was settled (Figure 5b), and the charge level reached a stationary point (see below).

(**a**) (**b**)

Figure 5. Powder configuration after gravity settling: (**a**) 25 mg (after 40 ms) and (**b**) 1 mg (after 200 ms). Carrier particles are shown in gray, and API particles are in purple.

Figure 6 shows the total charge acquired by the samples following the deposition simulations. The API and carrier curves are shown individually, as well as the total charge curve. Note that the charges are evaluated at very short time increments, and the symbols on the plot lines are only for reference. Contrary to the simulation with only lactose particles (Figure 4b), which became negatively charged, in this case, the carrier acquires a positive charge, while the salbutamol acquires a negative charge. Figure 6a shows some oscillations in the total charge of the carrier: first, it increases, then it decreases, and then it increases again, probably depending on the instantaneous ratio between carrier–wall and carrier–API interactions. Similar fluctuations can be observed in Figure 6b, which shows the result with the 1 mg sample and a different mixing ratio, but in a less pronounced way, suggesting that, in this case, the greater quantity of active principle and the reduced number of carrier particles promote API–carrier interactions at the expense of carrier–wall interactions. In both cases, the total net charge is negative and has a greater magnitude for the larger sample. A plateau in the contact charging process is observed more markedly in the 25 mg case.

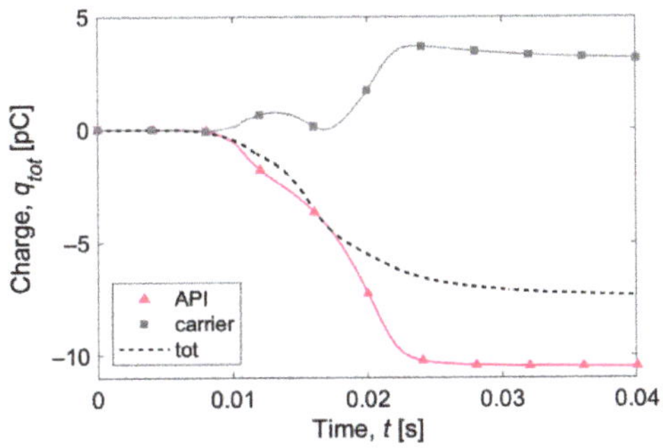

(a) (b)

Figure 6. Total charge acquired by (**a**) 25 mg and (**b**) 1 mg of powder after gravity settling in the capsule. The lines show instantaneous evaluation of the charges, and the symbols are only for references.

In Figure 7, the specific charge evolutions (expressed as the charge-to-surface ratio, CTS) in the first 40 ms of the two simulations for carrier and API particles are reported. Carrier particles in the 1 mg simulation acquire a higher specific charge than carrier particles in the 25 mg case. The opposite behavior is observed for API particles, for which the specific charge magnitude is higher for the 25 mg simulation.

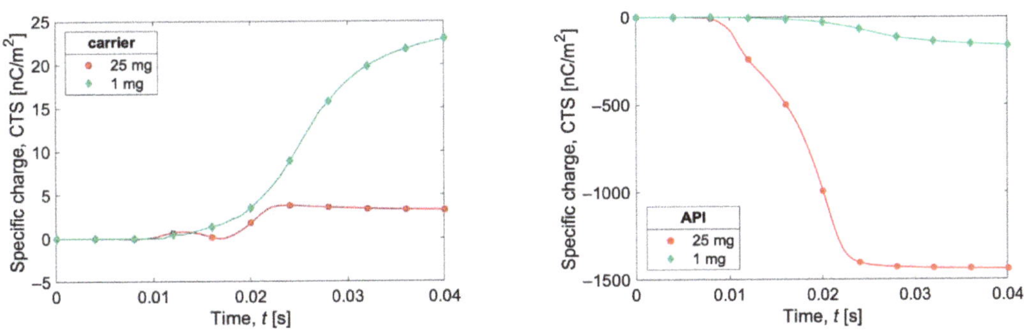

(a) (b)

Figure 7. Charge-to-surface ratio (CTS) acquired during gravity settling by (**a**) carrier and (**b**) API for the two simulations (1 mg and 25 mg). The lines show the instantaneous evaluation of the charges, and the symbols are only for references.

In Table 4, a summary of the net and specific charges after gravity settling is reported. The 25 mg test shows a higher net charge (four times higher), but the specific charge is higher for the 1 mg test, whether it is expressed as the CTS (charge-to-surface ratio) or as the CTM (charge-to-mass ratio). The numerical value of the CTS is equal to the CTM for API particles, since the mass/surface ratio is 1 g/m^2.

Table 4. Charging level after gravity settling.

		Net Charge (pC)	CTM (nC/g)	CTS (nC/m^2)
API	25 mg	−10.54	−1445	−1445
	1 mg	−4.77	−525	−525
Carrier	25 mg	3.09	0.13	3.17
	1 mg	2.95	2.67	66.67
Sample	25 mg	−7.46	−0.38	−7.60
	1 mg	−1.82	−1.64	−34.18

Compared to the already mentioned specific charge of 25 mg lactose powder stored overnight in a gelatine capsule reported by Chow et al. [19], CTM = −0.779 nC/g, a lower value is obtained with the simulations (−0.38 nC/g), suggesting that the presence of salbutamol as API mitigates the powder charging.

3.3. Tapping of the Capsule

The capsule was subjected to the vibrating motion presented in Section 3.1. in the presence of both API and carrier particles. The configuration obtained after gravity settling (Figure 5) was taken as the initial condition, considering the charge recorded after gravity settling. The simulation was carried out for 400 ms, corresponding to 48 taps, with a shaking frequency of 120 Hz.

Figure 8 shows the total net charge acquired by the 25 mg (Figure 8a) and 1 mg (Figure 8b) samples after capsule shaking. Now, the lactose carrier particles acquire an overall negative charge, as in the simulations with carrier particles only (Figure 4b). The total net charge is negative in both the 1 mg and 25 mg samples. The final charge is about −0.6 nC for the 1 mg sample and −0.8 nC for the 25 mg sample. The charge acquired by 25 mg of carrier particles (Figure 4b) shaken 50 times was more than twice, suggesting again the possible mitigatory effect [9,10] in the total charge acquisition due to the presence of salbutamol API particles.

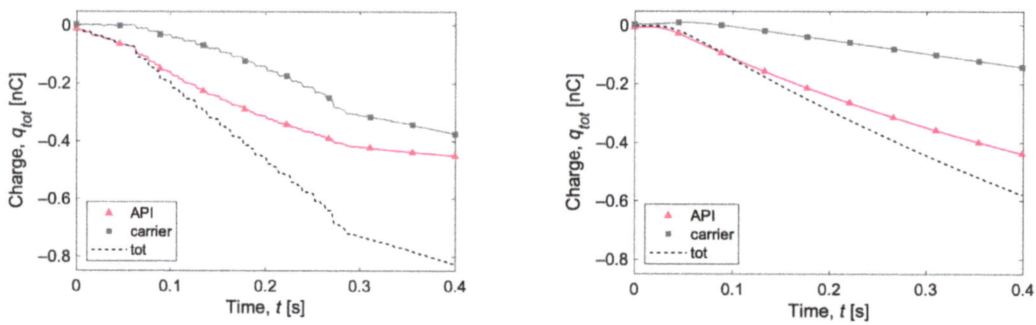

(**a**) (**b**)

Figure 8. Total net charge acquired by (**a**) 25 mg and (**b**) 1 mg of powder after capsule shaking. The lines show instantaneous evaluation of the charges, and the symbols are only for references.

The evolution of the specific charge is reported in Figure 9 as the charge-to-mass ratio (CTM) and in Figure 10 as the charge-to-surface ratio (CTS). The specific charge recorded for the API particles (Figures 9a and 10a) is substantial, exceeding −50,000 nC/g (or −50,000 nC/m^2) with both capsule loadings. The curves for the 25 mg case show a somewhat flattening trend at the end of the simulations, suggesting the beginning of a plateau phase.

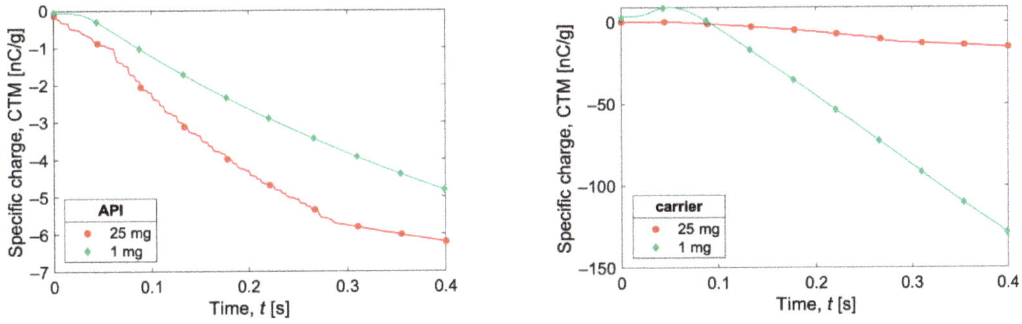

(a) (b)

Figure 9. Specific charge expressed as charge-to-mass ratio (CTM) during capsule shaking for (**a**) API and (**b**) carrier particles. The lines show instantaneous evaluation of the charges, and the symbols are only for references.

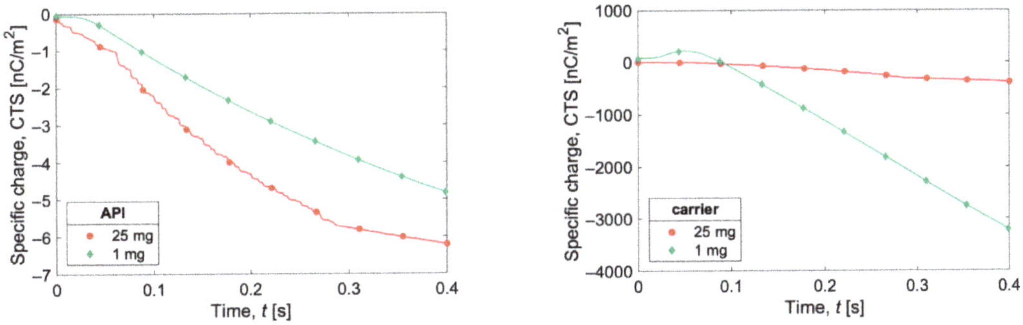

(a) (b)

Figure 10. Specific charge expressed as charge-to-surface ratio (CTS) during capsule shaking for (**a**) API and (**b**) carrier particles. The lines show instantaneous evaluation of the charges, and the symbols are only for references.

Observing the evolution of carrier particles' specific charge (Figures 9b and 10b), the curve for the 1 mg sample is of particular interest. An initial positive charge buildup is observed, followed by a decrease in the charge and a subsequent markedly linear trend, until it reaches a specific charge value of almost −130 nC/g (or −3000 nC/m^2). The specific charge for the 25 mg sample is more than one order of magnitude lower.

Figure 11 shows the individual charge distribution for API (top) and carrier (bottom) particles at different times: 50 ms, 200 ms, and 400 ms. The charge distribution of API particles becomes wider, going from 50 ms to 200 ms, slightly narrowing again at the end of the simulation (Figure 11c, top). The mean value shifts to the left, i.e., with an increase in charge with negative polarity.

The evolution of carrier particles' charge distribution is different for the 1 mg and 25 mg samples. With the 25 mg sample, more and more carrier particles acquire a positive charge over time, while the opposite tendency is observed for the 1 mg sample: after 50 ms, most of the particles carry a positive charge (Figure 11a, bottom), while no carrier particle is charged positively at the end of the simulation (Figure 11c, bottom).

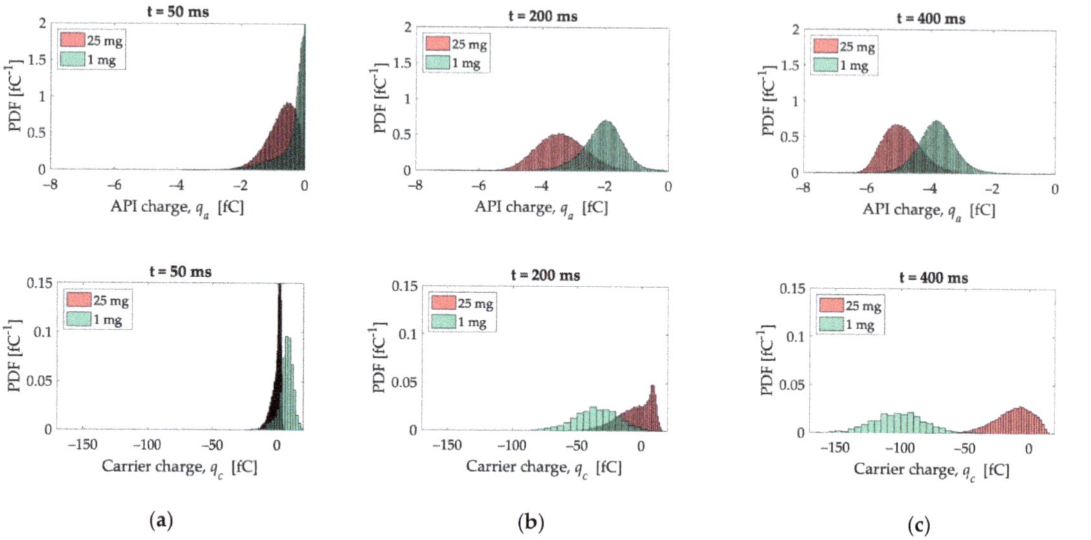

Figure 11. API (**top**) and carrier (**bottom**) charge distribution at different times: (**a**) 50 ms, (**b**) 200 ms, and (**c**) 400 ms.

Figure 12 shows a colored map of the surface charge density (σ) observed in the capsule after shaking in the x-direction for 60 ms, as estimated according to Equation (9). The charge is positive, as the wall has the lowest work function. Comparable charge density values are observed between the two different loadings. However, a wider and more scattered area is associated to the 1 mg sample case, probably due to the more chaotic movement of API particles, while the 25 mg case is characterized by higher values localized near the main impact locations (normal to the shaking direction). The charge density estimated for the capsule is about 10 times lower than the CTS of carrier particles in the 1 mg configuration (see Figure 10b), while the opposite trend is observed in the 25 mg simulation. On the other hand, in both configurations, the CTS calculated for API particles is significantly higher than the capsule surface charge density (see Figure 10a).

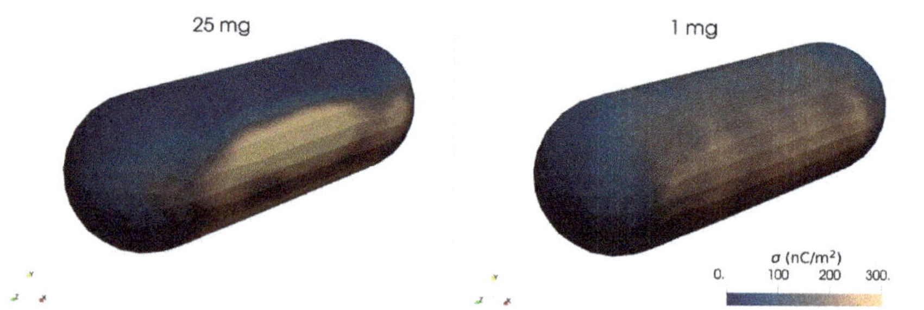

Figure 12. Capsule charge density due to charge exchange with the bottom of the capsule after shaking in the x-direction for 60 ms.

3.4. Collision Statistics

To interpret the charge acquisition data, it is useful to relate the evolution of the acquired charge to the events that generate the charge transfer, i.e., the collisions. To that purpose, the MFIX code was modified to track individual collision events in DEM separately for particle–particle and particle–wall cases.

Figure 13a shows the percentage of API particles in contact with the walls of the capsule as a function of time. In terms of the fraction of the total, the 1 mg case shows a higher value, but the time variation is not significant in both cases. Figure 13b shows the evolution of the carrier-to-API coordination number (CN), i.e., the number of API particles in contact with a carrier particle averaged over all carrier particles. Interestingly, the CN remains almost constant in the carrier-rich 25 mg simulation, while in the API-rich 1 mg simulation, the carrier CN goes from about 100 to 4 within the first 100 ms, meaning that collisions are strong enough to determine that detachment of most of the API particles from the carrier. This can be the explanation of the shift in the carrier charge distribution, in which the mean value goes from a positive value at 0.05 s (Figure 11a, bottom) to a progressively negative value (Figure 11b,c, bottom), as the carrier–API contact is the event that can give rise to a positively charged carrier particle.

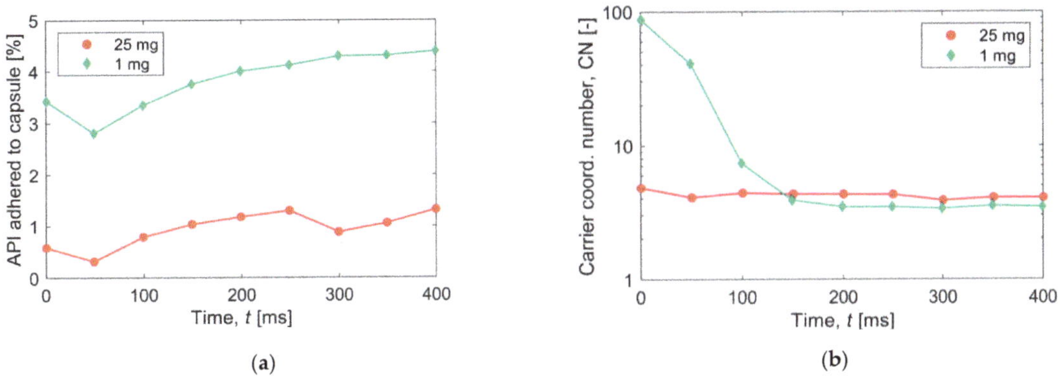

Figure 13. Collision statistics: (**a**) percentage of API particles adhered to the walls and (**b**) mean coordination number (CN) for carrier particles.

Figure 14a shows the instantaneous number of collision events during the last part of the simulation (from 300 to 400 ms); the data for particle–particle (PP) and particle–wall (PW) contacts are presented separately. In the 25 mg simulation, more than 10^4 particle–particle contacts are recorded, while the number of particle–wall collision events does not exceed 1000. The contact events in the 1 mg simulation stand in between, with more particle–wall contacts than particle–particle contacts, most probably due to the lower total solid mass. Figure 14b shows the average normal impact velocity; in this case, the velocity values are lower for the 25 mg simulations. Particle–wall contacts tend to occur with a higher impact velocity in both configurations.

The contact statistics data are completed in Figure 15, which shows how the collision events are distributed among the various phases involved. The majority of the collisions in the 25 mg simulations are low-energy carrier–carrier collisions, during which the charge exchanged is minimal since particles are made of the same material. The large number of carrier–carrier collisions could also be the reason why settling occurs much faster than in the 1 mg simulation, since all of these collisions dissipate the initial kinetic energy of the particles.

3.5. Estimate of Force Contributions

Given the significant value of the observed electrostatic charges (see, e.g., Figure 8), it is interesting to investigate whether or not the consequent electrostatic forces actually influence particle trajectories and to what extent.

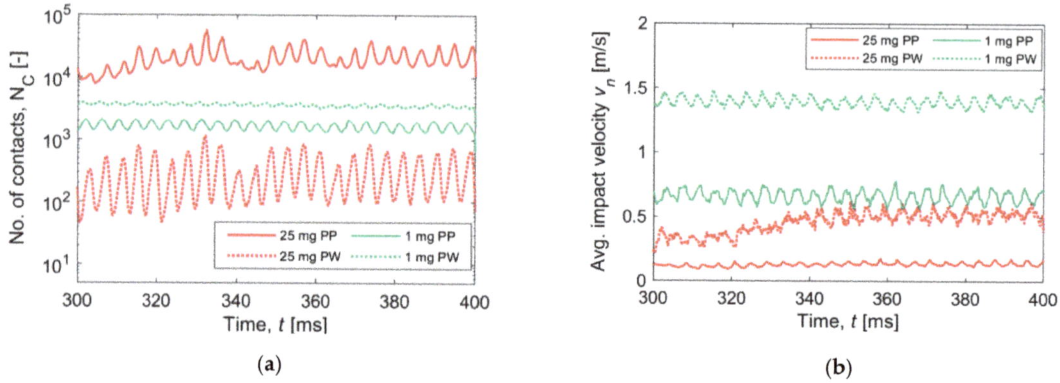

Figure 14. Collision statistics: (**a**) instantaneous number of contacts and (**b**) average impact velocity for the two simulations. Data for particle–particle (PP) and particle–wall (PW) contacts are presented separately.

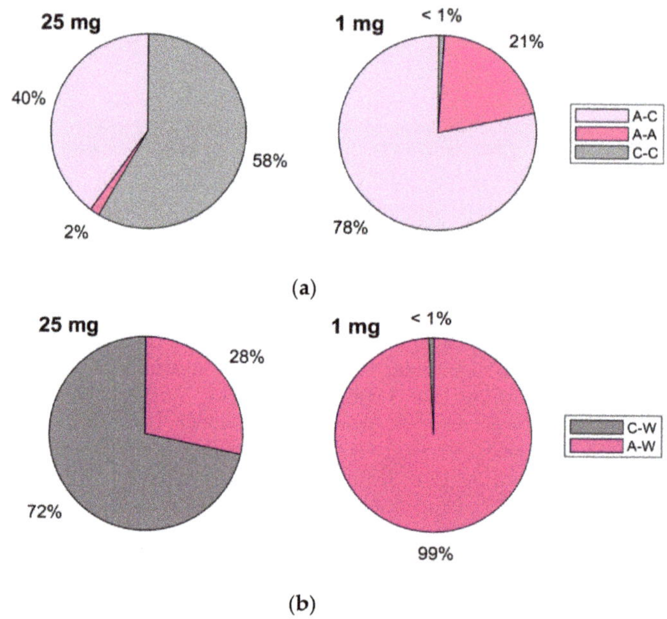

Figure 15. (**a**) Percentage of API–API (A-A), API–carrier (A-C) and carrier–carrier (C-C) collisions with respect to the total particle–particle contacts for the two simulations. (**b**) Percentage of API–wall (A-W) and carrier–wall (C-W) collisions with respect to the total particle–wall contacts for the two simulations.

By equating the weight force of the particle with the particle–wall electrostatic force calculated according to Equation (11), it is possible to estimate the charge magnitude necessary for the particle to remain attached to the wall, assuming for simplicity that the electrostatic attractive force acts in the opposite direction to the gravitational force:

$$|q_W| = 16\pi R^2\sqrt{\frac{\varepsilon_0 \rho R g}{3}} \quad (12)$$

Figure 16 shows $|q_W|$ as a function of particle diameter. The final charge magnitude distribution is also reported in the form of a scatter plot for the two systems investigated. Most API particles at the end of the simulation possess a charge level that is higher than $|q_W|$, while no carrier particle exceeds such a value (all points are below the blue curve). The results suggest that during capsule handling and shaking, a fraction of API particles are likely to end up retained inside the capsule, on its internal walls, due to the charge buildup, while this is much less likely to happen for a carrier particle.

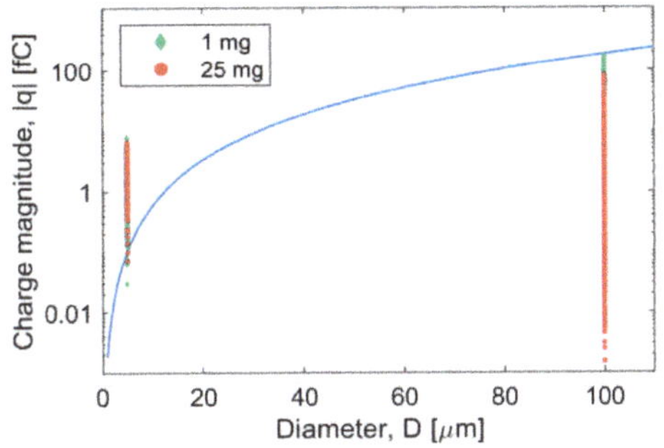

Figure 16. Charge at which P-W electrostatic force is equal to weight as a function of particle diameter (blue line). A scatter plot of the final charge magnitude is reported in the plot for 1 mg and 25 mg simulations, separately for API (5 μm diameter) and carrier particles (100 μm diameter).

It is useful to note that other cohesive/adhesive forces and inertial fictitious forces are at play (see, e.g., Table 2), so their relative contribution should be compared. An evaluation in terms of the maximum force values is discussed below.

The maximum electrostatic force was calculated with the Coulomb law, considering the maximum particle charge magnitude in both simulations, which is 7.7 fC for API particles and 170 fC for carrier particles. For the calculation of the distance of the Coulomb force, it is assumed that the two bodies are in contact; therefore, for example, in the case of a p-P-P interaction, the sum of the particle radii was considered.

To estimate the maximum cohesive and adhesive forces, the reference values of the pull-off force in the JKR model are considered, i.e., the values reported in Table 2.

The fictitious force associated with the shaking motion follows a sinusoidal trend in time. Its maximum value is given by the following:

$$F_{shk} = 4\pi^2 A f^2 \times m_p \tag{13}$$

where A is the shaking amplitude, f is the shaking frequency, and m_p is the mass of the particle. In the present study, this force reaches about 50 times the weight of the particle.

Figure 17 shows a comparison between the estimated maximum force contribution for carrier and API particles. As expected from the externally imposed motion, the fictitious force associated with shaking primarily determines the motion of the carrier particles, as it is higher than any other contributions. Electrostatic forces are lower than cohesive forces, with the API–carrier electrostatic interaction being the lowest (and lower than weight). On the other hand, the adhesive API–carrier force contribution is dominant in the case of API particles, the fictitious force appears to be irrelevant compared to the other forces at play, and the electrostatic forces are relevant.

It is interesting to note that the electrostatic API–API force is higher than the cohesive JKR reference force, suggesting that the repulsive Coulomb force between like-charged particles can overcome the attractive van der Waals interactions. In reality, with sufficient charge difference, even like-charged particles can be attracted to one another due to a mutual polarization of surface charge [57,58]. This is not accounted for in simple Coulombic interactions, as modeled here, and more sophisticated approaches would be necessary.

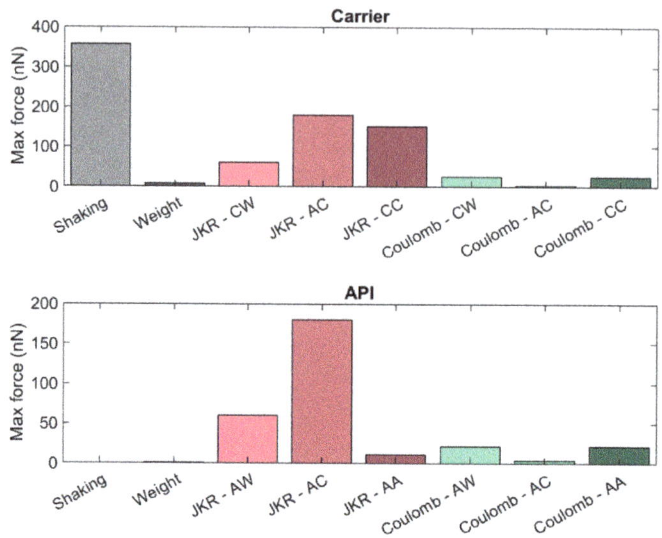

Figure 17. Comparison between estimated maximum force contributions for API and carrier particles. AW = API–wall; CW = carrier–wall; AA = API–API; AC = API–carrier; CC = carrier–carrier.

4. Conclusions

An extended DEM approach was applied to evaluate triboelectric charging of a lactose–salbutamol binary mixture in the gelatine capsule of a DPI. Two powder configurations were considered, with different loaded doses and mixing ratios. The selected triboelectric charging model and parameters were validated with the experimental data available in the literature. Unlike in previous works, dynamic charging and interparticle electrostatic interactions of carrier and API powders were detailed investigated, including the charge buildup on the capsule walls. The detailed analysis of the flow and collision behavior during tapping allowed us to elucidate the mechanisms leading to the final charge polarity of all three materials, a condition not predictable a priori.

Gravity settling simulations were performed to estimate the charge buildup due to operations such as capsule filling and storage. The total net charge after settling was found to be negative, despite the opposite charge polarity acquired by the carrier (positive) and API (negative). Then, the capsule was subjected to a vibrating motion to simulate routine handling operations and check whether such movements can give rise to a significant electrostatic charge. A consistently higher charge buildup was measured with the vibrating motion compared to the gravity settling simulations, with an overall negative charge also for carrier particles. The net charge was higher for the carrier-rich 25 mg formulation, but the specific charge was higher for the API-rich 1 mg formulation. The two different configurations also show a different charge distribution, with a more pronounced bipolar charging tendency in the case of the 25 mg dose. To interpret the charging dynamics, a detailed study of contact statistics was performed. The mechanical shaking movement promoted API detachment from the carrier, especially in the 1 mg configuration (with higher API dosage). With the 25 mg configuration, a high number of low-energy carrier-carrier collisions were recorded which tended to dissipate the kinetic energy of the particles. On the other hand, API–API and API–wall collisions were prevalent in the 1 mg configuration. Finally, a comparison of the forces at play revealed that electrostatic forces are relevant for API particles and might play a major role in cohesion and adhesion phenomena. The present work lays the foundation for new developments in the relations between particle-scale charge transfer and buildup and the macroscopic manifestations, particularly in relation to DPI performances.

Author Contributions: Conceptualization, F.O.A., A.D.R. and F.P.D.M.; methodology, F.O.A., A.D.R. and F.P.D.M.; software, F.P.D.M.; validation, F.O.A.; formal analysis, F.O.A.; investigation, F.O.A.; resources, A.D.R. and F.P.D.M.; data curation, F.O.A.; writing—original draft preparation, F.O.A.; writing—review and editing, A.D.R. and F.P.D.M..; visualization, F.O.A.; supervision, A.D.R. and F.P.D.M.; project administration, A.D.R. and F.P.D.M.; funding acquisition, A.D.R. and F.P.D.M. All authors have read and agreed to the published version of the manuscript.

Funding: The present research was conducted with the support of the ICSC National Centre for High Performance Computing, Big Data and Quantum Computing, funded by European Union—NextGenerationEU, through Piano Nazionale di Ripresa e Resilienza (PNRR), Missione 4, Componente 2, Investimento 1.4 (grant number CN00000013 and CUP H23C22000360005), which is gratefully acknowledged.

Data Availability Statement: The data presented in this study are available from the corresponding authors upon request.

Conflicts of Interest: The authors declare no conflict of interest.

References

1. Wong, J.; Chan, H.K.; Kwok, P.C.L. Electrostatics in Pharmaceutical Aerosols for Inhalation. *Ther. Deliv.* **2013**, *4*, 981–1002. [CrossRef]
2. Matsusaka, S.; Maruyama, H.; Matsuyama, T.; Ghadiri, M. Triboelectric Charging of Powders: A Review. *Chem. Eng. Sci.* **2010**, *65*, 5781–5807. [CrossRef]
3. Šupuk, E.; Zarrebini, A.; Reddy, J.P.; Hughes, H.; Leane, M.M.; Tobyn, M.J.; Timmins, P.; Ghadiri, M. Tribo-Electrification of Active Pharmaceutical Ingredients and Excipients. *Powder Technol.* **2012**, *217*, 427–434. [CrossRef]
4. Šupuk, E.; Hassanpour, A.; Ahmadian, H.; Ghadiri, M.; Matsuyama, T. Tribo-Electrification and Associated Segregation of Pharmaceutical Bulk Powders. *KONA Powder Part. J.* **2011**, *29*, 208–223. [CrossRef]
5. Lacks, D.J.; Shinbrot, T. Long-Standing and Unresolved Issues in Triboelectric Charging. *Nat. Rev. Chem.* **2019**, *3*, 465–476. [CrossRef]
6. Murtomaa, M.; Laine, E. Electrostatic Measurements on Lactose-Glucose Mixtures. *J. Electrostat.* **2000**, *48*, 155–162. [CrossRef]
7. Hoe, S.; Young, P.M.; Traini, D. A Review of Electrostatic Measurement Techniques for Aerosol Drug Delivery to the Lung: Implications in Aerosol Particle Deposition. *J. Adhes Sci. Technol.* **2011**, *25*, 385–405. [CrossRef]
8. Telko, M.J.; Kujanpää, J.; Hickey, A.J. Investigation of Triboelectric Charging in Dry Powder Inhalers Using Electrical Low Pressure Impactor (ELPI^{TM}). *Int. J. Pharm.* **2007**, *336*, 352–360. [CrossRef]
9. Naik, S.; Sarkar, S.; Hancock, B.; Rowland, M.; Abramov, Y.; Yu, W.; Chaudhuri, B. An Experimental and Numerical Modeling Study of Tribocharging in Pharmaceutical Granular Mixtures. *Powder Technol.* **2016**, *297*, 211–219. [CrossRef]
10. Hoe, S.; Traini, D.; Chan, H.K.; Young, P.M. The Contribution of Different Formulation Components on the Aerosol Charge in Carrier-Based Dry Powder Inhaler Systems. *Pharm. Res.* **2010**, *27*, 1325–1336. [CrossRef] [PubMed]
11. Wong, J.; Kwok, P.C.L.; Noakes, T.; Fathi, A.; Dehghani, F.; Chan, H.K. Effect of Crystallinity on Electrostatic Charging in Dry Powder Inhaler Formulations. *Pharm. Res.* **2014**, *31*, 1656–1664. [CrossRef] [PubMed]
12. Pinto, J.T.; Wutscher, T.; Stankovic-Brandl, M.; Zellnitz, S.; Biserni, S.; Mercandelli, A.; Kobler, M.; Buttini, F.; Andrade, L.; Daza, V.; et al. Evaluation of the Physico-Mechanical Properties and Electrostatic Charging Behavior of Different Capsule Types for Inhalation Under Distinct Environmental Conditions. *AAPS PharmSciTech* **2020**, *21*, 128. [CrossRef]
13. Peart, J. Powder Electrostatics: Theory, Techniques and Applications. *KONA Powder Part. J.* **2001**, *19*, 34–45. [CrossRef]
14. Mukherjee, R.; Gupta, V.; Naik, S.; Sarkar, S.; Sharma, V.; Peri, P.; Chaudhuri, B. Effects of Particle Size on the Triboelectrification Phenomenon in Pharmaceutical Excipients: Experiments and Multi-Scale Modeling. *Asian J. Pharm. Sci.* **2016**, *11*, 603–617. [CrossRef]
15. Lacks, D.J.; Sankaran, R.M. Triboelectric Charging in Single-Component Particle Systems. *Part. Sci. Technol.* **2016**, *34*, 55–62. [CrossRef]
16. Chowdhury, F.; Elchamaa, B.; Ray, M.; Sowinski, A.; Passalacqua, A.; Mehrani, P. Apparatus Design for Measuring Electrostatic Charge Transfer Due to Particle-Particle Collisions. *Powder Technol.* **2020**, *361*, 860–866. [CrossRef]
17. Hoe, S.; Young, P.M.; Traini, D. Dynamic electrostatic charge of lactose-salbutamol sulphate powder blends dispersed from a Cyclohaler®. *Drug Dev. Ind. Pharm.* **2011**, *37*, 1365–1375. [CrossRef]
18. Kaialy, W. A Review of Factors Affecting Electrostatic Charging of Pharmaceuticals and Adhesive Mixtures for Inhalation. *Int. J. Pharm.* **2016**, *503*, 262–276. [CrossRef] [PubMed]
19. Chow, K.T.; Zhu, K.; Tan, R.B.H.; Heng, P.W.S. Investigation of Electrostatic Behavior of a Lactose Carrier for Dry Powder Inhalers. *Pharm. Res.* **2008**, *25*, 2822–2834. [CrossRef]
20. Capecelatro, J.; Longest, W.; Boerman, C.; Sulaiman, M.; Sundaresan, S. Recent Developments in the Computational Simulation of Dry Powder Inhalers. *Adv. Drug Deliv. Rev.* **2022**, *188*, 114461. [CrossRef] [PubMed]

21. Ponzini, R.; Da Vià, R.; Bnà, S.; Cottini, C.; Benassi, A. Coupled CFD-DEM Model for Dry Powder Inhalers Simulation: Validation and Sensitivity Analysis for the Main Model Parameters. *Powder Technol.* **2021**, *385*, 199–226. [CrossRef]
22. Alfano, F.O.; Benassi, A.; Gaspari, R.; Di Renzo, A.; Di Maio, F.P. Full-Scale DEM Simulation of Coupled Fluid and Dry-Coated Particle Flow in Swirl-Based Dry Powder Inhalers. *Ind. Eng. Chem. Res.* **2021**, *60*, 15310–15326. [CrossRef]
23. Benque, B.; Khinast, J.G. Carrier Particle Emission and Dispersion in Transient CFD-DEM Simulations of a Capsule-Based DPI. *Eur. J. Pharm. Sci.* **2022**, *168*, 106073. [CrossRef] [PubMed]
24. Sulaiman, M.; Liu, X.; Sundaresan, S. Effects of Dose Loading Conditions and Device Geometry on the Transport and Aerosolization in Dry Powder Inhalers: A Simulation Study. *Int. J. Pharm.* **2021**, *610*, 121219. [CrossRef] [PubMed]
25. Alfano, F.O.; Di Maio, F.P.; Di Renzo, A. Deagglomeration of Selected High-Load API-Carrier Particles in Swirl-Based Dry Powder Inhalers. *Powder Technol.* **2022**, *408*, 117800. [CrossRef]
26. Korevaar, M.W.; Padding, J.T.; Van der Hoef, M.A.; Kuipers, J.A.M. Integrated DEM-CFD Modeling of the Contact Charging of Pneumatically Conveyed Powders. *Powder Technol.* **2014**, *258*, 144–156. [CrossRef]
27. Pei, C.; Wu, C.Y.; England, D.; Byard, S.; Berchtold, H.; Adams, M. Numerical Analysis of Contact Electrification Using DEM-CFD. *Powder Technol.* **2013**, *248*, 34–43. [CrossRef]
28. Kolehmainen, J.; Ozel, A.; Boyce, C.M.; Sundaresan, S. Triboelectric Charging of Monodisperse Particles in Fluidized Beds. *AIChE J.* **2017**, *63*, 1872–1891. [CrossRef]
29. Hu, J.; Pei, C.; Zhang, L.; Liang, C.; Wu, C.Y. Numerical Analysis of Frictional Charging and Electrostatic Interaction of Particles. *AIChE J.* **2022**, *68*, e17444. [CrossRef]
30. Zafar, U.; Alfano, F.; Ghadiri, M. Evaluation of a New Dispersion Technique for Assessing Triboelectric Charging of Powders. *Int. J. Pharm.* **2018**, *543*, 151–159. [CrossRef]
31. Alfano, F.O.; Di Renzo, A.; Di Maio, F.P.; Ghadiri, M. Computational Analysis of Triboelectrification Due to Aerodynamic Powder Dispersion. *Powder Technol.* **2021**, *382*, 491–504. [CrossRef]
32. Mukherjee, R.; Sansare, S.; Nagarajan, V.; Chaudhuri, B. Discrete Element Modeling (DEM) Based Investigation of Tribocharging in the Pharmaceutical Powders during Hopper Discharge. *Int. J. Pharm.* **2021**, *596*, 120284. [CrossRef] [PubMed]
33. Chowdhury, F.; Ray, M.; Passalacqua, A.; Mehrani, P.; Sowinski, A. Evaluating the Electrostatic Charge Transfer Model for Particle-Particle Interactions. *J. Electrostat.* **2021**, *112*, 103603. [CrossRef]
34. Naik, S.; Hancock, B.; Abramov, Y.; Yu, W.; Rowland, M.; Huang, Z.; Chaudhuri, B. Quantification of Tribocharging of Pharmaceutical Powders in V-Blenders: Experiments, Multiscale Modeling, and Simulations. *J. Pharm. Sci.* **2016**, *105*, 1467–1477. [CrossRef]
35. Zhu, Q.; Gou, D.; Li, L.; Chan, H.K.; Yang, R. Numerical Investigation of Powder Dispersion Mechanisms in Turbuhaler and the Contact Electrification Effect. *Adv. Powder Technol.* **2022**, *33*, 103839. [CrossRef]
36. Almeida, L.C.; Bharadwaj, R.; Eliahu, A.; Wassgren, C.R.; Nagapudi, K.; Muliadi, A.R. Capsule-Based Dry Powder Inhaler Evaluation Using CFD-DEM Simulations and next Generation Impactor Data. *Eur. J. Pharm. Sci.* **2022**, *175*, 106226. [CrossRef]
37. Mitani, R.; Ohsaki, S.; Nakamura, H.; Watano, S. Numerical Study on Particle Adhesion in Dry Powder Inhaler Device. *Chem. Pharm. Bull* **2020**, *68*, 726–736. [CrossRef]
38. Benque, B.; Khinast, J.G. Understanding the Motion of Hard-Shell Capsules in Dry Powder Inhalers. *Int. J. Pharm.* **2019**, *567*, 118481. [CrossRef]
39. Alfano, F.O.; Sommerfeld, M.; Di Maio, F.P.; Di Renzo, A. DEM Analysis of Powder Deaggregation and Discharge from the Capsule of a Carrier-Based Dry Powder Inhaler. *Adv. Powder Technol.* **2022**, *33*, 103853. [CrossRef]
40. Johnson, K.L.; Kendall, K.; Roberts, A.D. Surface Energy and the Contact of Elastic Solids. *Proc. R. Soc. Lond. A* **1971**, *324*, 301–313. [CrossRef]
41. Di Renzo, A.; Napolitano, E.S.; Di Maio, F.P. Coarse-Grain Dem Modelling in Fluidized Bed Simulation: A Review. *Processes* **2021**, *9*, 279. [CrossRef]
42. Di Renzo, A.; Di Maio, F.P. Comparison of Contact-Force Models for the Simulation of Collisions in DEM-Based Granular Flow Codes. *Chem. Eng. Sci.* **2004**, *59*, 525–541. [CrossRef]
43. Mindlin, R.D.; Deresiewicz, H. Elastic Spheres in Contact under Varying Oblique Forces. *J. Appl. Mech.* **1953**, *20*, 327–344. [CrossRef]
44. Ai, J.; Chen, J.-F.; Rotter, J.M.; Ooi, J.Y. Assessment of Rolling Resistance Models in Discrete Element Simulations. *Powder Technol.* **2011**, *206*, 269–282. [CrossRef]
45. Alfano, F.O.; Di, R.; Gaspari, A.; Benassi, R.; Di Maio, A.; Alfano, F.O.; Di Renzo, A.; Gaspari, R.; Benassi, A.; Paolo, F.; et al. Modelling Deaggregation Due to Normal Carrier-Wall Collision in Dry Powder Inhalers. *Processes* **2022**, *10*, 1661. [CrossRef]
46. Garg, R.; Galvin, J.; Li, T.; Pannala, S. Documentation of Open-Source MFIX–DEM Software for Gas-Solids Flows. 2012. Available online: https://mfix.netl.doe.gov/documentation/dem_doc_2012-1.pdf (accessed on 1 February 2023).
47. Santasusana, M.; Irazábal, J.; Oñate, E.; Carbonell, J.M. The Double Hierarchy Method. A Parallel 3D Contact Method for the Interaction of Spherical Particles with Rigid FE Boundaries Using the DEM. *Comput. Part Mech.* **2016**, *3*, 407–428. [CrossRef]
48. Matsusaka, S.; Ghadiri, M.; Masuda, H. Electrification of an Elastic Sphere by Repeated Impacts on a Metal Plate. *J. Phys. D Appl. Phys.* **2000**, *33*, 2311–2319. [CrossRef]
49. Matsuyama, T.; Yamamoto, H. Charge Relaxation Process Dominates Contact Charging of a Particle in Atmospheric Conditions. *J. Phys. D Appl. Phys.* **1995**, *28*, 2418–2423. [CrossRef]

50. Feynman, R.P.; Leighton, R.B.; Sands, M. *The Feynman Lectures in Physics, Mainly Electromagnetism and Matter*; The New Millennium, Ed.; Basic Books: New York, NY, USA, 2011; Volume 2, ISBN 978-0-465-07998-8.
51. van Wachem, B.; Thalberg, K.; Remmelgas, J.; Niklasson-Björn, I. Simulation of Dry Powder Inhalers: Combining Micro-Scale, Meso-Scale and Macro-Scale Modeling. *AIChE J.* **2017**, *63*, 501–516. [CrossRef]
52. Nguyen, D.; Remmelgas, J.; Björn, I.N.; van Wachem, B.; Thalberg, K. Towards Quantitative Prediction of the Performance of Dry Powder Inhalers by Multi-Scale Simulations and Experiments. *Int. J. Pharm.* **2018**, *547*, 31–43. [CrossRef]
53. Zellnitz, S.; Pinto, J.T.; Brunsteiner, M.; Schroettner, H.; Khinast, J.; Paudel, A. Tribo-Charging Behaviour of Inhalable Mannitol Blends with Salbutamol Sulphate. *Pharm. Res.* **2019**, *36*, 80. [CrossRef]
54. Trigwell, S.; Grable, N.; Yurteri, C.U.; Sharma, R.; Mazumder, M.K. Effects of Surface Properties on the Tribocharging Characteristics of Polymer Powder as Applied to Industrial Processes. *IEEE Trans. Ind. Appl.* **2003**, *39*, 79–86. [CrossRef]
55. Ibrahim, T.H.; Burk, T.R.; Etzler, F.M.; Neuman, R.D. Direct Adhesion Measurements of Pharmaceutical Particles to Gelatin Capsule Surfaces. *J. Adhes Sci. Technol.* **2000**, *14*, 1225–1242. [CrossRef]
56. Coates, M.S.; Fletcher, D.F.; Chan, H.-K.; Raper, J.A. The Role of Capsule on the Performance of a Dry Powder Inhaler Using Computational and Experimental Analyses. *Pharm. Res.* **2005**, *22*, 923–932. [CrossRef] [PubMed]
57. Lindgren, E.B.; Chan, H.K.; Stace, A.J.; Besley, E. Progress in the Theory of Electrostatic Interactions between Charged Particles. *Phys. Chem. Chem. Phys.* **2016**, *18*, 5883–5895. [CrossRef] [PubMed]
58. Qin, J.; Li, J.; Lee, V.; Jaeger, H.; de Pablo, J.J.; Freed, K.F. A Theory of Interactions between Polarizable Dielectric Spheres. *J. Colloid Interface Sci.* **2016**, *469*, 237–241. [CrossRef] [PubMed]

pharmaceutics

MDPI

Article

Impact of Vertical Blender Unit Parameters on Subsequent Process Parameters and Tablet Properties in a Continuous Direct Compression Line

Marius J. Kreiser [1,2], Christoph Wabel [1] and Karl G. Wagner [2,*]

[1] Product and Process Development, Pfizer Manufacturing Deutschland GmbH, 79108 Freiburg, Germany; marius.kreiser@pfizer.com (M.J.K.); christoph.wabel@pfizer.com (C.W.)
[2] Department of Pharmaceutical Technology and Biopharmaceutics, University of Bonn, 53121 Bonn, Germany
* Correspondence: karl.wagner@uni-bonn.de

Abstract: The continuous manufacturing of solid oral-dosage forms represents an emerging technology among the pharmaceutical industry, where several process steps are combined in one production line. As all mixture components, including the lubricant (magnesium stearate), are passing simultaneously through one blender, an impact on the subsequent process steps and critical product properties, such as content uniformity and tablet tensile strength, is to be expected. A design of experiment (DoE) was performed to investigate the impact of the blender variables hold-up mass (HUM), impeller speed (IMP) and throughput (THR) on the mixing step and the subsequent continuous manufacturing process steps. Significant impacts on the mixing parameters (exit valve opening width (EV), exit valve opening width standard deviation (EV SD), torque of lower impeller (T_L), torque of lower impeller SD (T_L SD), HUM SD and blend potency SD), material attributes of the blend (conditioned bulk density (CBD), flow rate index (FRI) and particle size (d_{10} values)), tableting parameters (fill depth (FD), bottom main compression height (BCH) and ejection force (EF)) and tablet properties (tablet thickness (TT), tablet weight (TW) and tensile strength (TS)) could be found. Furthermore, relations between these process parameters were evaluated to define which process states were caused by which input variables. For example, the mixing parameters were mainly impacted by impeller speed, and material attributes, FD and TS were mainly influenced by variations in total blade passes (TBP). The current work presents a rational methodology to minimize process variability based on the main blender variables hold-up mass, impeller speed and throughput. Moreover, the results facilitated a knowledge-based optimization of the process parameters for optimum product properties.

Keywords: continuous manufacturing; continuous mixing technology; vertical blender; direct compression; lubrication; material characterization

Citation: Kreiser, M.J.; Wabel, C.; Wagner, K.G. Impact of Vertical Blender Unit Parameters on Subsequent Process Parameters and Tablet Properties in a Continuous Direct Compression Line. *Pharmaceutics* **2022**, *14*, 278. https://doi.org/10.3390/pharmaceutics14020278

Academic Editor: Colin Hare

Received: 9 December 2021
Accepted: 20 January 2022
Published: 25 January 2022

1. Introduction

Continuous manufacturing lines are supplied by various vendors and distributed amongst the pharmaceutical industry. Next to the benefit of continuous manufacturing, the modular setup allows for an easier transfer amongst various production sites, as the setup can be more easily cloned from the pilot plant or launch site [1]. The PCMM (portable, continuous, modular and miniature) installed at the Pfizer site in Freiburg, Germany, consists of a GEA Compact Feeder, a vertical continuous blender (CMT—continuous mixing technology), a MODUL™ P tablet press equipped with an NIR (near infrared) probe installed in the feed frame and an at-line combi-tester to analyze tablet properties, such as thickness, weight and crushing strength (Figure 1).

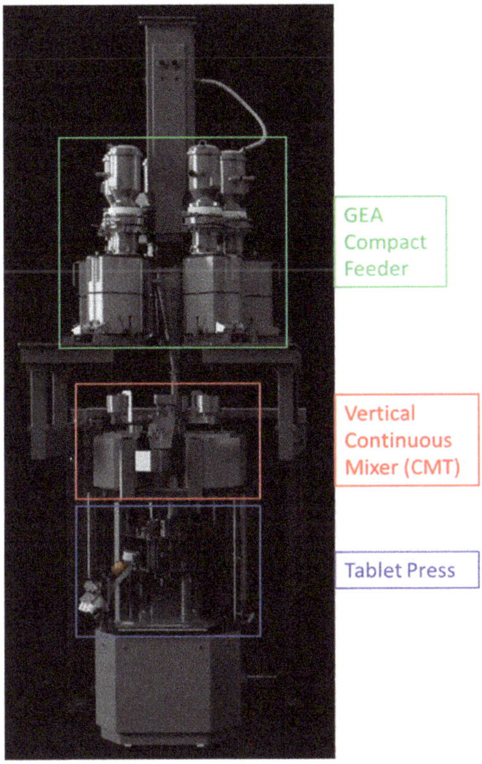

Figure 1. Overview of the direct compression line used for this trial.

Basically, for each raw material, the powder is transferred from a polyethylene bag via a top-up valve into an agitated hopper, where co-rotating screws supply the powder by the loss in weight (LiW) principle at a composition related feed rate. The continuous process demands a periodical refill of the hopper triggered by a defined refill level, performed by a rotating volumetric refill device with flexible volume inserts [2,3].

Since feeding is the first step within a continuous process, it is consequently one of the first critical control elements besides the material attributes. Accurate feeding is substantial for the quality of a continuous process to avoid deviations regarding the quality of blend and content uniformity of the tablets [4–8].

To provide low variability in feed rate, the optimal feeder design and the corresponding parameter settings, such as refill level, top-up volume, screw pitch, feed-factor array (governing dosing in volumetric mode during e.g., refill) and gearbox type (Figure 2), should be individually adjusted based on composition, throughput and powder attributes [3,4,7–13].

Several feeders supply each raw material separately, and the powder falls through the conical-shaped inlet hopper into the vertical continuous mixer. It is composed of two regions: the upper delumping region and the lower mixing region (Figure 3). In both, the impellers can be adjusted independently regarding speed, direction and vertical position, i.e., the gap between impeller and conical sieve. In the delumping region, a downstream sieve (d = 2.1 mm) is set to delump possible agglomerates. The powder leaves the upper region and arrives in the conical mixing region, where a second impeller is mounted. The whole setup of the CMT is attached to load cells, which monitor the weight of the powder within the mixer. This hold-up mass (HUM) is defined in the recipe and determines the mass, will always be mixed in the CMT continuously throughout the process.

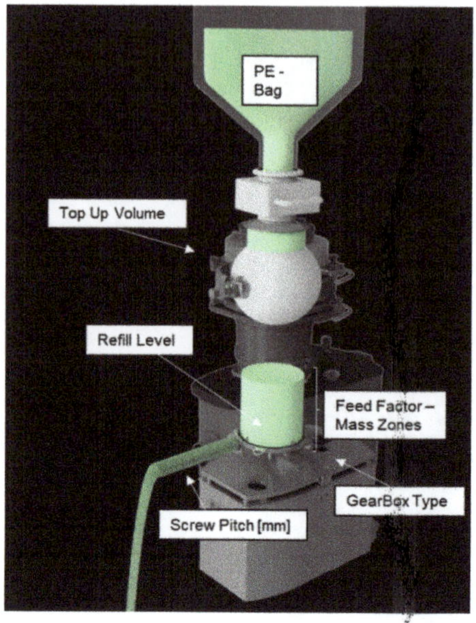

Figure 2. Overview of a GEA Compact Feeder and corresponding adjustment options.

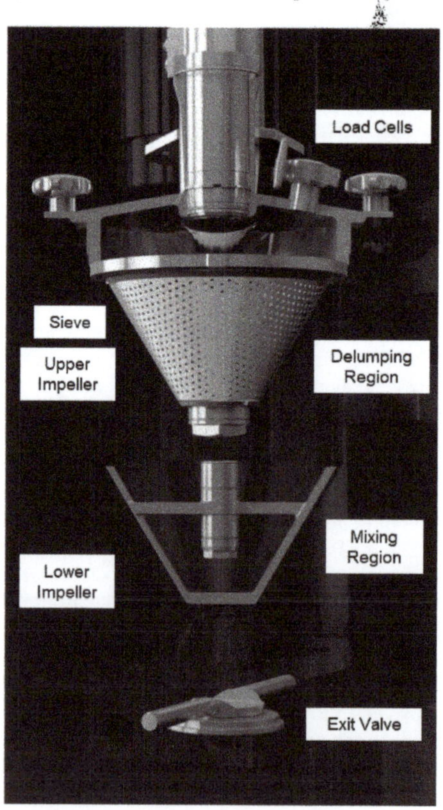

Figure 3. Overview of the CMT. View of the upper impeller is obstructed by the sieve.

Other papers focus on a horizontal continuous mixer, where the HUM is considered a function of flow rate and impeller speed and cannot be set individually [14,15]. In contrast, the HUM in a vertical continuous mixer remains constant, and various shear rates (impeller speeds) can be applied despite a constant residence-time distribution (RTD) [8]. As in this case the continuous DC line includes only one mixing step for all mixture components, including the lubricant. An impact especially on lubricant-sensitive mixtures, as well as on the blend uniformity of the mixture and, subsequently, content uniformity of the tablets, can be expected [16–18].

The exit valve is located at the bottom of the CMT. By means of a proportional–integral–derivative (PID) control loop, the exit valve opening width is adjusted automatically based on the current HUM value in order to keep the mass of the CMT constant. The controlled exit valve ensures that the same amount of mass entering the CMT will simultaneously leave the CMT ($mass_{in} = mass_{out}$). Feed fluctuations of each feeder and the respective variability in the mass flow can be balanced that way. Smaller exit valve opening widths are recommended so that newly entering raw materials can be properly mixed together with the blend that is already present in the blender. Otherwise, unmixed or poorly mixed material can pass by and leave the CMT without being blended, causing content-uniformity variability [8].

The mean residence time (MRT, Equation (1)) of a particle can be calculated based on the overall throughput and the HUM. It reflects the mixing period of that particle within the CMT [8].

$$\text{MRT}\ [\text{min}] = \frac{\text{HUM}\ [\text{kg}]}{\text{THR}\ \left[\frac{\text{kg}}{\text{h}}\right]} * 60\frac{\text{min}}{\text{h}} \quad (1)$$

The total blade passes (TBP, Equation (2)) reveals how often the impeller, on average, will pass a particle and show the intensity of the shear transmitted to the powder. With an increasing number of revolutions and respectively increased shear, a lubricant, such as magnesium stearate (MgSt), can be introduced more homogeneously into the blend or even filmed onto the particles, potentially resulting in decreasing tensile strength of tablets. Therefore, particular attention is paid to the single mixing step in the CMT, where the lubricant will be mixed right from the start, together with the remaining raw materials, potentially resulting in a narrow process window between a homogeneous and an over-lubricated blend. Hence, it is required to set a suitable combination for HUM and IMP to ensure that TS and disintegration, as well as dissolution time, are within specification [19–25]. Thresholds regarding HUM and IMP are, besides the mass balance model (MBM), part of the control strategy of the CMT. If the process values exceed the specific limits, an alarm occurs and the process stops. Furthermore, variations in HUM and IMP could also impact the exit valve opening width and, therefore, the mixing quality.

$$\text{TBP} = \text{MRT}\ [\text{min}] * \text{IMP}\ [\text{rpm}] \quad (2)$$

After the powder exits the CMT, it travels through the feed chute into the feed frame, where powder will be held up and be fed into the tablet press. Position sensors in the feed chute measure the filling levels. Using an internal feedback loop, we can control the turret speed of the tablet press according to the filling levels, thus preventing powder from backing up or the tablet press from running empty. An increasing feed-chute level results in an increased turret speed of the tablet press, i.e., increased powder demand, and vice versa.

The NIR probe in the feed frame is the first chemometric measurement in the process. It is therefore important to understand the impact of upstream settings and process states on the conformity of potency, as predicted by the NIR model.

The NIR probe measures a defined volume of the powder. The corresponding spectra are used to predict the API (active pharmaceutical ingredient) content. If inhomogeneity of the blend or variability in the upstream process units occurs, it can consequently be detected with NIR and is seen as a disturbance in the blend potency measured by NIR inside the feed frame [26].

Depending on the chosen control strategy, the impacted tablets can be diverted into the waste channel if the signals exceed the specification limits. As soon the signals are within specification limits again, the diverter switches back to the good product channel after a defined lead-lag time [27–29].

The tablet press was running with a control mode enabled, wherein the tablet weight control is based on the pre-compression displacement and the fill depth is adjusted accordingly. The bottom main compression height controls the thickness and compression force and, therefore, the crushing strength of the tablets. At the end of the tablet press, the tablets can be directed into the good channel, diverted into the waste channel or directed to the combi-tester, where at-line measurements regarding tablet properties can be performed in containment (see Section 2.7).

This paper assesses to what extent HUM, IMP and THR impact the downstream process of a direct compression mixture. It focuses on correlations and coherences and evaluates the predictability of process parameters based on the CMT settings, especially since the lubricant and all other formulation constituents are mixed simultaneously in one single mixing step. As a model formulation, Saccharin Monohydrate was used as API surrogate.

2. Materials and Methods

2.1. Materials

For this trial, saccharin sodium monohydrate (JMC, Ulsan, South Korea), microcrystalline cellulose (Avicel PH 102, FMC, Cork, Ireland), calcium di-phosphate (A-Tab, Innophos, Chicago Heights, IL, USA), sodium starch glycolate (Roquette, Lestrem, France) and magnesium stearate V (Mallinckrodt, St. Louis, MO, USA) were used.

2.2. DoE Settings

A central composite face design with a star points at the face of each side defined by a 2-level factorial design was conducted by using MODDE Pro 12.1 (Satorius Stedim Data Analytics AB, Umea, Sweden) (Table 2). A quadratic model was used, wherein the following parameters (Table 1) were considered:

Table 1. Overview of considered responses, where HUM, IMP and THR were adjusted as input variables.

Responses	
Mixing parameters	T_L, EV Blend potency as predicted by the NIR model
Material attributes of the blend	FRI Particle size (d_{10}) CBD
Tableting parameters	FD BCH EF
Tablet properties	TS TT TW

Compounds and composition remained constant over the entire experiment. In general, 17 runs, including 3 replicates of a center point, were performed (Table 2). After adjusting the new CMT parameters, a transition phase was initiated (3 × MRT) to wash out the powder mixed at the former setting. A compression-force profile was conducted by using 118, 157, 169, 236 and 275 MPa compression pressure for each phase. Subsequently, the process was run for at least 10 min in a steady-state phase.

Table 2. DoE settings, where phase 7, 9 and 11 are the replicates of the center point.

Phase	Throughput (kg/h)	Hold-Up Mass (g)	Impeller Speed (rpm)	MRT (min)	TBP (rev)
1	10	400	200	2.4	480
2	10	400	650	2.4	1560
3	10	600	425	3.6	1530
4	10	800	200	4.8	960
5	10	800	650	4.8	3120
6	20	400	425	1.2	510
7	20	600	425	1.8	765
8	20	600	200	1.8	360
9	20	600	425	1.8	765
10	20	800	425	2.4	1020
11	20	600	425	1.8	765
12	20	600	650	1.8	1170
13	30	400	200	0.8	160
14	30	400	650	0.8	520
15	30	600	425	1.2	510
16	30	800	650	1.6	1040
17	30	800	200	1.6	320

During the transition phase and the compression-force profile, the tablet press was operated in manual mode, without using the combi-tester, to analyze tablet properties. In manual mode, samples were taken and weighed manually to select the correct fill depth. During each steady-state phase, manual mode was switched to automatic mode, in which the NIR probe was active. For each steady state phase, 275 MPa compression pressure was set; a tablet sample was taken in the middle of the steady state phase, using the combi-tester; and a powder sample was withdrawn at the end of each steady state phase by opening the sampling port underneath the feed frame and collecting approximately 300 g of powder.

2.3. Feeder Settings

The continuous manufacturing line used was equipped with PID-controlled LiW feeders. To ensure consistent powder supply, the following feeder settings were used (Table 3).

Table 3. Feeder settings for each raw material.

	Microcrystalline Cellulose	Saccharin Sodium Monohydrate	Calcium Di-Phosphate	Sodium Starch Glycolate	Magnesium Stearate
Composition (%)	49.104	21.844	24.552	3	1.5
Top-Up Volume (L)	1.6	1.2	1.6	1.2	0.8
Gearbox Type	1 (63:1)	2 (235:1)	2 (235:1)	3 (455:1)	3 (455:1)
Screw Pitch (mm)	20	10	20	10	20
Refill Level (dm^3)	0.5	0.74	0.3	0.25	1.5

2.4. Bulk and Tapped Density

Bulk and tapped density were measured by using an Erweka SVM 222 (ERWEKA GmbH, Langen, Germany) according to Ph.Eur. A 250 mL graduated flask was used and filled with an appropriate amount of powder of each raw material and blend. The initial volume and the volume (V_0) after 750 and 1250 taps were noted. Each sample was analyzed in triplicate. Hausner Ratio and Carr's Index were calculated as shown in Equations (3) and (4) and interpreted as shown in Table 4.

$$\text{Hausner Ratio} = \frac{\rho_{\text{tapped}}}{\rho_{\text{bulk}}} \quad (3)$$

$$\text{Carr Index} = \frac{\rho_{\text{tapped}} - \rho_{\text{bulk}}}{\rho_{\text{tapped}}} * 100 \tag{4}$$

Table 4. Classification of Carr Index [30].

Flowability	Carr's Index
Excellent	<15
Correct	15–25
Poor	>25

2.5. *Freeman Powder Rheometer FT4*

The FT4 Powder Rheometer (Freeman Technology Inc., Worcestershire, UK) was used to characterize flow properties of powders and granulates. For this trial, 3 methods (stability and variable flow rate, powder compressibility and shear cell) were used to analyze the impact of CMT parameters on the flowability of the resulting blends.

2.5.1. Stability and Variable Flow Rate

In this trial, a cylindrical 25 mm × 25 mL split vessel was used. After an initial condition cycle, the powder was split to obtain a defined amount of powder to ensure reproducible measurements. The actual testing consisted of seven alternating conditioning and test cycles where the blade was inserted in the powder bed and was moved downward, with a rotational blade tip speed of 100 mm/s, to remove history and operator influence. Subsequently, 4 cycles with decreasing blade tip speed (100, 70, 40 and 10 mm/s) were performed. The required energy is based on the resistance of the blade to flow in the downward motion [31].

The basic flow energy (BFE) is defined by the required energy to move the blade downward at test-cycle 7. The specific energy (SE) represents the energy that is required during an upward traverse at the same test cycle. The stability index (SI) is calculated by the ratio of the energy at test-cycle 7 and test-cycle 1. The flow-rate index (FRI) reflects the results of the reducing blade-tip speed, where the energy of the lowest rotational speed and the highest is set in ratio.

$$\text{FRI} = \frac{\text{energy test 11 } \left(10\ \frac{\text{mm}}{\text{s}}\right)}{\text{energy test 8 } \left(100\ \frac{\text{mm}}{\text{s}}\right)} \tag{5}$$

Basically, at higher flow rates, less energy is required, since the entrained air acts as a lubricant. At lower flow rates, the powder in front of the blades is more likely to be consolidated, due to the absence of entrained air; therefore, the interlocking of particles is more probable. Consequently, higher FRI values are common for cohesive powders. In this study, FRI values < 1 are shown; they are typical for powders or blends containing lubricants. The conditioned bulk density (CBD) was measured after the initial conditioning cycle and the split of the powder, where agglomerates and air inclusions could be evened to ensure reproducible measurements [11,30,32–34].

2.5.2. Powder Compressibility

The compressibility method was used to investigate how the density of the measured powder changes with increasing normal stress. A split vessel (25 mm × 10 mL) was used for this trial. After three conditioning cycles, the powder was split, and the blade was changed for a vented piston. In total, 8 compression steps were performed (1, 2, 4, 6, 8, 10, 12, and 15 kPa) and were held for 60 s at each force. In this work, only compressibility (change in volume after compression (%)) was used. Low compressibility values occurred for powders with a low amount of entrained air where particles are packed compactly. High compressibility values were seen if voids within the powder occurred. This was likely with cohesive powders [30,35].

2.5.3. Shear Cell

A shear cell test was performed by using the FT4. For this method, a 25 mm × 10 mL split vessel was used. As normal stress, 7, 6, 5, 4 and 3 kPa were adjusted, and the initial consolidation stress was 9 kPa.

For this method, a $\tau - \sigma$-diagram can be obtained, where one pre-shear point and five yield points can be observed. Using a Mohr circle analysis, a linearized yield locus can be obtained, where the τ-axis intersection is interpreted as cohesion and presents the obtained shear stress during powder deformation when no normal stress is applied [11,36–39].

2.6. Particle Size Distribution

For particle size measurements, a Sympatec QicPic (Sympatec GmbH, Clausthal-Zellerfeld, Germany) was used. It is a dynamic high-speed image-analysis system with a LED pulse-light source and high-resolution high-speed camera. An M7 lens was used that covers particles between 4.2 and 2888 µm. Dispersion pressure was set to 1 bar for all raw materials and blends to maintain comparability. A dry dispersion line RODOS/L with VIBRI attachment was in place, and the sample size remained constant for each material (5 mL). To determine the particle size, the EQPC method was used, where d_{10}, d_{50} and d_{90} values were obtained (see Supplementary Table S34).

2.7. Tableting

A MODUL™ P tablet press (GEA Pharma Systems, Courtoy™, Halle, Belgium) was implemented at the end of the continuous manufacturing line. Mode 2 (Courtoy dual-control force method) was selected, where the tablet weight control is based on pre-compression displacement measurements adjusting the fill depth, accordingly [40].

Round convex tablets with an 11 mm diameter and 1.12 mm cup height were manufactured. During steady state, a target compression pressure of 275 MPa was set.

The target tablet weight of 600 mg, tablet crushing strength and tablet thickness were tested periodically in the middle of each steady state, using the at-line combi-tester (Kraemer Elektronik GmbH, Darmstadt, Germany).

The feed chute level was controlled to a constant level at 40%, and the paddle speed remained constant at 45 and 40 rpm. Turret speed set-points and speed tolerances of the tablet press were adapted to the respective mass throughput (11 rpm ± 2.2 rpm; 21 rpm ± 4.2 rpm and 32 rpm ± 6.4 rpm).

Tensile Strength

The tensile strength of the convex round tablets was calculated based on the following equation [41]:

$$\text{Tensile Strength} = \frac{10P_s}{\pi D^2}\left(2.84\frac{t}{D} - 0.126\frac{t}{W} + 3.15\frac{W}{D} + 0.01\right)^{-1} \quad (6)$$

where P_s = tablet core crushing strength, D = tablet core diameter, t = tablet core thickness and W = cylinder length. Tablet-crushing strength was measured by using the combi-tester, which is directly connected to the continuous manufacturing line.

2.8. Blend Potency

To analyze the impact of the CMT settings on the blend potency, an NIR spectrometer (SentroProbe DR LS NIR 170C ATEX, Sentronic GmbH, Dresden, Germany) was installed in the feed frame, with an insertion depth of 1 mm. Using PharmaMV 5.3 (Perceptive Engineering, Daresbury, UK), we recorded a spectrum every 4 s. Approximately 150–200 mg of the blend was measured during one measurement cycle. The collected data were preprocessed by first applying the Savitzky–Golay filter and then the standard normalize variate method (SNV). After that, the data were processed by a partial least square (PLS) regression model. The integration time was 9 ms, with 133 average scans.

2.9. *Software*

2.9.1. MODDE

The DoE was designed by using MODDE Pro 12.1. A multiple linear regression (MLR) model was used to evaluate the significance of the input factors on the responses. Furthermore, MODDE was used to obtain model equations to predict the responses.

2.9.2. Osi Pi

A considerable benefit of the PCMM is the implementation of OsiPi (OsiSoft, San Leandro, CA, USA), which enables access to all essential process values. All data generated by the PCMM are continuously monitored and stored by using OsiPi.

Pi Vision is a web-based tool wherein process data can be visualized in real time. Since the process data are stored in the PI Server, PiVision also can visualize previous batches if process states need to be evaluated retrospectively. For this trial, all process-related data were gathered by using PiDataLink, which is an Add-In to Excel (Microsoft Corporation, Redmond, Washington, USA) that enables data to be imported from the PI Server.

2.9.3. GraphPad Prism

GraphPad Prism 9 (GraphPad Software, Inc., San Diego, CA, USA) was used to generate the figures and to calculate the correlations (Pearson correlations) between the process parameters, including the p-values. All correlation coefficients are shown Supplementary Figure S57. To evaluate the size of the correlation, the following thumb rule was used (Table 5):

Table 5. Interpretation of Pearson correlation coefficients [42].

Correlation Coefficient	Interpretation
0.9 to 1.0 (−0.9 to −1.0)	Very high correlation
0.7 to 0.9 (−0.7 to −0.9)	High correlation
0.5 to 0.7 (−0.5 to −0.7)	Moderate correlation
0.3 to 0.5 (−0.3 to −0.5)	Low correlation
0.0 to 0.3 (−0.0 to −0.3)	Negligible correlation

3. Results and Discussion

3.1. *DoE Results*

The DoE reveals to what extent the input variables throughput (THR), hold-up mass (HUM) and impeller speed (IMP) affect the response parameters, such as the exit valve opening width and SD, torque of the lower impeller and corresponding SD, HUM SD and blend potency uniformity, as measured by NIR, in regard to the mixing step. Furthermore, the impact on material attributes of the blend (FRI, CBD and d_{10} values), tablet press parameters (FD, BCH and EF) and tablet properties (TS, TW, TT and corresponding standard deviation) are presented. A visualization where responses are expected is shown in Figure 4. The data were fitted by using an MLR model, wherein significant model terms are identifiable when error bars (=95% confidence interval) do not cross the zero-line. Corresponding-fit statistics are shown in Supplementary Tables S2–S33. In this paper, models with $Q^2 > 0.500$ (=estimate of prediction precision) and $R^2 \geq 0.800$ (=model fit) are considered good models, indicating a significant correlation between input variables and responses.

3.1.1. Mixing Parameters

For each presented response regarding mixing quality, impeller speed is a significant model term (Figure 5). The DoE showed that the influence on the exit valve opening width is driven by THR and IMP, resulting in higher opening widths if throughput and impeller speed are high as well. Regarding variability in EV, torque and blend potency, the impeller speed is the only significant model term. For torque values, HUM and IMP seem to share the same extent of deflection. With regard to HUM SD values, all three input factors and HUM*IMP were significant.

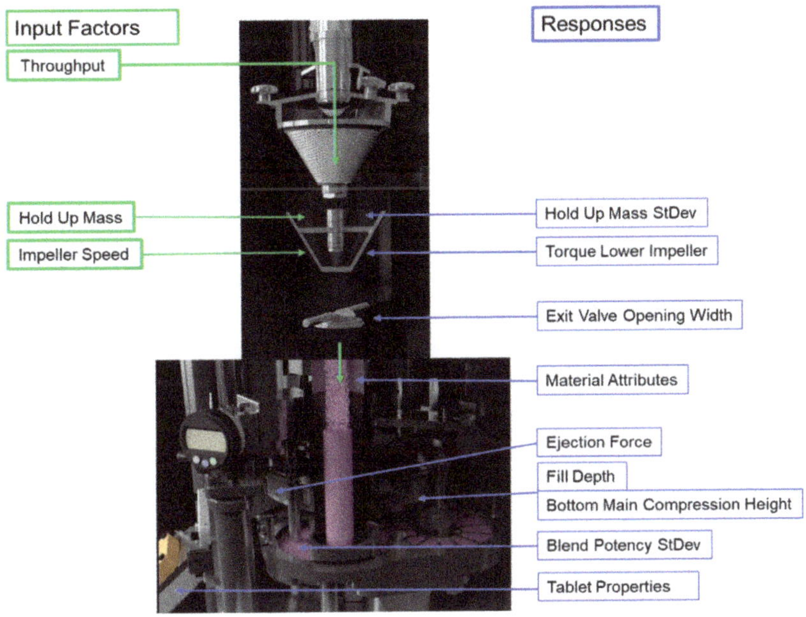

Figure 4. Process overview of input factors (green, **left** side) and observed responses (blue, **right** side).

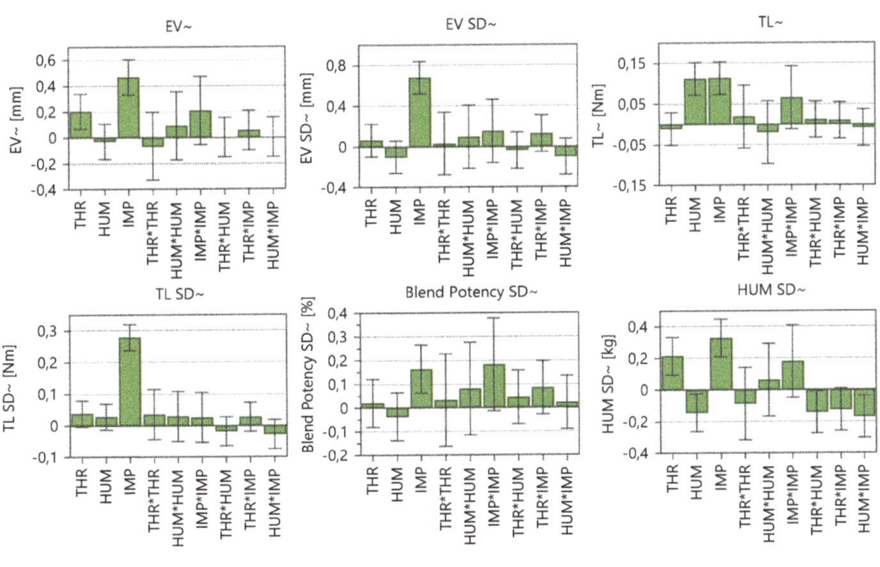

Figure 5. Coefficients plot of the impact of input variables on responses regarding the blending unit and uniformity of the blend. The 95% confidence interval is displayed as an error bar.

As shown in Table 6, Q2 and R2 imply, that exit valve opening width (+SD) and torque (+SD) can be considered good models. As the variabilities of the responses were not linearly distributed a logarithmic data transformation was conducted (as shown in the corresponding chapters and Supplementary Figures S7 and S11).

Table 6. Overview of fit statistics regarding mixing parameters after removing non-significant model terms.

Response Factor	Data Transformation	Q^2	R^2	Adjusted R^2
Exit Valve Opening Width	Logarithmic	0.860	0.905	0.883
Exit Valve Opening Width SD	Logarithmic	0.822	0.933	0.893
Torque Lower Impeller	Logarithmic	0.851	0.916	0.896
Torque Lower Impeller SD	Logarithmic	0.882	0.949	0.933
Blend Potency SD	Logarithmic	0.491	0.669	0.622
HUM SD	Logarithmic	0.428	0.727	0.664

Further details regarding fit statistics and model equations are shown in Supplementary Section A, "Summary of Fit: Mixing Parameter".

3.1.2. Material Attributes of the Blend

The conditioned bulk density (CBD), flow-rate index (FRI) and d_{10} values of the blend were evaluated (Figure 6). THR, HUM and IMP show a similar impact on CBD and d_{10} values of the powder. In contrast, the coefficients regarding FRI show a positive impact of THR and a negative influence by IMP and THR*THR.

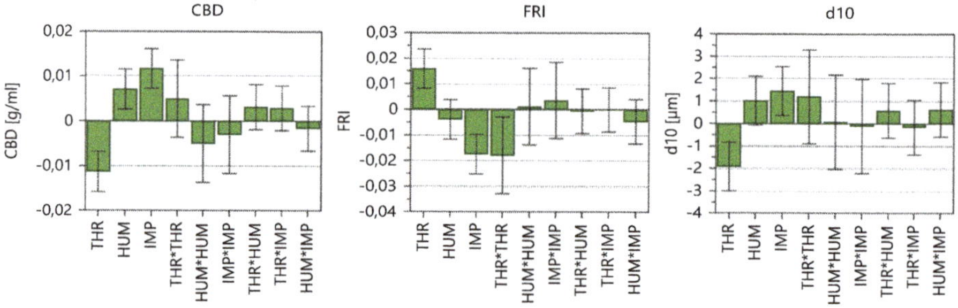

Figure 6. Coefficients plot of model terms regarding material attributes of the blends. The 95% confidence interval is displayed as an error bar.

Table 7 shows the fit statistics after removing non-significant model terms, whereby models regarding CBD, FRI and d_{10} can be considered good models. For further details, see Supplementary Section B, "Summary of Fit: Material Attributes of the Blend".

Table 7. Overview of fit statistics regarding material attributes of the blend.

Response Factor	Data Transformation	Q^2	R^2	Adjusted R^2
Conditioned Bulk Density	-	0.735	0.850	0.816
Flow Rate Index	-	0.800	0.896	0.848
Particle Size (d_{10})	-	0.587	0.842	0.747

3.1.3. Tableting Parameters

Regarding tableting parameters, the fill depth, bottom main compression height and ejection force were evaluated, wherein throughput and impeller speed are significant model terms for all three parameters (Figure 7). That means, these input factors have a statistically significant impact on all three tableting parameters. For example, higher throughput and lower impeller speed result in lower TBP and, therefore, in lower lubrication, leading to lower powder densities, higher required fill depths and higher ejection forces.

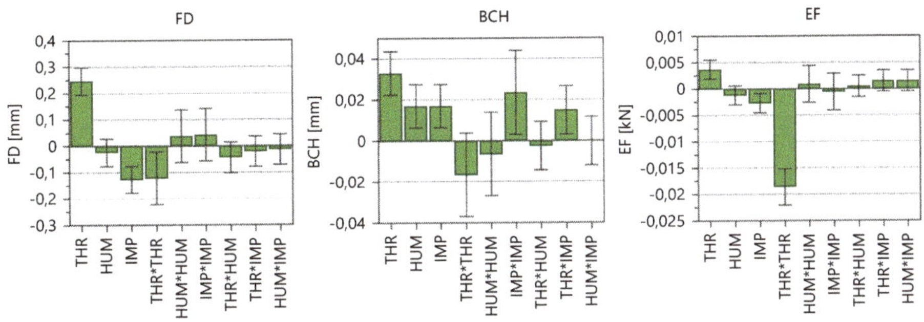

Figure 7. Coefficients plot of model terms regarding tablet press parameters. The 95% confidence interval is displayed as an error bar.

Furthermore, the fill depth and ejection force share the same deflection of the three significant model terms, namely THR, IMP and THR*THR.

Table 8 shows the fit statistics after removing non-significant model terms. All three parameters show high values regarding Q^2 and R^2. For further information regarding fit statistics and model equations, see Supplementary Section C, "Summary of Fit: Tablet Press Parameters".

Table 8. Overview of fit statistics regarding tablet-press parameters.

Response Factor	Data Transformation	Q^2	R^2	Adjusted R^2
Fill Depth	-	0.873	0.941	0.914
Bottom Main Compression Height	-	0.774	0.928	0.885
Ejection Force	-	0.892	0.944	0.931

3.1.4. Tablet Properties

To investigate the impact of CMT parameters on the tablet properties, the tensile strength (TS), tablet weight (TW) and tablet thickness (TT) obtained during steady state at 275 MPa compression pressure were evaluated.

In Figure 8, it can be observed that, besides the three input factors, namely THR, HUM and IMP, THR*THR, THR*IMP and HUM*IMP are significant model terms for tensile strength. That means, higher THR, lower HUM and lower IMP resulted in lower TBP and, therefore, lower lubrication, which increased the tensile strength of the tablets. Further explanations regarding TBP and tablet properties can be seen in the paragraph "Tensile Strength".

On the other hand, the tablet weight and thickness are both influenced by similar input variables. As the tablet-weight variance was always within control limits, an automatic weight adjustment did not occur. Consequently, TW was impacted by the density of the blends and FD. Considering the MLR, the high IMP, high IMP^2 and high THR*IMP resulted in higher TBP and higher densities. Since the FD adjustments only occurred occasionally when the displacement at the pre-compression exceeded internal limits at which the calculated weights are too high/low, higher powder density resulted in higher TW. Regarding variability in tablet properties, throughput has the highest impact on tablet weight and thickness standard deviations, whereas no significant model term regarding TS SD could be found.

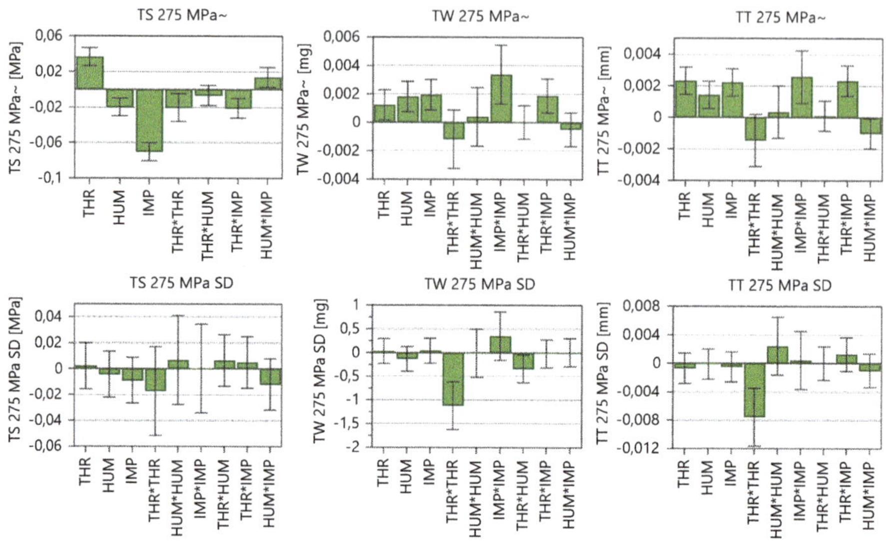

Figure 8. Model terms regarding tensile strength (TS), tablet weight (TW), tablet thickness (TT) and corresponding standard deviation. The 95% confidence interval is displayed as an error bar.

According to Table 9, tensile strength, tablet weight and tablet thickness can be considered good models. Again, as the variabilities of the responses were not linearly distributed, a logarithmic data transformation was conducted. Corresponding figures, model equations and fit statistics are shown in Supplementary Section D, "Summary of Fit: Tablet Properties".

Table 9. Overview of fit statistics regarding tablet properties.

Response Factor	Data Transformation	Q^2	R^2	Adjusted R^2
Tensile Strength	Logarithmic	0.907	0.976	0.958
Tensile Strength SD	-	−0.090	0.283	0.117
Tablet Weight	Logarithmic	0.641	0.904	0.847
Tablet Weight SD	-	0.472	0.856	0.770
Tablet Thickness	Logarithmic	0.718	0.953	0.917
Tablet Thickness SD	-	0.395	0.694	0.592

3.2. *Response Factors*

For recapitulation, Figure 9 demonstrates the relationships between all parameters obtained and evaluated within this DoE. Starting from the CMT settings, the flowchart depicts the downstream process parameters where correlations are expected to be found.

3.2.1. Mixing Parameters

Exit-Valve-Opening Width

As presented in Figure 5, the throughput and impeller speed were the significant model terms for the exit valve opening width (Q^2 = 0.860 and R^2 = 0.905). The low model validity observed was due to the extremely low variability seen in the replicated center points, and, hence, it is not a cause for concern.

In this regard, Figure 11a shows the exit valve opening width dependent on overall mass throughput, wherein increasing the throughput led to an increasing opening width. Furthermore, all EV at 650 rpm were higher than 10 mm. Figure 11b shows the exit valve depending on impeller speed. It confirms that the high impeller speed was an important reason for an increasing EV, while variations in HUM seemingly did not impact the exit valve (0.042 $p = 0.874$). Furthermore, a contour plot is used to demonstrate the significance of both model terms throughput and impeller speed (Figure 10). To determine suitable CMT settings based on this plot, small exit valve opening widths (<5 mm) are preferable, which is in line with the findings of Toson [8]. Additionally, data regarding blend potency SD confirmed the maximum of 5 mm opening value of the exit valve (further details below).

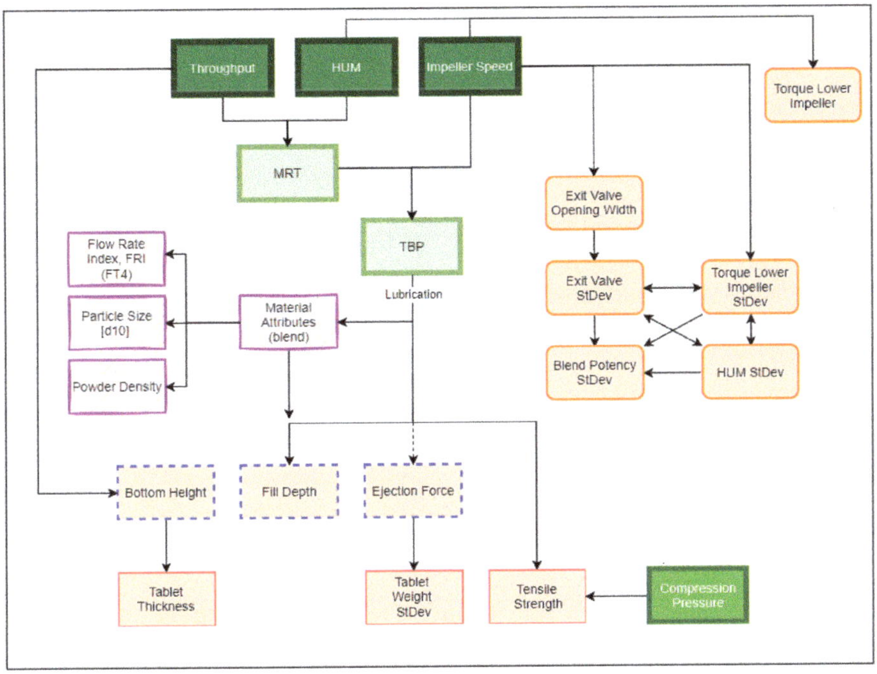

Figure 9. Qualitative overview of process parameter connections and correlations. Input factors are marked in dark green (thick borders), confounding input parameters are marked in light green and the considered response parameters are shown in light orange. The color/shape of the borders classifies the responses into mixing parameters (orange line, rounded corners), material attributes of the blend (purple, striped background), tableting parameters (blue, dotted borders) and tablet properties (red, thin borders). Compression pressure (green) is considered an independent input factor of the tablet press.

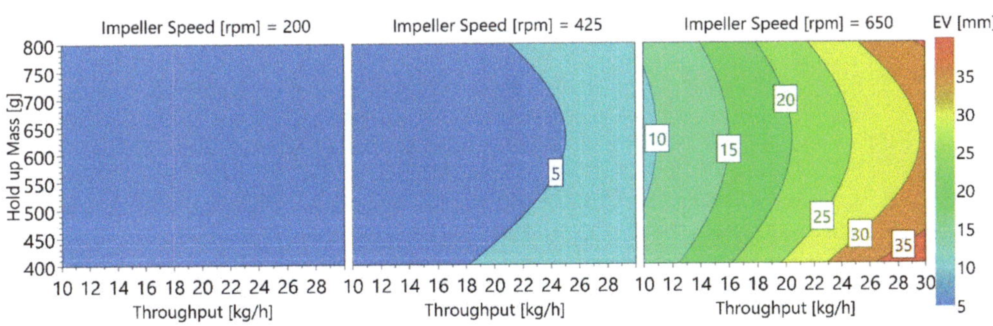

Figure 10. Contour plot of exit valve opening width in dependence of THR, HUM and IMP.

Regarding EV SD, Table 6 reveals that impeller speed was the only significant model term ($Q^2 = 0.822$ $R^2 = 0.933$). Furthermore, Figure 11c shows the EV standard deviation as a function of the EV opening width (0.785 $p = 0.0002$). This correlation leads to the conclusion that higher EV values increased the risk of a fluctuating opening width, impacting the variability of the blend potency values (0.952 $p < 0.0001$) and subsequently affecting content uniformity of the tablets. A correlation matrix with downstream parameters concerning the EV is shown in Figure 12.

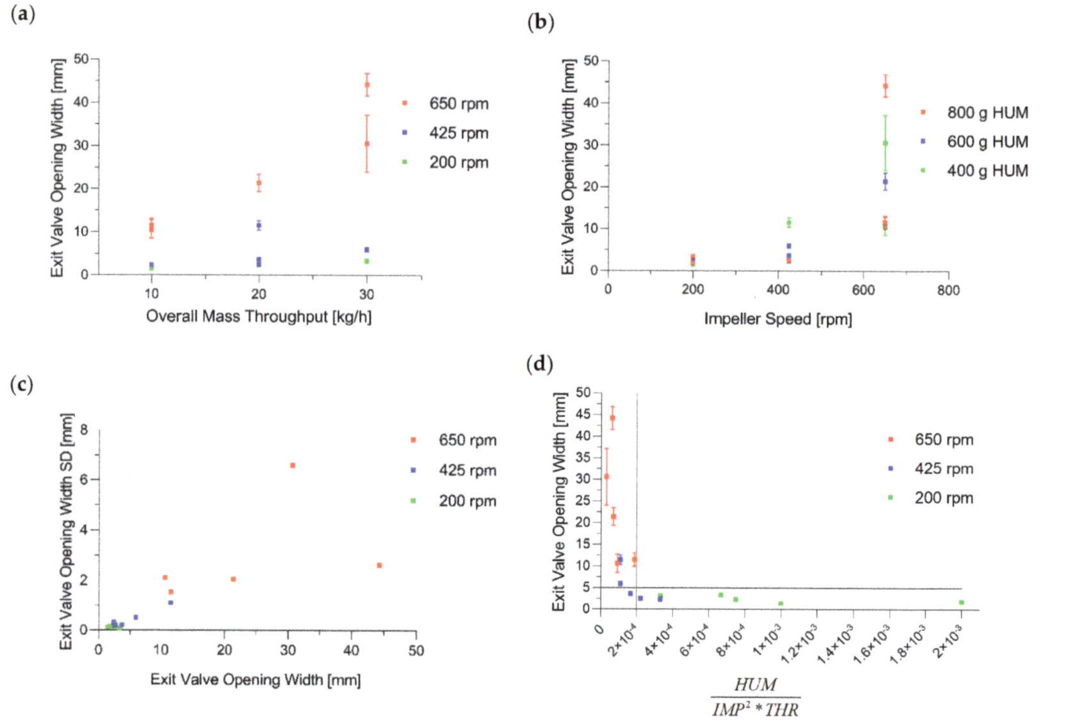

Figure 11. (**a**) Exit valve opening width vs. throughput (kg/h) in relation to varying impeller speeds. (**b**) EV in dependence of impeller speed. (**c**) EV SD vs. EV. (**d**) EV as function of $\frac{\text{HUM [g]}}{\text{IMP}^2\ [\text{rpm}^2] * \text{THR}\ \left[\frac{\text{kg}}{\text{h}}\right]}$ where x-values higher than 2×10^{-4} result in EV below 5 mm.

Figure 11d shows a correlation of the EV with the ratio of $\frac{\text{HUM [g]}}{\text{IMP}^2\ [\text{rpm}^2]\ * \text{THR}\ \left[\frac{\text{kg}}{\text{h}}\right]}$. This empirically found normalization revealed good processing for values exceeding 2×10^{-4}.

As the decreased impeller speed proved to have the highest impact on reducing the exit valve opening width, it is certainly the primary parameter for reducing the EV value below 5 mm. However, one needs to consider the impact on the powder attributes of the blend, because a decrease in impeller speed will decrease the TBP and, therefore, the amount of lubrication. As described in the following sections, the TBP impacts the CBD, FRI, d_{10} values, FD and TS.

HUM SD

HUM is an essential variable in MRT and TBP (Equations (1) and (2)), and this is why it is crucial to choose suitable blender parameters to maintain a consistent process. Accordingly, the fluctuation in HUM led to variabilities in the MRT and TBP.

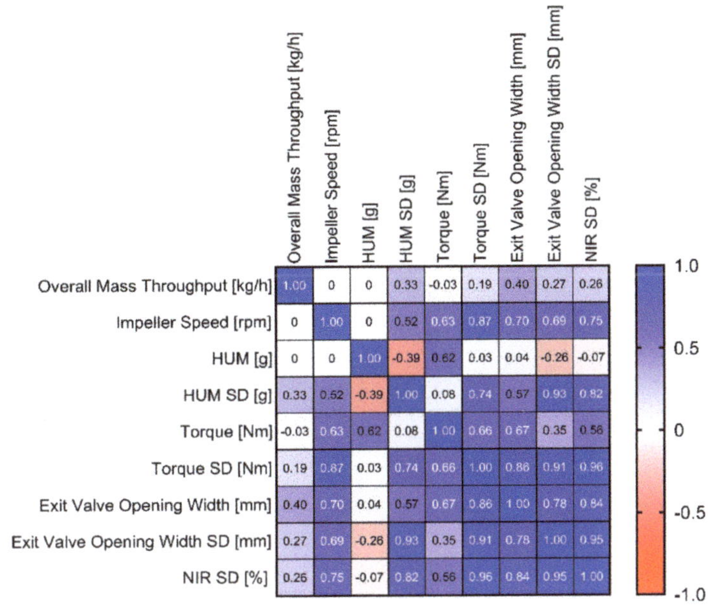

Figure 12. Correlation matrix of input variables and mixing parameters.

Figure 13a shows the HUM SD as a function of impeller speed (0.514 $p = 0.035$). It reveals that the HUM standard deviations were not directly impacted by throughput. However, throughput is a significant model term, since comparatively low HUM standard deviations were obtained at low throughputs. On the other hand, higher impeller speeds tended to result in a larger span of HUM SD, and this could be caused by an unfavorable powder bed shape, due to higher centrifugal forces, as described by Toson et al. [8].

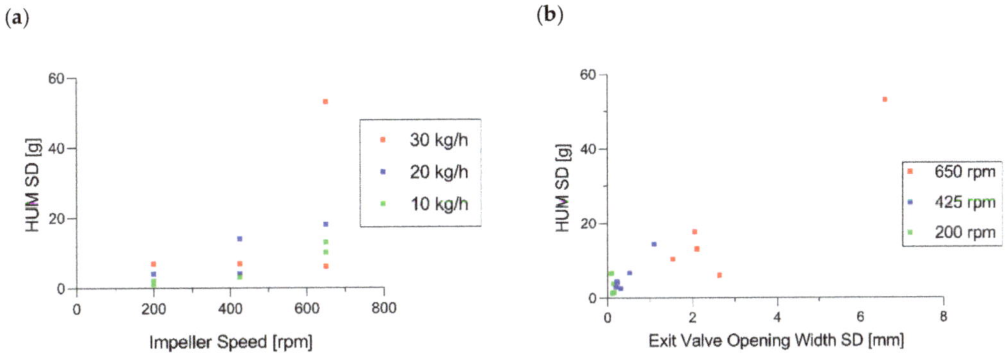

Figure 13. (**a**) HUM standard deviation as a function of impeller speed. (**b**) Dependencies between SD in EV and HUM.

Figure 13b shows that the previously mentioned EV SD correlated with HUM SD (0.929 $p < 0.001$). That could be traced back to the PID control loop between HUM and EV, where EV is a function of HUM process values in order to maintain $mass_{in} = mass_{out}$. Therefore, if variability could be observed in the HUM, then it occurred in EV, as well. To avoid those fluctuations, we can rely on the previous section, where impeller speed is the recommended parameter to control the corresponding process parameters. A detailed example is given in Supplementary Section E, "Additional Demonstration of HUM SD".

Torque of Lower Impeller

Regarding Table 6, the models for T_L ($Q^2 = 0.851$ and $R^2 = 0.916$) and T_L SD ($Q^2 = 0.882$ and $R^2 = 0.949$) could be considered good models. The low model validity for torque SD is, again, caused by low variability in the replicated center points, and, therefore, it is not a cause for concern. Basically, the torque represents the required energy to turn the impeller within the CMT and can be used to monitor the mixing process [43].

Since the model terms in Figure 5 showed similar coefficients of HUM and impeller speed, the torque could be seen as a function of the sum of both factors (0.888 $p < 0.0001$) (Figure 14a).

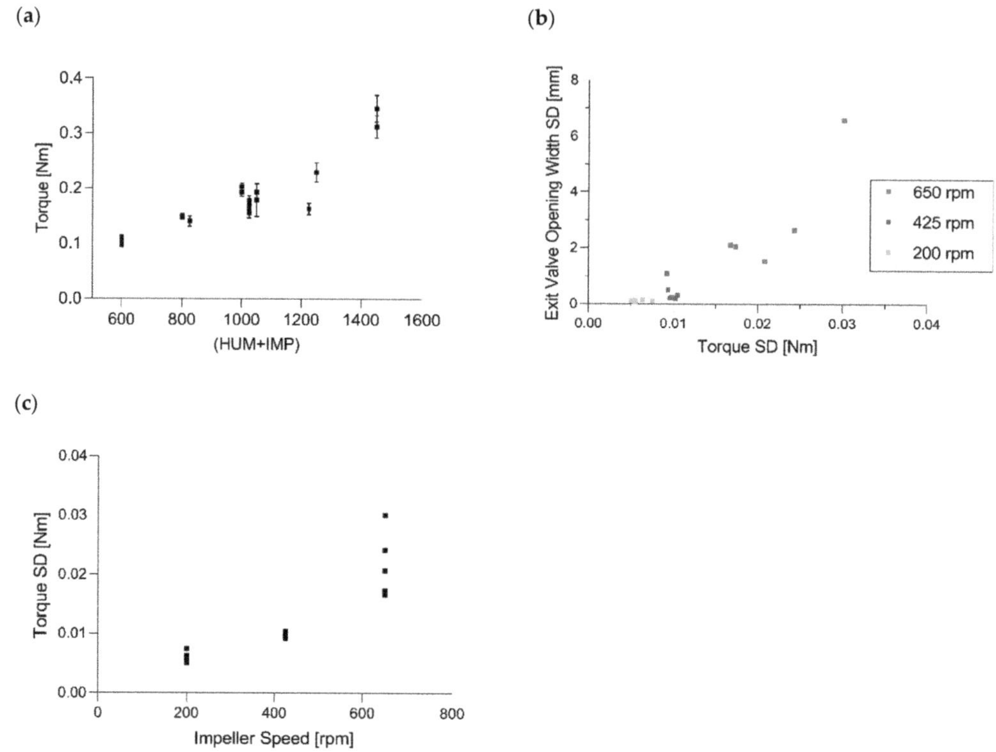

Figure 14. (**a**) Torque of the lower impeller as a function of the sum of HUM and IMP. (**b**) Correlation between variability in torque and exit valve opening width. (**c**) Impact of impeller speed on the torque values.

Figure 14b demonstrates the linearity between T_L SD and EV SD (0.906 $p < 0.0001$). The correlation between these standard deviations is based on the impact of impeller speed (IMP—T_L SD: 0.874 $p < 0.0001$), wherein the higher impeller rotation resulted in higher variabilities in both parameters (Figures 5 and 14c).

Since standard deviations in both the torque and exit valve were strongly correlated, it is recommended to only focus on the EV values if monitoring is required.

Blend Potency SD

Reflecting previously described process parameters, the correlations between blend potency SD and EV (0.843 $p < 0.0001$), EV SD (0.952 $p < 0.0001$), HUM SD (0.817 $p < 0.0001$), T_L SD (0.965 $p < 0.0001$) and IMP (0.753 $p = 0.0005$) could be observed (Figure 12).

Higher exit valve opening widths implicate that the powder bed was not entirely closed at the bottom of the CMT and that particles newly entering the CMT could exit unmixed [8]. Consequently, blend potency SDs and, therefore, blend inhomogeneities could be explained by insufficient mixing based on the structure of the powder bed within the blend. Figure 15a shows the blend potency standard deviation as a function of impeller speed, wherein all values at 650 rpm were above 2.5%. This observation could also be confirmed by using Figure 5, wherein IMP was the significant model term. Thus, to reduce blend potency SDs and, therefore, improve blend homogeneity and content uniformity of the tablets, reduction of the impeller speed is again proposed.

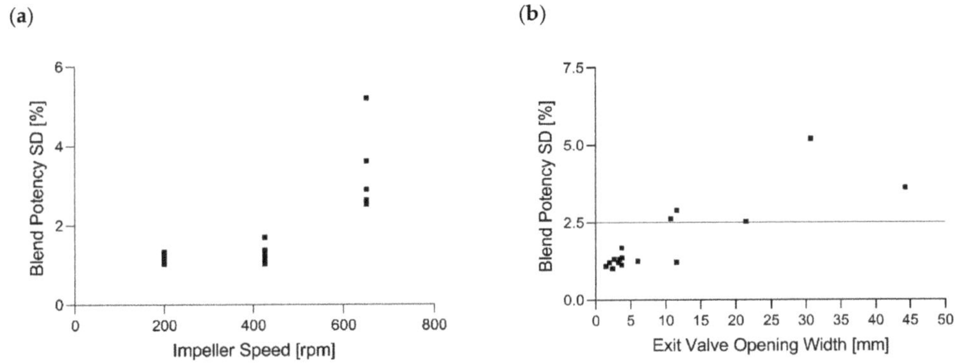

Figure 15. Blend Potency SD as a function of (**a**) impeller speed and (**b**) exit valve opening width.

Figure 15b shows that all blend potency values obtained at exit valve opening widths below 10 mm were smaller than 2.5%. To minimize the risk of a higher blend potency SD, the presented results confirm maximum EV values below 5 mm.

Furthermore, independent of the blender variables, a potential risk for blend potency inhomogeneity could be adhesion of API at the walls due to electrostatic charging of particles [44].

3.2.2. Material Attributes of the Blend

During continuous mixing with a vertical blender, TBP is the decisive factor in describing the impact on material attributes since magnesium stearate will be mixed simultaneously throughout the entire mixing process. It is the combination of impeller speed and blend time (MRT) (Equation (2)), governing shear and mixing intensity of the lubricant into the blend. For improvements regarding EV position and HUM uniformity following changes in material attributes must be considered:

Higher TBP represents more contact between impeller and powder particles and hence, it is implied that lubricant can be distributed more homogeneously into the blend with the potential risk of film formation. This would impact the tablet tensile strength and will be further discussed in section Tensile Strength.

Powder Density

With more impeller revolutions, more cavities of particles and granules can be filled and a layer around the particles can be built. On one hand, that increases the weight without increasing the volume and on the other hand, it is reducing particle-particle frictions due to the reduced friction of magnesium stearate filmed particles. Particles can now arrange more compactly, increasing the powder density [45,46].

Figure 16a demonstrates an exponential relationship between TBP and CBD asymptotically reaching a value of 0.598 $\frac{g}{mL}$ at 1560 TBP. At extreme values, such as 3120 revolutions, powder density will not increase any further and a maximum seemed to be reached, which led to the conclusion that increasing TBP only affected the material up to a certain limit.

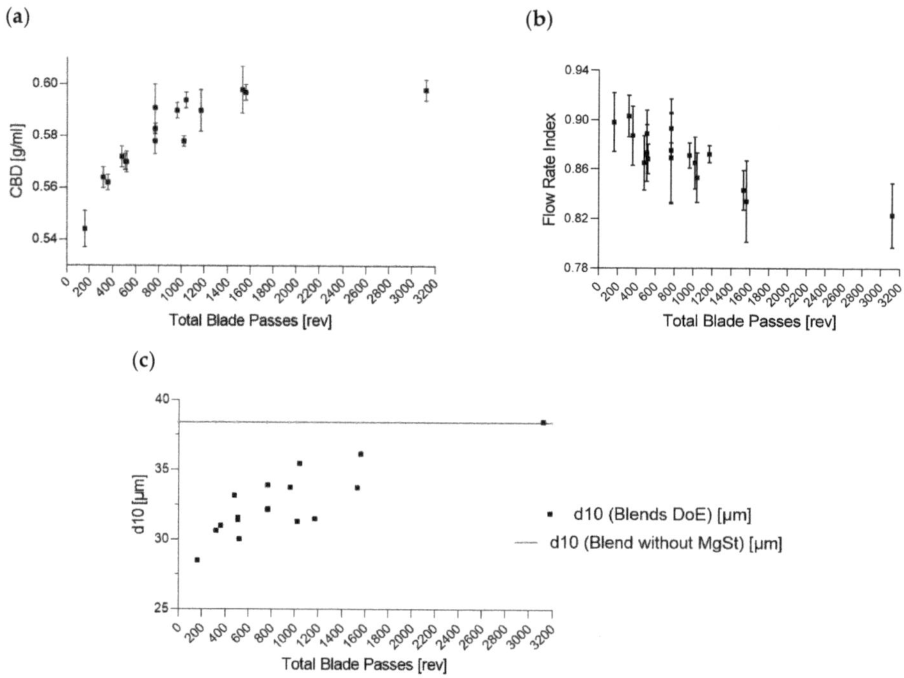

Figure 16. (**a**) Conditioned bulk density (CBD) (g/mL), (**b**) flow rate index (FRI) and (**c**) particle size (d_{10}) (µm) as a function of total blade passes.

Considering Figure 6 and the equation for TBP (Equation (2)), the significant model terms of the DoE revealed the same information, where higher values in HUM and impeller speed increased CBD, and higher throughputs decreased the powder density ($Q^2 = 0.735$ and $R^2 = 0.850$).

Flow Rate Index

Figure 16b shows the flow rate index (FRI) as a function of TBP (-0.846 $p < 0.0001$). In contrast to CBD, the FRI decreased with the rising TBP. Due to an increasing lubrication effect at the higher TBP, less energy was needed to move the blade through the powder bed, since the required energy is based on the resistance at the downward motion. Again, a plateau could be observed wherein the increasing TBP did not necessarily impact the FRI any further. The MLR analysis showed a model fit of $Q^2 = 0.800$ and $R^2 = 0.896$.

Particle Size

The description of density changes based on TBP also applies to the particle size (d_{10}) (Figure 16c). At a high TBP, more magnesium stearate adhered to the particles, leading to a lower amount of the remaining free MgSt particles within the blend, and thus increasing the d_{10} values (0.836 $p < 0.0001$). As a reference, a blend without magnesium stearate was mixed by using a Turbula blender (Willy A. Bachofen AG, Muttenz, Switzerland), where a d_{10} value of 38.38 µm was obtained (Figure 16c).

Therefore, the appearance of smaller particle sizes in the blend could be traced back to MgSt. As seen at 3120 revolutions, the d_{10} value was similar to the blend without MgSt, implicating that the fine fraction of MgSt was almost completely attached to the remaining raw materials at the higher TBP. Moreover, particle-size changes due to destruction of particles could be ruled out. In this case, the d_{10} values would have decreased with a higher shear.

Regarding the DoE results in Figure 6, a good model for d_{10} values could be obtained (R^2 = 0.842 and Q^2 = 0.587). Particle sizes of raw materials and blends are shown in Supplementary Table S34.

3.2.3. Tableting Parameters

Fill Depth

Higher powder density (CBD) will result in lower fill depths to fulfill the weight requirements, which could be confirmed in this paper (-0.844 $p < 0.0001$). As described above, the density of the blend was a function of TBP; that was why the fill depth was adjusted according to changes in TBP (-0.775 $p < 0.0001$), as well. Figure 17a shows the comparison between CBD and fill depth in dependence of TBP, where the increasing TBP resulted in higher density values and therefore in lower required fill depths. According to the TBP in Equation (2) and DoE Results in Figure 7, this observation could be confirmed since impeller speed was shown as negative and throughput as positive model term on fill-depth values.

As already described, after a specific amount of revolutions, neither CBD nor FD values showed further changes with increasing TBP.

Figure 17b shows the fill depth as a function of particle size (d_{10}). In general, smaller particle sizes are considered to decrease essential flowability, impacting a complete fill of the dies [47].

Regarding the die-filling process described by Xie and Puri [48], for powders with smaller particles, it is more challenging to lose entrained air due to cohesion during filling. Therefore, more volume and higher fill depths are required.. In this work, the correlation could be traced back again to lubrication, as described before, and not to cohesion (-0.224 $p = 0.387$).

Osorio and Muzzio [49] showed that higher powder compressibility values increase weight variability during capsule filling. Additionally, capsule weight decreased as powder compressibility increased. The same principle applies for die filling in this study, where higher powder compressibility led to higher fill depth values (0.703 $p < 0.002$) (Figure 17c). This observation may also be helpful if a capsule machine were used instead of a tablet press in continuous downstream processing.

Ejection Force

The ejection force is the required force to eject the tablet from the die and depends on the friction between the tablet and the die walls. Consequently, the reduction in ejection force is mainly influenced by the lubrication of the powder [50]. Usually, high ejection forces are accompanied by tableting problems and may cause damages to the tooling [51,52].

Regarding this dataset, the model-terms throughput and impeller speed shared the same deflections as for the fill depth (Figure 7); that means, it is indicated that a higher TBP results in a higher lubrication and lower ejection forces. However, although a strong correlation between ejection force and TBP was expected, only a correlation between ejection force and tablet-weight variability could be found (0.787 $p = 0.0002$). Nevertheless, a robust model regarding ejection force could be obtained by an MLR analysis (Q^2 = 0.892 and R^2 = 0.944). For further explanation regarding TBP and ejection force, see Supplementary Section G, "Ejection Force".

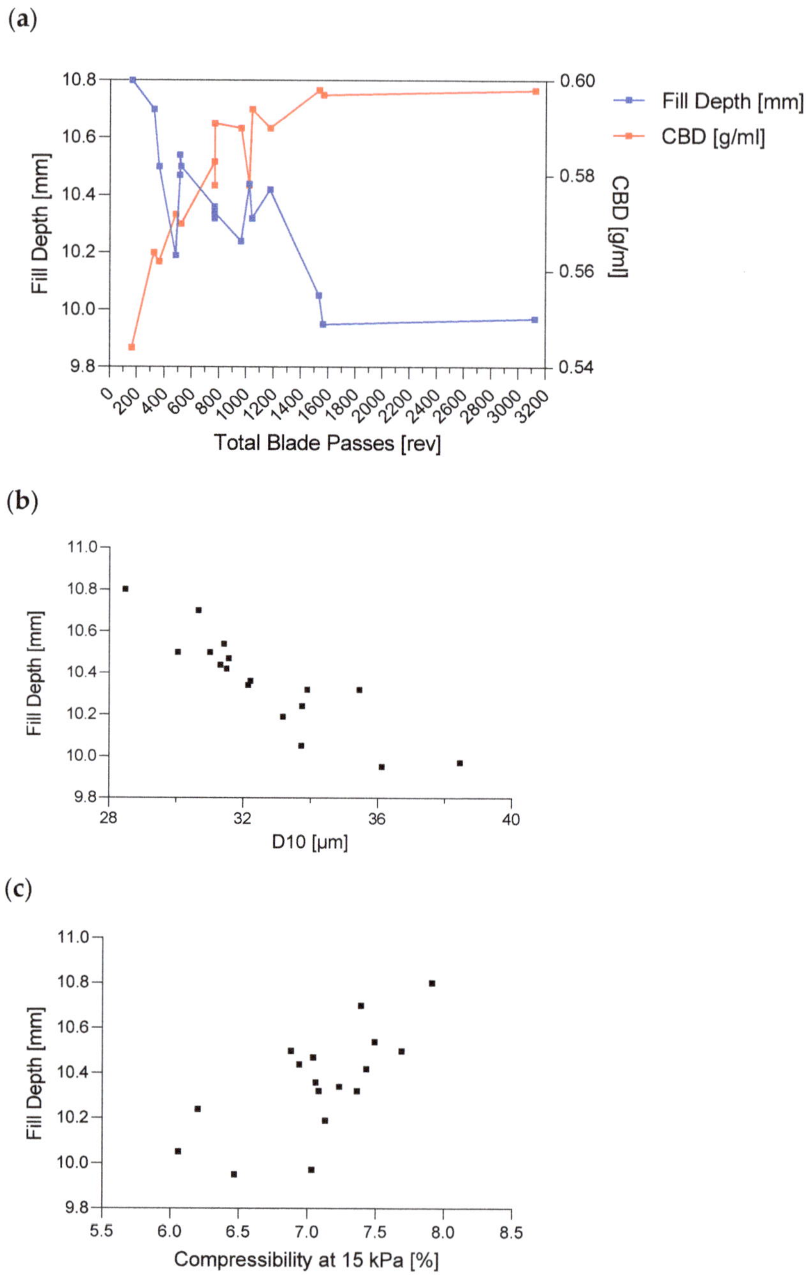

Figure 17. (**a**) Fill depth as function of total blade passes compared to bulk density. (**b**) Linearity between fill depth and d_{10} values. (**c**) Impact of compressibility on fill depth.

3.2.4. Tablet Properties

Even if good models for TS, TW and TT could be found, only few correlations regarding TW and TT could be obtained (see paragraph "Ejection Force").

Tensile Strength

Figure 18 demonstrates the tensile strength (TS) as a function of TBP (−0.704 p = 0.002), wherein a higher TBP resulted in lower tensile strength at the same compression pressure (275 MPa), due to increased lubrication efficiency. According to the DoE results in Figure 8, the significant model terms corresponded to the TBP Equation (2), where a higher throughput, lower HUM and lower impeller speed result in a lower TBP and, therefore, in higher tensile strengths of the tablet. If previous process states need to be optimized by adapting CMT parameters, a similar TBP should be maintained to ensure the correct TS.

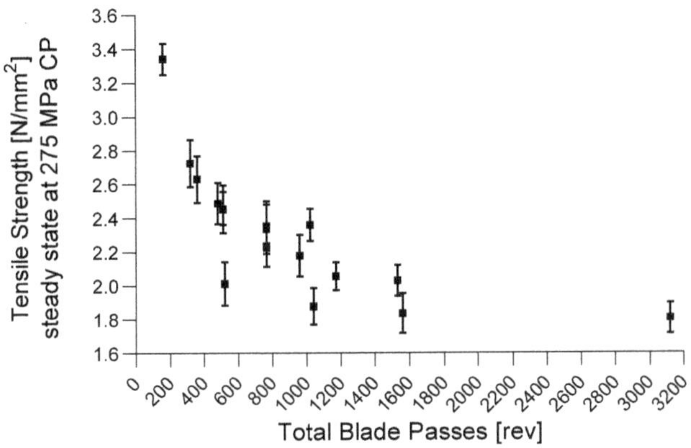

Figure 18. TS as a function of TBP at 275 MPa compression pressure.

Again, after 1560 revolutions, a plateau was reached, and no further reduction in tensile strength could be noticed with the increasing TBP.

Compression-Force Profile

A compression-force profile was conducted by using 118, 157, 169, 236 and 275 MPa compression pressure. During phase 16, no compression-force profile could be performed, because HUM increased from 0.8 to ~1.1 kg and the exit valve opened up to 45 mm, without any chance of decreasing. Thus, a consistent process flow could not be reached, and the correct setting of the FD and compression pressure was not possible.

Figure 19 includes the TS as a function of the corresponding compression pressure and TBP. Figure 19a demonstrates the profiles of each phase as a function of compression pressure, wherein the lowest TBP showed the highest values. Figure 19b reflects the TS as a function of TBP, where higher compression pressure led to profiles with higher values. Table 10 shows the fit statistics regarding tensile strengths obtained during the compression-force profiles.

Table 10. Overview of fit statistics regarding tensile strengths obtained during compression-force profiles.

Response Factor	Data Transformation	Q^2	R^2	Adjusted R^2
Tensile Strength at 118 MPa	Logarithmic	0.905	0.958	0.942
Tensile Strength at 157 MPa	Logarithmic	0.877	0.963	0.944
Tensile Strength at 169 MPa	Logarithmic	0.870	0.940	0.918
Tensile Strength at 236 MPa	Logarithmic	0.923	0.978	0.964
Tensile Strength at 275 MPa	Logarithmic	0.927	0.975	0.963

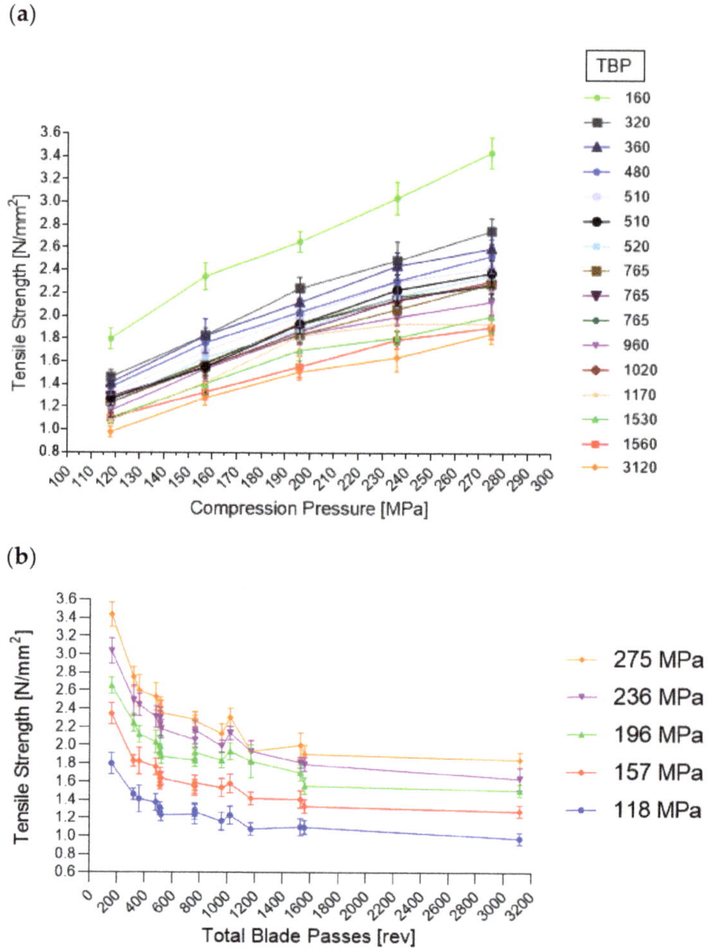

Figure 19. (**a**) Overview of all phases (TBP) regarding compression pressure and tensile strength. (**b**) Overview of all compression pressures and the corresponding tensile strength based on the lubrication (TBP).

4. Sweet Spot

By using MODDE, it is possible to detect a sweet spot where several criteria are met. For this paper, exit valve opening width (1–5 mm), blend potency SD (0–3%), tensile strength (2–3 MPa) and tablet-weight variability (0–2.5 mg) are considered critical parameters. In brackets, the favorable process values are shown. Figure 20 shows a visualization of a combination of input variables (throughput, hold-up mass and impeller speed) in which all criteria are met (light green). At an impeller speed of 650 rpm, no sweet spot could be achieved. With reducing impeller speeds, sweet spots at low throughputs are possible at 425 and 200 rpm. At 200 rpm, sweet spots could be achieved at low throughputs independently of HUM. Since it is preferred to run the process with higher throughputs, an optimal setting for this formulation can be observed at a combination of high throughputs and high HUM values at 200 rpm impeller speed.

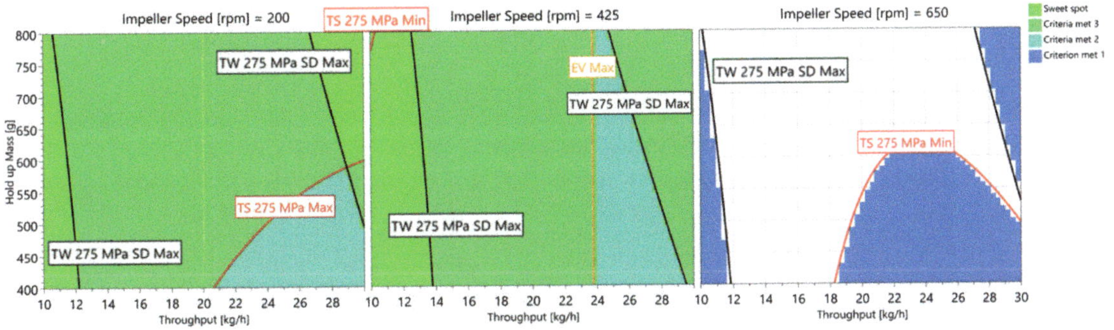

Figure 20. Sweet spot (light green) reveals the combination of the DoE input variables in which the criteria are met. The color of the borders indicate which criterion is not met anymore. Black borders = TW SD, red borders = TS and orange boarders = EV.

5. Conclusions

This paper showed the evaluation of the downstream process states based on throughput, hold-up mass and impeller speed in a continuous direct compression line, including a single blending step, in a vertical blender (CMT). For all settings in the performed DoE, the same composition and compounds were used, so that the initial material attributes and lubrication sensitivity remained constant.

In this study, the model terms of the process states based on the CMT parameters were evaluated by means of a MLR analysis. Corresponding fit statistics are shown in Table 11.

Table 11. Overview of the models obtained in this study.

Responses	Q^2	R^2	Adjusted R^2
Exit valve opening width	0.860	0.905	0.883
Exit valve opening width SD	0.822	0.933	0.893
Torque of lower impeller	0.851	0.916	0.896
Torque of lower impeller SD	0.882	0.949	0.933
Conditioned bulk density	0.735	0.850	0.816
Flow rate index	0.800	0.896	0.848
Fill depth	0.873	0.941	0.914
Bottom main compression height	0.774	0.928	0.885
Ejection force	0.892	0.944	0.931
Tablet thickness	0.718	0.953	0.917
Tablet weight	0.642	0.904	0.847
Tensile strength	0.907	0.976	0.958

Furthermore, the connections between the parameters were evaluated. Regarding mixing parameters, it has been shown that the exit valve opening width and variability in exit valve, in hold-up mass, in torque and in blend potency are significantly correlated and can all be controlled mainly by impeller speed. If the improvement of these parameters is required, it needs to be considered that changes in impeller speed will also lead to changes in TBP.

With higher TBP, more shear is transmitted to the powder and more magnesium stearate will adhere to the remaining particles, leading to more lubrication and higher variation in material attributes. Hence, TBP significantly correlated with the blend's material attributes (density, d_{10} values and flow-rate index), the fill depth and the tensile strength of the tablets.

Target criteria (exit valve opening width (1–5 mm), blend potency SD (0–3%), tensile strength (2–3 MPa) and tablet-weight variability (0–2.5 mg)) could generally be found at impeller speeds between 200 and 425 rpm and at throughputs between 10 and 12 kg/h independent of HUM. To run the process as fast as possible, high throughput, high HUM and 200 rpm IMP are required to fulfill the target criteria and, therefore, represent the optimal setting for this formulation.

Supplementary Materials: The following supporting information can be downloaded at https://www.mdpi.com/article/10.3390/pharmaceutics14020278/s1.

Author Contributions: Conceptualization, M.J.K. and K.G.W.; methodology, M.J.K. and K.G.W.; software, M.J.K.; formal analysis, M.J.K. and K.G.W.; investigation, M.J.K.; data curation, M.J.K.; writing—original draft preparation, M.J.K..; writing—review and editing, M.J.K., C.W. and K.G.W.; visualization, M.J.K.; supervision, K.G.W. and C.W.; project administration, K.G.W. and C.W. All authors have read and agreed to the published version of the manuscript.

Funding: This research was funded by Pfizer Manufacturing Deutschland GmbH.

Acknowledgments: We would like to thank the Pfizer MCT colleagues in Sandwich, UK, for supporting the powder analysis. We also want to acknowledge the work of the Pfizer R&D colleagues in Groton, USA, and Sandwich, UK, who carefully read and reviewed the manuscript.

Conflicts of Interest: The authors declare no conflict of interest.

Abbreviations

API	active pharmaceutical ingredient
BCH	bottom main compression height
BFE	basic flow energy
CBD	conditioned bulk density
CMT	continuous mixing technology
CP	compression pressure
DC	direct compression
DoE	design of experiment
EF	ejection force
EV	exit valve opening width
EV SD	exit valve opening width standard deviation
FD	fill depth
FRI	flow rate index
HUM	hold-up mass
IMP	impeller speed
LiW	loss in weight
MBM	mass balance model
MCC	microcrystalline cellulose
MgSt	magnesium stearate
MLR	multiple linear regression
MRT	mean residence time
NIR	near infrared
PCMM	portable, continuous, modular, miniature
PID	proportional–integral–derivative
PLS	partial least square
PV	process value
RTD	residence-time distribution
SE	specific energy
SI	stability index
SNV	standard normal variate
SD	standard deviation
TBP	total blade passes
THR	throughput

T_L	torque lower impeller
T_L SD	torque lower impeller standard deviation
TS	tensile strength
TT	tablet thickness
TW	tablet weight

References

1. Blackwood, D.O.; Bonnassieux, A.; Cogoni, G. Continuous direct compression using portable continuous miniature modular & manufacturing (PCM&M). In *Chemical Engineering in the Pharmaceutical Industry*; John Wiley & Sons, Ltd.: Hoboken, NJ, USA, 2019; pp. 547–560. [CrossRef]
2. Hsiao, W.-K.; Hörmann, T.R.; Toson, P.; Paudel, A.; Ghiotti, P.; Stauffer, F.; Bauer, F.; Lakio, S.; Behrend, O.; Maurer, R.; et al. Feeding of particle-based materials in continuous solid dosage manufacturing: A material science perspective. *Drug Discov. Today* **2020**, *25*, 800–806. [CrossRef] [PubMed]
3. Nowak, S.; K-Tron, C. *Three Ways to Improve Continuous Loss-in-Weight Feeding Accuracy*; CSC Publishing: St Paul, MN, USA, 2016.
4. Engisch, W.E.; Muzzio, F.J. Feedrate deviations caused by hopper refill of loss-in-weight feeders. *Powder Technol.* **2015**, *283*, 389–400. [CrossRef]
5. Gao, Y.; Muzzio, F.; Ierapetritou, M. Characterization of feeder effects on continuous solid mixing using fourier series analysis. *AIChE J.* **2011**, *57*, 1144–1153. [CrossRef]
6. Hanson, J. Control of a system of loss-in-weight feeders for drug product continuous manufacturing. *Powder Technol.* **2018**, *331*, 236–243. [CrossRef]
7. Tahir, F.; Palmer, J.; Khoo, J.; Holman, J.; Yadav, I.K.; Reynolds, G.; Meehan, E.; Mitchell, A.; Bajwa, G. Development of feed factor prediction models for loss-in-weight powder feeders. *Powder Technol.* **2019**, *364*, 1025–1038. [CrossRef]
8. Toson, P.; Siegmann, E.; Trogrlic, M.; Kureck, H.; Khinast, J.; Jajcevic, D.; Doshi, P.; Blackwood, D.; Bonnassieux, A.; Daugherity, P.D.; et al. Detailed modeling and process design of an advanced continuous powder mixer. *Int. J. Pharm.* **2018**, *552*, 288–300. [CrossRef]
9. Engisch, W.E.; Muzzio, F.J. Loss-in-Weight Feeding Trials Case Study: Pharmaceutical Formulation. *J. Pharm. Innov.* **2015**, *10*, 56–75. [CrossRef]
10. Engisch, W.E.; Muzzio, F.J. Method for characterization of loss-in-weight feeder equipment. *Powder Technol.* **2012**, *228*, 395–403. [CrossRef]
11. Escotet-Espinoza, M.S.; Moghtadernejad, S.; Scicolone, J.; Wang, Y.; Pereira, G.; Schäfer, E.; Vigh, T.; Klingeleers, D.; Ierapetritou, M.; Muzzio, F.J. Using a material property library to find surrogate materials for pharmaceutical process development. *Powder Technol.* **2018**, *339*, 659–676. [CrossRef]
12. Hopkins, M. Loss in Weight Feeder Systems. *Meas. Control.* **2006**, *39*, 237–240. [CrossRef]
13. Wang, Y.; Li, T.; Muzzio, F.J.; Glasser, B.J. Predicting feeder performance based on material flow properties. *Powder Technol.* **2017**, *308*, 135–148. [CrossRef]
14. Gao, Y.; Vanarase, A.; Muzzio, F.; Ierapetritou, M. Characterizing continuous powder mixing using residence time distribution. *Chem. Eng. Sci.* **2011**, *66*, 417–425. [CrossRef]
15. Marikh, K.; Berthiaux, H.; Gatumel, C.; Mizonov, V.; Barantseva, E. Influence of stirrer type on mixture homogeneity in continuous powder mixing: A model case and a pharmaceutical case. *Chem. Eng. Res. Des.* **2008**, *86*, 1027–1037. [CrossRef]
16. Lee, K.T.; Kimber, J.A.; Cogoni, G.; Brandon, J.K.; Wilsdon, D.; Verrier, H.M.; Grieb, S.; Blackwood, D.O.; Jain, A.C.; Doshi, P. Continuous Mixing Technology: Characterization of a Vertical Mixer Using Residence Time Distribution. *J. Pharm. Sci.* **2021**, *110*, 2694–2702. [CrossRef]
17. Mehrotra, A.; Llusa, M.; Faqih, A.; Levin, M.; Muzzio, F.J. Influence of shear intensity and total shear on properties of blends and tablets of lactose and cellulose lubricated with magnesium stearate. *Int. J. Pharm.* **2007**, *336*, 284–291. [CrossRef]
18. Swaminathan, V.; Kildsig, D.O. Effect of Magnesium Stearate on the Content Uniformity of Active Ingredient in Pharmaceutical Powder Mixtures. *Aaps Pharmscitech* **2002**, *3*, 27. [CrossRef]
19. Johansson, M.E. Granular magnesium stearate as a lubricant in tablet formulations. *Int. J. Pharm.* **1984**, *21*, 307–315. [CrossRef]
20. Ketterhagen, W.R.; Mullarney, M.P.; Kresevic, J.; Blackwood, D. Computational approaches to predict the effect of shear during processing of lubricated pharmaceutical blends. *Powder Technol.* **2018**, *335*, 427–439. [CrossRef]
21. Kushner, J. Incorporating Turbula mixers into a blending scale-up model for evaluating the effect of magnesium stearate on tablet tensile strength and bulk specific volume. *Int. J. Pharm.* **2012**, *429*, 1–11. [CrossRef]
22. Kushner, J.; Moore, F. Scale-up model describing the impact of lubrication on tablet tensile strength. *Int. J. Pharm.* **2010**, *399*, 19–30. [CrossRef]
23. Kushner, J.; Schlack, H. Commercial scale validation of a process scale-up model for lubricant blending of pharmaceutical powders. *Int. J. Pharm.* **2014**, *475*, 147–155. [CrossRef] [PubMed]
24. Portillo, P.M.; Ierapetritou, M.G.; Muzzio, F.J. Characterization of continuous convective powder mixing processes. *Powder Technol.* **2008**, *182*, 368–378. [CrossRef]
25. Wang, J.; Wen, H.; Desai, D. Lubrication in tablet formulations. *Eur. J. Pharm. Biopharm.* **2010**, *75*, 1–15. [CrossRef] [PubMed]

26. Vanarase, A.U.; Alcalà, M.; Rozo, J.I.J.; Muzzio, F.J.; Romañach, R.J. Real-time monitoring of drug concentration in a continuous powder mixing process using NIR spectroscopy. *Chem. Eng. Sci.* **2010**, *65*, 5728–5733. [CrossRef]
27. De Leersnyder, F.; Peeters, E.; Djalabi, H.; Vanhoorne, V.; Van Snick, B.; Hong, K.; Hammond, S.; Liu, A.Y.; Ziemons, E.; Vervaet, C.; et al. Development and validation of an in-line NIR spectroscopic method for continuous blend potency determination in the feed frame of a tablet press. *J. Pharm. Biomed. Anal.* **2018**, *151*, 274–283. [CrossRef] [PubMed]
28. Peeters, M.; Peeters, E.; Van Hauwermeiren, D.; Cogoni, G.; Liu, Y.; De Beer, T. Determination and understanding of lead-lag between in-line NIR tablet press feed frame and off-line NIR tablet measurements. *Int. J. Pharm.* **2021**, *611*, 121328. [CrossRef]
29. Van Hauwermeiren, D.; Peeters, M.; Peeters, E.; Cogoni, G.; Yang, L.A.; De Beer, T. Development of a tablet press feed frame lead lag determination model using in-line and off-line NIR measurements. *Int. J. Pharm.* **2021**, *612*, 121284. [CrossRef]
30. Madian, A.; Leturia, M.; Ablitzer, C.; Matheron, P.; Bernard-Granger, G.; Saleh, K. Impact of fine particles on the rheological properties of uranium dioxide powders. *Nucl. Eng. Technol.* **2020**, *52*, 1714–1723. [CrossRef]
31. Freemann Technology. *Instruction Documents: W7013 Stability and Variable Flow Rate*; Freemann Technology: Tewkesbury, UK, 2007.
32. Freeman, R. Measuring the flow properties of consolidated, conditioned and aerated powders—A comparative study using a powder rheometer and a rotational shear cell. *Powder Technol.* **2007**, *174*, 25–33. [CrossRef]
33. Freemann Technology. *Instruction Documents: W7012 Variable Flow Rate*; Freemann Technology: Tewkesbury, UK, 2007.
34. Freemann Technology. *Instruction Documents: W7030 Basic Flowability Energy*; Freemann Technology: Tewkesbury, UK, 2007.
35. Freemann Technology. *Instruction Documents: W7008 Compressibility*; Freemann Technology: Tewkesbury, UK, 2007.
36. Freemann Technology. *Instruction Documents: W7018 Shear Cell*; Freemann Technology: Tewkesbury, UK, 2007.
37. Freemann Technology. *Instruction Documents: W7050 1ml Shear Cell*; Freemann Technology: Tewkesbury, UK, 2007.
38. Freeman, R.E.; Cooke, J.R.; Schneider, L.C.R. Measuring shear properties and normal stresses generated within a rotational shear cell for consolidated and non-consolidated powders. *Powder Technol.* **2009**, *190*, 65–69. [CrossRef]
39. Wang, Y.; Koynov, S.; Glasser, B.J.; Muzzio, F.J. A method to analyze shear cell data of powders measured under different initial consolidation stresses. *Powder Technol.* **2016**, *294*, 105–112. [CrossRef]
40. Peeters, E. *Investigation of the Tableting Process in Continuous Production: Influence of Feeding and Extended Dwell Time during Compression on Dependent Process Variables and Tablet Properties*; Ghent Univeristy: Ghent, Belgium, 2014.
41. Pitt, K.G.; Newton, J.M.; Stanley, P. Tensile fracture of doubly-convex cylindrical discs under diametral loading. *J. Mater. Sci.* **1988**, *23*, 2723–2728. [CrossRef]
42. Mukaka, M.M. Statistics corner: A guide to appropriate use of correlation coefficient in medical research. *Malawi Med. J.* **2012**, *24*, 69–71. [PubMed]
43. Knight, P.; Seville, J.; Wellm, A.; Instone, T. Prediction of impeller torque in high shear powder mixers. *Chem. Eng. Sci.* **2001**, *56*, 4457–4471. [CrossRef]
44. Moghtadernejad, S.; Escotet-Espinoza, M.S.; Oka, S.; Singh, R.; Liu, Z.; Román-Ospino, A.D.; Li, T.; Razavi, S.; Panikar, S.; Scicolone, J.; et al. A Training on: Continuous Manufacturing (Direct Compaction) of Solid Dose Pharmaceutical Products. *J. Pharm. Innov.* **2018**, *13*, 155–187. [CrossRef]
45. Morin, G.; Briens, L. The Effect of Lubricants on Powder Flowability for Pharmaceutical Application. *AAPS PharmSciTech* **2013**, *14*, 1158–1168. [CrossRef]
46. Razavi, S.M.; Gonzalez, M.; Cuitiño, A.M. Quantification of lubrication and particle size distribution effects on tensile strength and stiffness of tablets. *Powder Technol.* **2018**, *336*, 360–374. [CrossRef]
47. Goh, H.P.; Heng, P.W.S.; Liew, C.V. Comparative evaluation of powder flow parameters with reference to particle size and shape. *Int. J. Pharm.* **2018**, *547*, 133–141. [CrossRef]
48. Xie, X.; Puri, V.M. Uniformity of Powder Die Filling Using a Feed Shoe: A Review. *Part. Sci. Technol.* **2006**, *24*, 411–426. [CrossRef]
49. Osorio, J.G.; Muzzio, F.J. Effects of powder flow properties on capsule filling weight uniformity. *Drug Dev. Ind. Pharm.* **2012**, *39*, 1464–1475. [CrossRef]
50. Uzondu, B.; Leung, L.Y.; Mao, C.; Yang, C.-Y. A mechanistic study on tablet ejection force and its sensitivity to lubrication for pharmaceutical powders. *Int. J. Pharm.* **2018**, *543*, 234–244. [CrossRef] [PubMed]
51. Anuar, M.; Briscoe, B. The elastic relaxation of starch tablets during ejection. *Powder Technol.* **2009**, *195*, 96–104. [CrossRef]
52. Dun, J.; Osei-Yeboah, F.; Boulas, P.; Lin, Y.; Sun, C.C. A systematic evaluation of poloxamers as tablet lubricants. *Int. J. Pharm.* **2020**, *576*, 118994. [CrossRef] [PubMed]

pharmaceutics MDPI

Article

Process Modeling and Simulation of Tableting—An Agent-Based Simulation Methodology for Direct Compression

Niels Lasse Martin [1,2,*,†], Ann Kathrin Schomberg [2,3,*,†], Jan Henrik Finke [2,3], Tim Gyung-min Abraham [1,2], Arno Kwade [2,3] and Christoph Herrmann [1,2]

1 Institute of Machine Tools and Production Technology (IWF), Technische Universität Braunschweig, Langer Kamp 19b, 38106 Braunschweig, Germany; t.abraham@tu-bs.de (T.A.); c.herrmann@tu-bs.de (C.H.)
2 Center of Pharmaceutical Engineering (PVZ), Technische Universität Braunschweig, Franz-Liszt-Str. 35A, 38106 Braunschweig, Germany; jan.finke@tu-bs.de (J.H.F.); a.kwade@tu-bs.de (A.K.)
3 Institute for Particle Technology (iPAT), Technische Universität Braunschweig, Volkmaroder Str. 5, 38104 Braunschweig, Germany
* Correspondence: n.martin@tu-bs.de (N.L.M.); a.schomberg@tu-bs.de (A.K.S.); Tel.: +49-531-391-7693 (N.L.M.)
† These authors contributed equally to this work.

Abstract: In pharmaceutical manufacturing, the utmost aim is reliably producing high quality products. Simulation approaches allow virtual experiments of processes in the planning phase and the implementation of digital twins in operation. The industrial processing of active pharmaceutical ingredients (APIs) into tablets requires the combination of discrete and continuous sub-processes with complex interdependencies regarding the material structures and characteristics. The API and excipients are mixed, granulated if required, and subsequently tableted. Thereby, the structure as well as the properties of the intermediate and final product are influenced by the raw materials, the parametrized processes and environmental conditions, which are subject to certain fluctuations. In this study, for the first time, an agent-based simulation model is presented, which enables the prediction, tracking, and tracing of resulting structures and properties of the intermediates of an industrial tableting process. Therefore, the methodology for the identification and development of product and process agents in an agent-based simulation is shown. Implemented physical models describe the impact of process parameters on material structures. The tablet production with a pilot scale rotary press is experimentally characterized to provide calibration and validation data. Finally, the simulation results, predicting the final structures, are compared to the experimental data.

Keywords: agent-based modeling and simulation; process modeling and simulation; tableting; product structures and characteristics

Citation: Martin, N.L.; Schomberg, A.K.; Finke, J.H.; Abraham, T.; Kwade, A.; Herrmann, C. Process Modeling and Simulation of Tableting—An Agent-Based Simulation Methodology for Direct Compression. *Pharmaceutics* **2021**, *13*, 996. https://doi.org/10.3390/pharmaceutics13070996

Academic Editors: Colin Hare and Anne Marie Healy

Received: 23 April 2021
Accepted: 22 June 2021
Published: 30 June 2021

Publisher's Note: MDPI stays neutral with regard to jurisdictional claims in published maps and institutional affiliations.

1. Introduction

The landscape of pharmaceutical processes and their control is and will further be changing in the coming decades. The adoption of quality by design (QbD) concepts and process analytical technology (PAT) approaches will be fostered, and classical batch production will be superseded by continuous pharmaceutical processes as well. The latter is getting more and more established in pharmaceutical industries due to its high potential in agility, flexibility, cost, and robustness and therefore ensuring a continuous supply of high-quality drugs [1]. Nevertheless, continuous as well as batch processes can be represented in simulation models, allowing the investigation of parameter changes and the resulting impact on the products. At the moment, and in the transition towards a continuous production in pharmaceutics, the process chains consist of combined batch and continuous processes, where each process has a significant impact on intermediate and final product structures and characteristics. Therefore, process chain models may need to represent batch as well as continuous processes.

Until now, the verification of final product quality is mainly ensured by offline batch acceptance sampling after nearly every process step [2]. In case of process deviations, methodologies forecasting the final product quality or even feed-forward oriented control strategies for process parameter adaption are mostly missing. Hence, this gap often causes batch loss and imposes high cost.

Dynamic simulation approaches allow for the forecasting of processes and their process chains and, when combined with real world data, data-driven, knowledge-based control of these processes [3]. In the planning phase, forecasting the resulting product structures and characteristics of the planned process chain is possible. In the operation phase and based on proper simulation approaches, deviations can be compensated for by adjusting the parameters of upcoming process steps, adapting treatment times, or repeating or adding further steps. Furthermore, simulation approaches allow the analysis of complex interdependencies and are therefore useful to generate process knowledge. Four typical simulation paradigms can be distinguished in dynamic simulation modeling [4] and are shown in Figure 1. Depending on the abstraction level of the model and the change of process parameters over time (continuous or discrete), different paradigms are more likely to be chosen. For example, models with a high degree of detail and a continuous parameter change are likely to be modelled using the dynamic systems paradigm.

Discrete event simulations (DES) are widely used for modeling complex networks such as Petri nets, which do not require continuous parameter changes but are very helpful for the determination of key performance indicators of process chains. System dynamics (SD) models describe the system behavior with few details of the entities on a high abstraction level, and are often used for complex social or political systems [5].

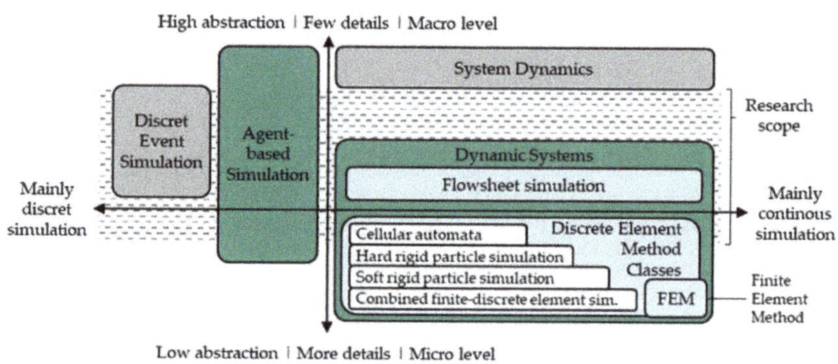

Figure 1. Typical dynamic simulation approaches for the tableting process mapped on the simulation modeling paradigms (following [4–6]).

For simulating a tableting process or sub-process, the common approach is the use of dynamic systems simulations such as flowsheet simulations (FSS). Furthermore, discrete element methods (DEM) are used, which can further be classified depending on their level of detail. This scale can reach from cellular automata to combined finite-discrete element simulation as illustrated in Figure 1. A detailed description for those methods can be found in [6]. Simulations built upon the finite element method (FEM) are used to model particle deformation behavior under external stresses and the propagation of stresses within particles. Cellular automata approaches model the interaction with neighboring particle groups rather than the particles themselves, while combined finite-discrete element approaches model particle deformation, interaction with other particles, form and size in detail. FSS are mainly used to describe the change of properties of continuous phases (whereas solids are also considered continuous phases) and are well established in the process industry. Similar to the DES models, networks of entities and transitions are formed, representing process chains and aggregates, for example. The entities of a FSS network, so called nodes, are used to calculate output streams of given input streams according to physical circumstances. Due to the more complex description of solids, the use of FSS models for solid processes has only recently been investigated [7–9].

Those dynamic systems simulation approaches have already proven their value in simulating production processes and chains or detailed process behaviors, respectively. However, considering the different abstraction levels of the approaches, FSS lack the ability to determine distributed product properties and their influence on the process. In contrast to that, discrete element simulations lack the ability to simulate whole process chains due to high computational efforts [10]. In past research, there have been efforts to adapt FSS especially for particulate processes in order to model detailed product characteristics to overcome this gap. This resulted in a modular open source system covering diverse process units by using multidimensional distributed product structures [11]. Additionally, reduced order models allow the integration of originally highly detailed models in FSS approaches at low computational expense [12].

In comparison to the common dynamic systems modeling approaches, agent-based (AB) models allow both the ability to represent process chain behaviors and the ability to determine heterogenic product properties as well as their interdependencies with the processes and the consideration of different detail levels [4,13]. Therefore, AB models could be used to simulate intermediate product changes and the effects of those changes on the process and vice versa. Furthermore, AB models are able to represent cellular automata [13]. This enables the simulation of interrelations between neighboring materials and processes, a typical discrete element simulation characteristic [6]. Therefore, AB simulations helps to gain a better understanding of process chains, sub-processes as well as product properties, and represents a practical alternative to the existing simulation approaches. Especially for considering not only single processes, but the complete process chain of a pharmaceutical production, AB approaches have advantages in traceability and the transfer of product characteristics for specific entities.

The overall aim in this research is to evaluate processes and entire process chains in the planning phase and to support process and process chain improvements during operation with respect to the quality of pharmaceutical products using simulations. AB models are capable of integrating the advantages of the common dynamic systems approaches. Consequently, an AB model on the tableting process in a rotary press is introduced, providing the basis for further research. The process step of tableting is chosen to demonstrate the applicability of the approach, which is, however, a generic approach that can be applied virtually on other process steps in the process chain. It highlights the necessity of careful process description and analysis (Section 2), setup of an AB model framework for a rotary press (Section 3), model identification, development and integration into this simulation approach, and its application and comparison to experimental results (Section 4), demonstrating its capability of determining product characteristics from parameter settings as well as tracing discrete product entities and material structures.

2. Tableting and Its Simulation

The diversity of particulate pharmaceutical products is high, reaching from powders to suspensions with numerous therapeutical applications and production processes. Due to its market share, the tablet is probably the most prominent particulate pharmaceutical product and there are several process routes for the production of tablets.

Direct compression, only comprising blending of raw materials and tableting, is the preferred process route for the tableting process, as very few process steps and handling are required and the powder materials are exposed to low temperatures, low moisture stresses, and short process times. The process parameters and the product structures for the direct compression process chain are illustrated and highlighted in Figure 2. This process chain covers several of the above-mentioned challenges, such as the combination of batch and continuous processes to produce a discrete final product or the variation of intermediate product structures resulting in critical quality attributes (CQAs) of the final product.

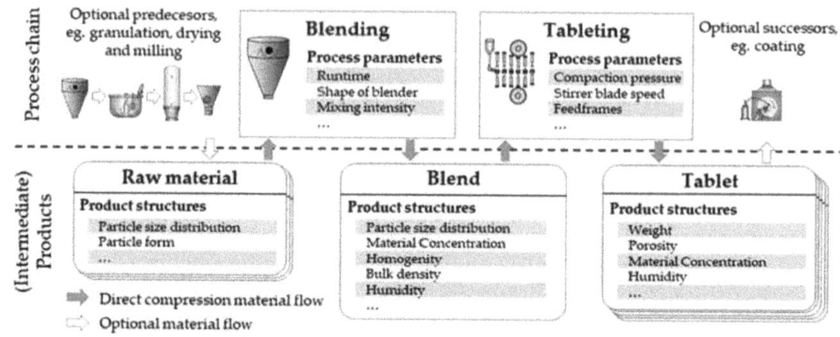

Figure 2. Direct compression process chain with exemplary (intermediate) product structures.

In addition, the applicability of direct compression is limited due to the high demands on the formulation regarding good flowability and low segregation tendencies. Depending on the properties and structure of API particles, the basic process chain of direct compression needs necessarily to be extended by different possible process steps, which are for example different granulation processes or subsequent tablet coating. Accordingly, the combined simulation of all processes for direct compression and the inclusion of granulation and coating will be the subject of future research. This research treats the direct compression process chain with a special focus on the tableting process.

2.1. Tableting Process

In the addressed tableting process, the process parameters and the preliminary product properties determine the intermediate product structures during the tableting process and consequently the final tablet properties. The tablet quality is often described by CQAs. As this case study focuses on the tablet structure depending on the excipient composition, the mass fraction of the respective excipient, tablet weight, porosity, and tensile strength are selected as CQAs. The porosity influences the disintegration time of the tablet and depends on the compression stress and the excipient properties. Additionally, a sufficient tensile strength is required to enable the handling and further processing of the tablets. The tableting process step exerts the highest influence on these tablet CQAs for the direct compression process chain. Therefore, the tablet production on a rotary press in pilot-scale was selected to be the simulative and experimental setup for this study. In Figure 3, the relevant components of a rotary press are shown. It consists of

1. a hopper containing the blend,
2. a filling pipe transporting the blend into the feed frame,
3. a feed frame equipped with one to three rotating paddle wheels with several stirring blades, transporting and filling the blend into the dies,
4. dies (and punches), passing the pre and main compression roller and the ejection mechanism. The main compression roller performs the compression of the powder in the die, leading to the formation of a tablet and the latter enables the ejection of the tablet from the die and out of the press.

The computational modeling of real-world processes allows better analysis and/or control of the processes. The modeling of the tableting process has already been described in the literature using different computational approaches, such as artificial intelligence [14,15], stochastic [16,17] and different simulation approaches (e.g., [18]), of which the existing FSS and DEM approaches are described in further detail below.

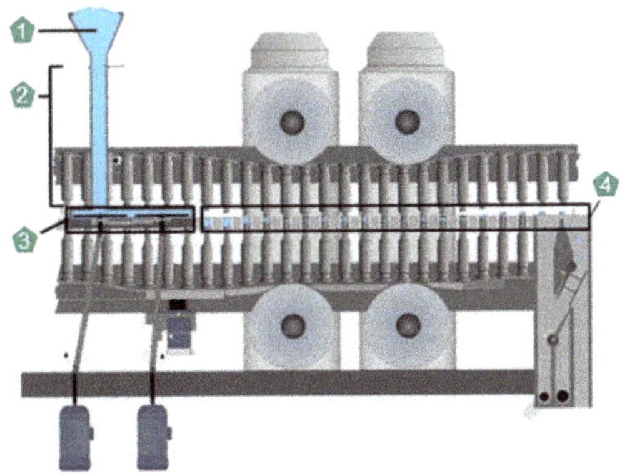

Figure 3. Rotary press components (provided by Korsch AG).

2.2. Existing Simulation Approaches and Agent-Based Simulation

Only a few approaches consider the complete direct compression scope. Simulation approaches using dynamic flow sheet approaches and considering the tableting process step in a direct compression are listed in Table 1. Compared to the low number of FSS approaches, several DEM and FEM approaches that consider the sub-process of the compression process exist.

Table 1. Dynamic FSS modeling and simulation approaches to pharmaceutical tableting.

Authors	Considered Process Scope			Considered CQA for Tableting Process
	Blending	Granulation	Tableting	
Boukouvala et al., 2012	X	X	X	▪ Tablet porosity ▪ Composition of material ▪ Relative standard deviations of input materials
Boukouvala et al., 2013	X	X	X	▪ Tablet dissolution ▪ Particle size distribution
Rogers et al., 2013	X		X	▪ Tablet hardness ▪ Friability ▪ Composition of material ▪ Dissolution performance

2.2.1. Dynamic Flowsheet Simulation Modeling

Boukouvala et al. use FSS modeling for the simulation of continuous tablet production via direct compression and compression after dry granulation. The tablet press is modelled in a very simplified manner using the popular Heckel equation in order to determine the porosity of the tablets. Therefore, the porosity of the final tablet is predominantly determined by the compaction force of the tablet press and material properties. Sub-processes such as the hopper, the filling pipe, or the feed frame are explicitly not considered, even though the conducted experiments show segregation in the feed frame. The input variables are varied, and a Monte Carlo simulation is applied to achieve multiple output evaluations to assess the sensitivity of the input parameters towards the output variable. This is carried out for different stages of the tablet production. The change of input variables over time is not explicitly considered [2].

In order to consider further CQAs, Boukouvala et al. have extended their first model with the aim of determining the tablet dissolution. The tablet dissolution is mainly determined by the tablet porosity, as well as composition and particle size distribution. For the tableting model, they used a Kawakita equation for modeling the final tablet porosity and the final tablet hardness is modelled as a function of the tablet relative density according to Kuentz and Leuenberger [19,20].

Rogers et al. extend the work of Boukouvala et al. [2,20], enabling the simulation of further CQAs such as the tensile strength [21], while Singh et al. even describe a control strategy for a multi-purpose continuous processing of pharmaceutical processes using the FSS model of Boukouvala et al. [9,20]. The dynamic FSS modeling approaches allow a good understanding and overview over the process chains and their parameters. Those parameters can even be controlled. However, the crucial components of FSS are the models for the individual processes and sub-processes. Those processes can be modeled to a certain degree of detail using semi-empirical or physical models. Comprehensive physical models that need microscale simulation, such as DEM, are not considered within the dynamic FSS, and therefore the determination of parameters for single products or product groups remains vague and could be improved [8].

2.2.2. Discrete and Finite Element Modeling

The discrete element modeling approaches are rather used to describe specific sub-processes such as the compression [22–25] or the die filling [26–32]. The particles in a typical DEM approach are modeled rigidly and are mainly used for explanatory models of physical behavior rather than for engineering purposes, which is to some extend the approach of the Finite Element Method (FEM), allowing the consideration of stresses within the machine design [22,33,34]. Baroutaji et al. describe the development of a FEM simulating the tablet compression to analyze, for example, the density distributions and stress maps in tablets or the tooling geometry [22]. Lewis et al. describe the development of an efficient combined finite-discrete element method (CFDEM) in order to simulate the powder compression of the tableting process [23]. The authors modeled the particles involved in the compression, considering the particles' individual shape, size, and size distribution. These approaches are modeled with a high level of detail of the particles and are useful for gaining a deeper process understanding, but due to time-consuming calculations, they are not appropriate for the simulation and control of process chains.

2.2.3. Agent-Based Modeling

The structure of a typical AB model consists of agents, their relationships to other agents, and their environment. The agent's own structure is shown in Figure 4.

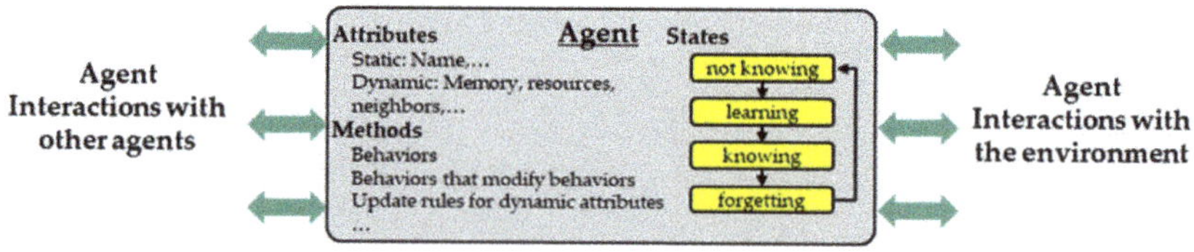

Figure 4. Typical agent structure and interactions (following [13]).

The attributes of an agent can be static or dynamic, and each agent has a behavior that might change depending on rules adapting the behaviors. Such rules can be, for example, physical models such as the Heckel equation or they can depend on the state of the agent. The states are connected via transitions that follow specific rules before the next state is reached. Depending on the logic of the state, those transitions switch their behavior time dependently, at specific ratios, or via trigger. Those triggers can occur from the interactions with other agents or the agent's environment. These and other aspects as well as use cases of AB modeling and simulation are well described in the literature [13,35,36].

The AB simulation paradigm was intended to simulate the dynamics of complex systems consisting of populations of autonomous, interacting agents or components. Nowadays, AB approaches are more and more common in production engineering [36]. AB simulation in particular can be used to describe the interactions of machines and (intermediate) products represented by individual agents [37].

To the authors' knowledge, AB simulation has not been used for pharmaceutical processes generally or tableting specifically. AB simulation enables the analysis of the development for specific product characteristics over the entire process chain, which is an advantage in comparison to the DEM and FSS. Here, processes can be adapted in the process of product development. Furthermore, the product quality can be controlled during the production phase. In addition, discrete and continuous processes and sub-processes can be considered together, which enhances the process understanding and is essential in the transition to continuous pharmaceutical production.

3. Agent-Based Simulation Model for the Tableting Process

For the development of an AB model of the tableting process, a thorough understanding of the sub-processes is required. Process models that describe the relationship between process parameters and product structures are necessary. A typical approach of modeling agents is outlined by Macal and North [35], consisting of the following three interlinked steps:

I. the identification of agents, agent groups, and their attributes,
II. the specification of the agent's behavior and
III. identifying the agent's interactions.

Following these three steps, the AB model was derived. For the derivation of the model, the pilot-scale rotary press XL 100 (KORSCH AG, Berlin, Germany) was considered, although the derived AB model is able to simulate any rotary press. A feed frame with one rotating paddle wheel with twelve stirring blades was used for the model setup as described below.

3.1. *Step I | Identification of Agents, Agent Groups and Their Attributes*

Two different agent types are introduced and shown in Figure 5; process agents and material agents. Process agents can be identified by determining the processes that have an individual behavior and a main impact on the final product structure. Considering the above mentioned rotary press, the process agents can be identified as (A) the hopper, (B) the filling pipe, (C) the feed frame and (D) the die (and punches).

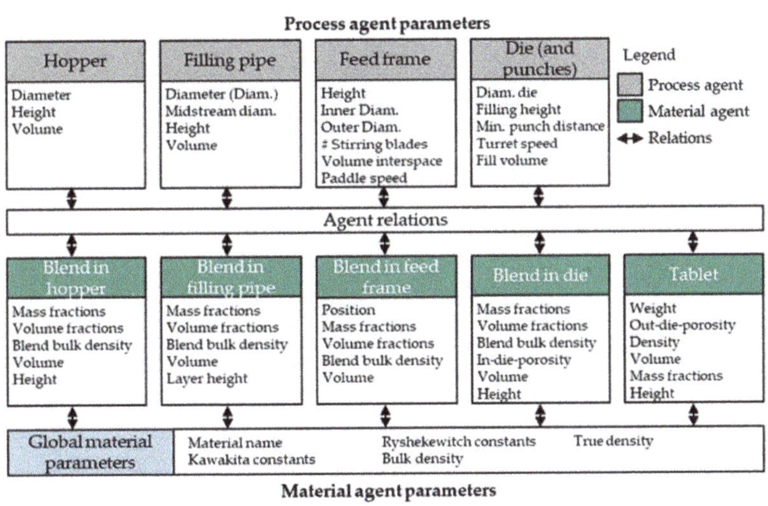

Figure 5. Existing model agents and their required parameters.

The identified material agents are the blend in the four different sub-process steps and especially the tablets at the end of the sub-process chain. Four different blend material agents are considered so as to better model the process-material behavior, the intermediate products, and the final tablet structures. The process parameters of the agents are listed in Figure 5. For the ease of understanding and the distinction of original processes and modeled agents, in the following all agents are written in *italics*.

As the CQAs of the final product are in the focus and the material flow rate is determined by the number and weight of tablets that are produced per time, the process chain and its parametrization is derived starting from the last material agent (*tablet*) and ending at the first process agent (*hopper*) to reasonably define the sizes and interactions. The agents in the agent group *tablet* are modeled as a passive data box, where the product structures and properties for each tablet are collected. Even though several dies are filled simultaneously on a rotary press, only one die is compressed at a time. Therefore, the authors choose to model the *blend in die* as a single agent representing alternatingly all dies of the rotary press. The twelve rotating stirrer blades on the paddle wheel in the feed frame transport the powder inside their respective interspaces. These interspaces are modeled as twelve compartments, each represented by one *blend in feed frame* agent. In order to simulate the powder flow within the filling pipe, a group of agents is defined as the *blend in filling pipe*. A flow profile inside the pipe with slower flow close to the wall compared to the middle is expected. Therefore, the agent group consists of two agent types–the *midstream*, representing the higher velocities in a central circular region, and the *outer ring* (Figure 6), in which velocities are lower. Both types of agents possess the same height in every vertical position, forming layers within the *filling pipe*, each containing a *midstream* and an *outer ring*. The *midstream* agents travel faster than the outer ring agents by a constant velocity difference.

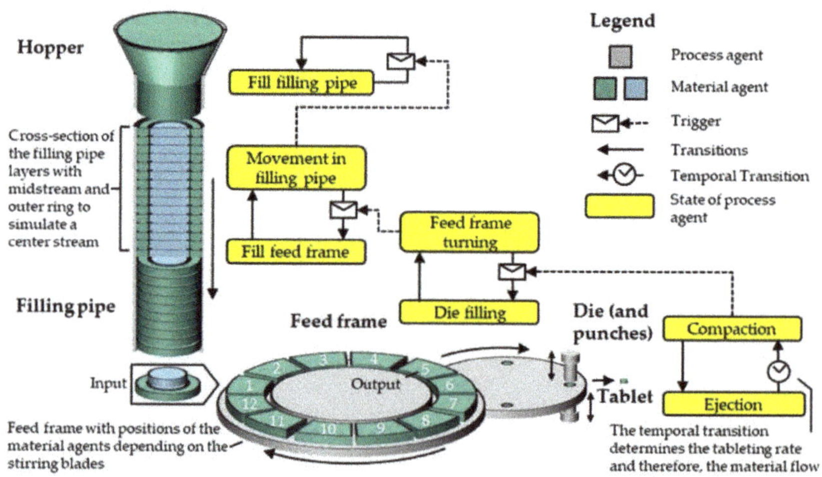

Figure 6. Schematic visualization of the process and material agents with the relevant states of the process agents.

In the state of equilibrium, the inflowing and outflowing powder streams in the feed frame and the filling pipe are equal in volume. Thus, the volume of the powder agents entering the feed frame from the filling pipe equals the powder volume filled to the die—the dosing volume. Therefore, the summarized volumes of one *outer ring* and a defined number of *midstream blend in filling pipe* agents, dependent on the velocity difference, are equal to the dosing volume of the *die*. The *hopper* serves as a feed for the blend in this model and the *blend in hopper* volume is reduced according to the remaining blend in the hopper.

3.2. Step II | Specification of the Agents' Behavior

For the specification of the agents' behavior, the different states of the process agents were identified and are shown in a schematic overview for the tableting process in Figure 6. Furthermore, process knowledge of each state of the agent groups needs to be implemented in the AB model according to the agents' behavior. In order to get the desired information, the process-structure relationships for the five process agents are described below.

3.2.1. Hopper

The hopper is used as a reservoir for the blend, which is initially introduced to the tableting process. The *blend in hopper* agent's structure is homogeneously distributed. Within the 'fill filling pipe' state, the structure of the *blend in hopper* is transferred into the newly generated *blend in filling pipe* agents. Thus, the mass fraction of the components inside the filling pipe is determined by the feed exiting the hopper. Segregation phenomena that may occur already in the hopper (e.g., during filling) will be the focus of future research work and are not included in this paper.

3.2.2. Filling Pipe

Due to the flow profile inside the pipe, the composition changes over the height of the pipe as well as over time, resulting in a time-dependent composition entering the feed frame. The *blend in filling pipe* agents neither interact with agents in the horizontal nor in the stream wise direction. However, the AB approach is prepared to incorporate such segregation and distribution phenomena in future research. The transferred material of the *blend in filling pipe* agents is mixed with the remaining powder of the *blend in feed frame* agents.

3.2.3. Feed Frame

The *feed frame* has twelve defined positions to which the *blend in feed frame* agents are clearly assigned (Figure 6). Inside the feed frame, a rotating paddle wheel transports the powder. In position 1, the interspaces are filled with powder from the *blend in filling pipe* agents. In the output position (5), the blend is filled into the die. The die is crossed by several feed frame paddles, so powder is filled from different *blend in feed frame agents* into the die with a descending proportion over time. Therefore, the product structures of the *blend in die* needs to be calculated according to the structures of the *blend in feed frames* and their filling proportions. The volume of the *blend in feed frame* is reduced accordingly. After each die filling, the *feed frame* agent rotates the *blend in feed frame* agents into a new position depending on the paddle speed. The mean bulk density during die filling can be calculated from the mass fraction x_i of the individual *blend in feed frame* agents entering the die and the respective bulk density of the powder $\rho_{bulk,i}$:

$$\frac{1}{\rho_{fill,mix}} = \sum_i \frac{x_i}{\rho_{bulk,i}} \tag{1}$$

Based on the powder weight filled into the die, the tablet weight m_T can be calculated considering the dosing height h_{dos}:

$$m_T = \pi \times \left(\frac{D_{die}}{2}\right)^2 \times h_{dos} \times \rho_{fill,mix} \tag{2}$$

3.2.4. Die (and Punches)

During the compression, the filling height is reduced to the compression height h_{min}. For the simulations as well as for the experiments the compression height is set to a constant value and thus determines the developing compression stress. Consequently, the density of the *blend in die* agent $\rho_{comp,mix}$ changes during compression, which can be calculated from the minimal in-die height h_{min} and the tablet weight m_T:

$$\rho_{comp,mix} = \frac{m_T}{\pi \times \left(\left(\frac{D_{die}}{2}\right)^2\right) \times h_{min}} \tag{3}$$

The compression curve (porosity over compression stress) can be applied to determine the resulting stress (and calculate the respective force) that is applied to achieve the bespoke compression density $\rho_{P,max,mix}$. A wide range of models was developed, describing such compressibility curves by mathematical equations [38,39]. The most prominent compression model is the model of Heckel [40]. Further compressibility models were developed by Kawakita, Gurnham, Cooper, and Eaton, as well as Wünsch et al. [39,41–43]. However, for describing the compressibility behavior of powder blends with known but varying composition, an approach with an appropriate mixing rule is necessary. Busignies et al. presented a volume-additive approach to predict the tablet density of a formulation using the Kawakita model [44]. They transformed the classical equation into a model of the tablet density where $\rho_{0,i}$ describes the density at low pressure (here 20 MPa) and a_i and b_i are constants that need to be calibrated for the respective material:

$$\frac{1}{\rho_{P,max,mix}} = \sum_i \frac{x_i}{\rho_{0,i}} \left(\frac{1 + (1 - a_i) \times b_i \times P}{(1 + b_i \times P)}\right) \tag{4}$$

In this study, Equation (4) is converted to equal the compression stress P and solved via the *pq*-formula to gain the compression stress of a blend with two components.

From a process agent's point of view, the ejection state is used to create a new *tablet* agent. After ejection, the tablet density changes due to the elastic recovery. Hirschberg et al. developed an approach to predict the out-die density $\rho_{out\text{-}die}$ based on the data of two tablets compressed at high and low compression stresses [45]. Therefore, the measured densities at maximum stress $\rho_{P,max}$, at zero axial stress $\rho_{P,0}$, and out-of-die $\rho_{out\text{-}die}$ are used. The change in density due to the instantaneous elastic recovery ($\Delta\rho_{in\text{-}die} = \rho_{P,max} - \rho_{P,0}$) is expected to increase linearly with the compression stress while the change due to the slow elastic recovery ($\Delta\rho_{slow} = \rho_{P,0} - \rho_{out\text{-}die}$) is expected to be constant:

$$\Delta\rho_{in-die} = k_{\Delta\rho_{instant}} \times P + l_{\Delta\rho_{instant}} \tag{5}$$

$$\Delta\rho_{slow,\,o} = \frac{\Delta\rho_{slow}(P_1) + \Delta\rho_{slow}(P_2)}{2} \tag{6}$$

$$\rho_{out-die} = \rho_{P,\,max} - \Delta\rho_{in-die} - \Delta\rho_{slow,o} \tag{7}$$

In order to predict the out-die density for mixtures of the excipients with varying compositions, a mixing rule has to be introduced. For the mixtures, only the mass fraction and the in-die density at the apparent stress $\rho_{P,\,max,mix}$ are known, so the elastic recovery has to be described based on the data of the pure substances. Therefore, according to the volumetric approach of Busignies et al., the densities at the different states ($\rho_{P,max}$, $\rho_{P,0}$ and $\rho_{out\text{-}die}$) for the high and low compression stress, investigated for the pure substances, are calculated for the mixtures (according to Equation (1)). Following this, the change in in-die density and out-die density (Equations (5) and (6)) can be calculated. Finally, the predicted out-die density results from the apparent in-die density at maximum stress $\rho_{P,max}$ minus the corrective function $\Delta\rho_{in\text{-}die}$ at the respective compression stress and the corrective value $\Delta\rho_{slow,ø}$ (Equation (7)). Finally, the out-die density is corrected using the difference of the predicted values and the measured out-die densities, as suggested by Hirschberg et al.

The final tensile strength of the tablets dominantly depends, besides other influences, on the out-die porosity. In general, the lower the out-die porosity $\varepsilon_{out\text{-}die}$, the higher the tensile strength σ. Ryshkewitch and Duckworth developed an exponential model to empirically describe this relationship [46]:

$$\sigma = \sigma_0 \times e^{-k_b \times \varepsilon_{out-die}} \tag{8}$$

The coefficient σ_0 describes the strength of a nonporous body of the same material, while k_b corresponds to the slope of the ln σ vs. $\varepsilon_{out\text{-}die}$ curve.

3.3. *Step III | Identifying the Agents' Interactions*

As described in Section 2.2, interactions in AB models can occur between agents (1–4) and between agents and their environment (A,B) as depicted in Figure 7. The latter, i.e., the environmental impacts on the process and the process impacts on the environment, have not been further considered even though aspects such as the humidity can have a significant impact on the product output. This research focuses on the interactions between agents. These interactions are modeled for connecting agents with the agents in their defined neighborhood and the dynamics that arise through this connection. Such interactions are mostly triggers (compare Section 2.1) for the transitions described in Step II.

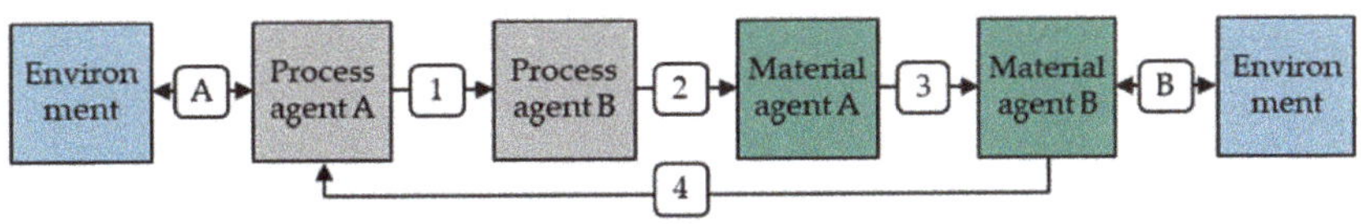

Figure 7. Exemplary agent interactions for this simulation model.

As illustrated in Figure 7, interactions between process agents (1), between process and material agents (2), as well as vice versa (4), and between material agents (3) are considered in this model.

The process-process interactions have already been shown in Figure 6 and are marked via triggers. These triggers align the processes behaviors. Therefore, only if a trigger is sent, the trigger-receiving transition of the process agent is able to open the transition. This enables the agent to switch into the next state connected by the transition and show its specific behavior. One purpose of a trigger could be, that a process can send a trigger informing other processes that a specific behavior of a state has been completed. This occurs for example after the creation of a tablet. At that moment the *punch and die* agent interacts with the *feed frame* agent to allow the refill of the next die. Several process-material interactions are implemented in the model, e.g., the material agents' volume depends on process agents' parameters such as the volume of the *blend in pipe* agents, which is dependent on the volume of the *die* and the geometries of the *filling pipe*. The material-material interactions are necessary to transfer information from one material agent to another. This occurs every time a new material agent is created and obtains its attributes from corresponding agents. The same applies when material agents merge with other material agents and the product structures of both agents need to be assigned and calculated, respectively. Interactions between material agents occur in the filling processes of the filling pipe, the feed frame compartments, and the die. In those cases, a new blend density and mass fractions are calculated. The material-process interactions are relevant for the correct result of the final tablet properties. One of those cases is the calculation of the punch distance, which differs slightly from h_{min} due to the elastic deformation of the rotary press depending on the applied compression stress resulting from the resistance against compression of the blend in the die.

An overview of the agents' states, their respective behavior as well as interactions and the required calculations are listed in Table 2. Important process input parameters for the simulation are the minimal in-die height h_{min}, the dosing height h_{dos}, and the turret speed. The paddle speed does not change in this case.

Table 2. Overview of the states of rotary press process agents, their behaviors, and their interactions.

Agent	State	Behavior	Interactions	Calculations
Hopper	Fill filling pipe	Determine/specify materials for the transmission in filling pipe	Filling the filling pipe with blend from hopper	
Filling pipe	Movement in filling pipe	Blend in pipe is transported towards feed frame, realize flow profile inside the pipe	Triggering hopper to fill voids	Material fractions
	Fill feed frame		Filling the feed frame compartments with blend in filling pipe	Material fractions
Feed frame	Turning	Turning the compartments according to paddle speed	Feed die with blend and triggering filling pipe	
	Die filling		Filling of die out of different feed frame compartments	Material fractions Mixed density (Equation (1)) Tablet weight (Equation (2))
Die (and punches)	Compaction	Applying the compression stress on the powder blend according to minimal in-die height h_{min} under consideration of the elastic deformation of the tablet press	Creating a stress based on product agent properties, creating tablet, triggering feed frame	Compression stress (Equations (3) and (4)) Tablet in-die density (Equation (3))
	Ejection	Waiting according to turret speed		Tablet out-die porosity (Equations (5)–(7)) Tablet tensile strength (Equation (8))

The whole AB model was implemented using the modeling and simulation software AnyLogic®. In the following, the simulation results are compared with experimental results.

4. Exemplary Application of the Agent-Based Methodology on a Rotary Press

For this case study, the effect of the change in particle size of the excipient anhydrous dicalcium phosphate in the feed is investigated. For illustration, the considerable change in particle size from one commercial quality with x_{50} = 167 µm to another with x_{50} = 64 µm and back is applied. Although changes in particle size are not expected to be as high during production as in this example, they may, however, occur due to different batches or fluctuations in the prior processes of the process chain. Due to the deformation behavior of anhydrous dicalcium phosphate, this change has a significant impact on the structural parameters as well as on the properties of the final tablet. Therefore, this case is well suited to demonstrate the capabilities of an AB simulation. In the following, the materials and the experimental setup are described.

4.1. Materials

Anhydrous dicalcium phosphate (DCPA) in two grades (DI-CAFOS® A60, DI-CAFOS® A150, Chemische Fabrik Budenheim KG, Budenheim, Germany) were used as model materials. The characteristic particle sizes of DCPA A60 and DCPA A150, determined by laser diffraction (Mastersizer 3000, Malvern Panalytical, Kassel, Germany), differ over the whole distribution (Table 3, Figure 8). To enable their processability, DCPAs were mixed with 1 wt% magnesium stearate (MgSt, Magnesia GmbH, Lüneburg, Germany) in a cube blender (ERWEKA GmbH, Langen, Germany) for five minutes at 30 rpm. The bulk ρ_b and tapped density ρ_t were determined according to the Ph. Eur. 9.3 2.9.34 using a 100 mL cylinder and a volumetric analyzer (Erich Tschacher Laboratoriumsbedarf, Bielefeld, Germany). As the two grades consist of the same chemical material, the solid density ρ_s, measured in triplicate with the helium pycnometer Ultrapyc 1200e (Quantachrome Instruments, Boynton Beach, FL, USA), are practically identical (Table 3).

Table 3. Characteristic particle sizes, solid, bulk, and tapped density for DCPA A60 and A150.

Material	x_{10} (μm)	x_{50} (μm)	x_{90} (μm)	ρ_s (g/cm^3)	ρ_b (g/cm^3)	ρ_t (g/cm^3)
DCPA A60	34	64	116	2.849	1.33	1.51
DCPA A150	96	167	263	2.842	0.68	0.75

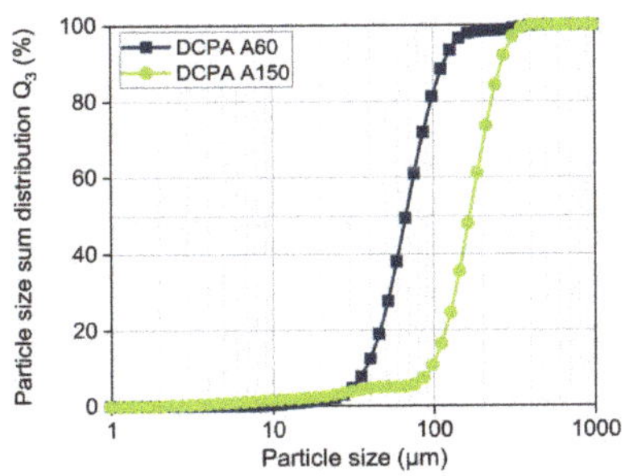

Figure 8. Particle size distribution of the DCPAs.

4.2. Experimental Methods

The pilot scale rotary press XL 100 (Korsch AG, Berlin, Germany) was used to investigate the tablet production process. The rotary press was equipped with four flat-faced round Euro-D punches with a diameter of 11.28 mm as a mixed rotor was used. A feed frame with one rotating paddle wheel with twelve stirring blades was used. The fill cam used had a fill depth of 10 mm, while the dosing depth was set to 7 mm. The minimal punch distance for compression was set to 3.2 mm, which is subsequently an input parameter for the simulation. Since the setting on the XL 100 does not take the elastic deformation of the press itself into account, the actual punch distance during compression was determined beforehand by comparison with the results of a compaction simulator (Styl'One, Medel'Pharm, Beynost, France).

For the process experiment, the rotary press was filled to the upper end of the filling pipe with DCPA A150. A constant filling level was ensured by continuously adding new powder manually. In order to guarantee complete filling of the dies, which might be a challenge at certain process parameter settings, the paddle speed was set to 60 rpm while the turret speed was 20 rpm [47]. Tablets were taken every minute, except between minute 8–18, where tablets were taken every two minutes. After one minute, the change in particle size was induced by subsequently filling about 1 kg DCPA A60 into the filling pipe while keeping its fill level approx. constant over a period of twelve minutes. Following, DCPA A150 was filled again until the end of the experiment. The tablet weight, height and diameter were determined (n = 6) after 24 h. The out-die porosity as well as the tensile strength were calculated according to Fell and Netwon [48].

In order to investigate the mass fraction of the two excipients in the tablets, DCPA A60 was dyed prior to the experiments. Methylene blue was sprayed as an aqueous solution onto DCPA A60 and subsequently dried in a fluidized bed using Solidlab 2 (Syntegon Technology GmbH, Waiblingen, Germany). The tablets were dissolved in 1 M hydrochloric acid (HCl) to solve the dye, and the methylene blue content was determined by UV-Vis spectroscopy (Specord 210 Plus, Analytik Jena GmbH, Jena, Germany) at a wavelength of 669 nm. Under consideration of the dye loading of DCPA A60, the corresponding mass fraction of DCPA A60 in the tablets was calculated (n = 3).

4.3. Calibration of the Process Models

The process models presented in Section 3 were calibrated for the excipients used. Therefore, tablets (n = 10) were compressed at certain compression stresses (30, 50, 100, 200, 300, 400 MPa) using the compaction simulator. In order to gain the coefficients a, b, and ρ_0

for the Kawakita model, the compressibility curves of the pure substances A60 and A150 were fitted in the range of 20 to 350 MPa. The starting pressure of 20 MPa was used, as the Kawakita model is generally not suitable for very low pressure ranges [49]. The coefficients can be found in Table 4.

Table 4. Coefficients for the Kawakita model.

Excipient	ρ_0	a	b	R^2
DCPA A150	1.22954	0.4981	0.0072	0.9962
DCPA A60	1.62302	0.4891	0.0041	0.9959

In order to calibrate the model of Hirschberg, the densities in different states ($\rho_{P,max}$, $\rho_{P,0}$ and $\rho_{out\text{-}die}$) of ten tablets of each component, DCPA A60 and A150, compressed at 30 and 400 MPa were used. The model of Ryshkewitch–Duckworth was calibrated using the tensile strengths of tablets of all six compression stresses. In order to determine the coefficients k_b and σ_0 and to develop a mixing rule for them, the compactability curves of the pure substances and the 50:50 blend were fitted with the model-equation (Equation (8)). The resulting progress of σ_0 presents an exponential development with rising mass fraction of DCPA A150 x_{A150}. Thus, for the mixtures σ_0 is calculated as follows:

$$\sigma_0 = c_1 \times e^{x_{A150} \times c_2} + c_3 \tag{9}$$

The coefficients c_1, c_2, and c_3 can be found in Table 5. Furthermore, k_b shows a linear increase with rising mass fraction of DCPA A150 x_{A150}. The value for k_b can be calculated for the respective composition of the materials by Equation (10). The coefficients m and n can be found in Table 5 as well.

$$k_b = m \times x_{A150} + n \tag{10}$$

Table 5. Coefficients to calculate σ_0 and k_b.

Coefficients to Determine σ_0		Coefficients to Determine k_b	
c_1	19.99	m	3.80
c_2	5.56	n	15.09
c_3	28.32		

4.4. Configuration of Simulation Setup and Error Analysis

Besides the above-mentioned agent definitions, further configurations to the AB model and model assumptions need to be made. Two events are modeled in order to change the raw material in the hopper from A150 to A60. Those time-dependent events are modeled according to the experimental set-up, i.e., the change in particle structures of the *blend in hopper* agent is scheduled after 1 min and resettled after 12 min.

The mean absolute value of the relative error f according to the final tablet weight between the experiment and the simulation was calculated according to Equation (11). This equation was applied for all measuring points k that have been made in the experiment exp_k (compare Section 4.2) and those equivalent measuring points in the simulation sim_k.

$$f = \sum_{k=1}^{n} \frac{\left|\frac{exp_k - sim_k}{sim_k}\right|}{n} \forall\, k = experimental\ measuring\ points \tag{11}$$

5. Results and Discussion

5.1. Calibration of the Simulation Model

The sensitivity of the system towards two parameters is characterized and their calibration is necessarily conducted based on comparison with experimental data, as these parameters are not directly assessable by experiments, so far. Firstly, the effect of the theoretical diameter of the midstream, assuming a constant velocity ratio of 2:1 to the outer ring of the filling pipe, is investigated within the simulation. Secondly, the filling pattern from the feed frame fragments to the die, defining what fraction of the whole die volume is filled by successive feed frame compartments passing over the die during one filling event is also evaluated within the simulation.

The sensitivity of the simulation towards deviation of the repeated midstream diameter increasing by always 6 mm shows that the relative error between the midstream diameter of 36 mm and 30 mm has the lowest relative errors (Figure 9a). Therefore, the mean of those two values (33 mm) is chosen as the midstream diameter. This proved well suited, as its f value of 1.32% is the lowest in this data set (see Table 6).

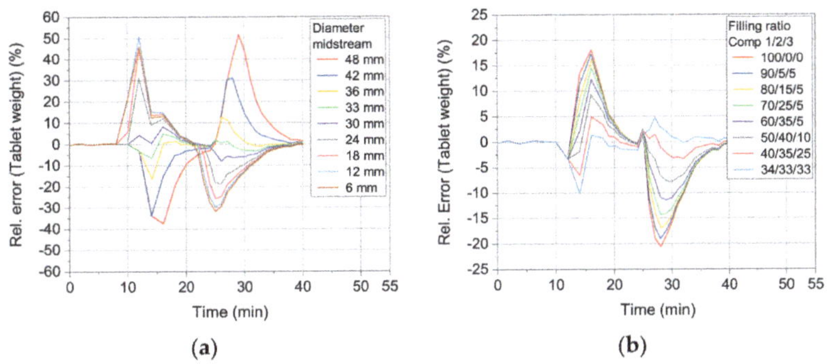

Figure 9. Relative error between simulation and experimental results for (**a**) different midstream diameter settings and (**b**) die filling ratio settings in relation to the tablet weight.

Table 6. Relative error of midstream model configuration regarding the tablet weight.

Midstream Diameter (mm)	42	36	33	30	24	18	12	6
Mean Absolute Value of the Relative Error (f) (%)	6.02	1.86	1.32	2.23	5.19	7.27	8.82	9.05

The sensitivity of the distribution of filling over three compartments was studied by reducing the fraction filled by the first compartment in steps of 10%-points, distributing the rest of the filling reasonably over the second and the third compartment with the assumption that the ratio declines over the compartment number.

The lowest values are found for the ratio 40/35/25 and 34/33/33 (Table 7). This finding shows that a distribution of the die filling is better described by filling from more than one compartment of the feed frame. However, it must be born in mind that this value most likely depends on the flowability of the formulation as well as on the process parameters of turret and paddle speed. A descending pattern of 40/35/25 was used for the case study.

Table 7. Relative error of die filling ratio configuration regarding the tablet weight.

Filling ratio (%)	1. Compartment	100	90	80	70	60	50	40	34
	2. Compartment	0	5	15	25	35	40	35	33
	3. Compartment	0	5	5	5	5	10	25	33
Mean Absolute Value of the Relative Error (%)		5.35	5.04	4.64	4.11	3.38	2.46	1.32	1.24

5.2. Validation of the Sub-Process Models

The behavior of the agents in the sub-processes plays an important role in the simulation, as they determine the interim results and thus the input of the subsequent sub-process models. Therefore, a careful prior validation of the underlying mathematical models is necessary and useful to investigate their accuracy and to identify possible sources of error for the simulation results. In the following, the three empirical models regarding the compressibility curve, the elastic recovery and the tensile strength are compared to experimental data examined by the compaction simulator. Subsequently, the results of the simulation are compared with the experimental data and discussed.

According to Busignies et al., the compressibility curves of DCPA A60 and DCPA A150 are fitted by the model of Kawakita (Figure 10a, dashed lines) to gain the material-specific coefficients a, b and ρ_0. Using Equation (4) and the coefficients, the compressibility curves of the mixtures with a mass fraction of 25, 50 and 75 wt.% DCPA A60 were calculated. The comparison of the predicted and the measured curves present errors between highly negative deviation (lower values of experimental data) for all materials at a very low compression stress < 50 MPa, displaying that the lower stress range is less well described by the Kawakita model.

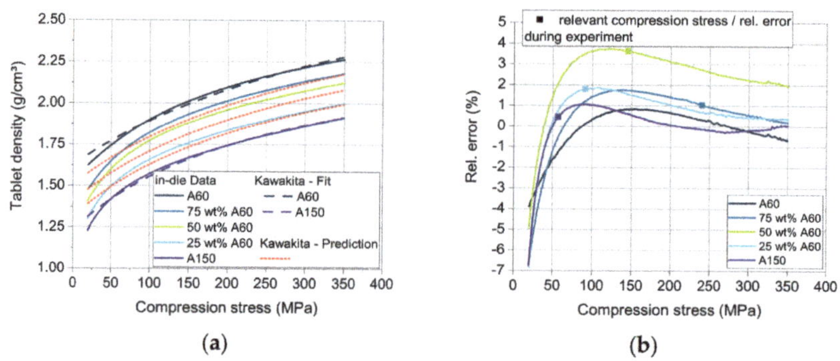

Figure 10. (**a**) Measured (solid lines), fitted (dashed lines), and calculated (dotted lines) compressibility curves of DI-CAFOS A60, A150, and mixtures, and (**b**) the relative error between the measured and the predicted data.

In the validation experiment as well as in the simulation, only specific compression stresses were reached for each blend composition, as the compression height stays approximately constant. These compression stresses and the corresponding relative errors are marked with a square (Figure 10b). The maximum positive deviation (4%) occurs for the blend with 50 wt.% DCPA A60 at about 100–150 MPa (Figure 10b), which was most likely to occur as it is the furthest from the support points of the two pure materials. The higher relative error for the blend with 50 wt.% DCPA A60 over the whole compression stress range may result from the volume-additive approach. The higher measured tablet density probably results from a volume contraction of the mixed powders in contrast to the additive volume of the single components. The small particles of DCPA A60 and their fragments may fill the pores between particles and fragments of DCPA A150, resulting in a higher tablet density than calculated. This trend can also be observed for the blends with 25 and 75 wt.% DCPA A60, although the relative error is smaller.

After the stress maximum during compaction and after ejection of the tablets, the elastic recovery takes place. As proposed by Hirschberg et al., the out-die porosity is calculated based on Equations (5)–(7) with consideration of the solid density. The relative error between the measured and the predicted out-die porosities of all substances and blends is considerably low with maximal 4% for 75 wt.% DCPA A60 and minimal −6% for 50 wt.% DCPA A60 (Figure 11b). The relevant compression stresses and the related relative errors, which occur in the validation experiment as well as in the simulation, are marked with a hollow symbol (Figure 11b).

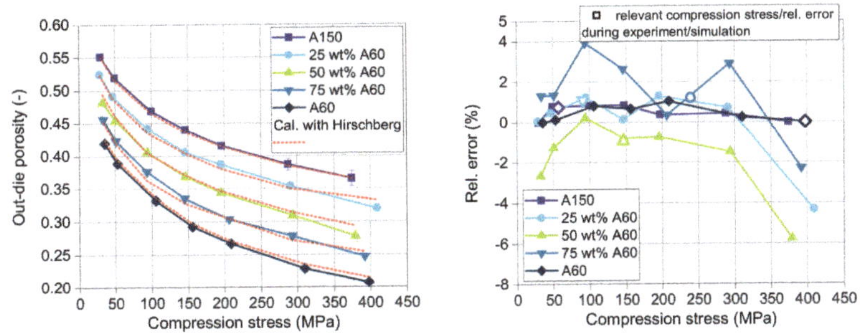

(**a**) (**b**)

Figure 11. (**a**) Measured (solid lines) and calculated (dotted lines) out-die porosities of DI-CAFOS A60, A150, and mixtures, and (**b**) the relative error between the measured and the calculated data.

Finally, the tablet tensile strength was estimated in dependence on the out-die porosity using the model of Ryshkewitch and Duckworth. As described in Section 4.3, a new mixing rule was developed to predict the tensile strength of tablets consisting of two different components. The coefficients of the model are calculated based on the mass fraction of DCPA A150 (Equations (9) and (10)). The compactability curves of A150 and the mixture 25 wt.% A60 (Figure 12a) show a strong increase in tensile strength with decreasing out-die porosity. With increasing mass fraction of A60, the curve flattens. The correct mathematical description of the curves is challenging due to high changes in tensile strength for low differences in the out-die porosity. This causes the high relative error between the measured and the predicted tensile strength with about −30 up to 20% (Figure 12b). Although, the calculated tensile strength curves fit the experimental values very well, as it can be seen in a first approximation in Figure 12a. Nevertheless, it has to be taken into account that the data for each composition is only relevant in a specific compression stress range and thus a porosity range, as the stress changes with the composition and so does the out-die porosity. The relevant relative errors of the tensile strength are marked for each blend composition with hollow symbols (Figure 12b).

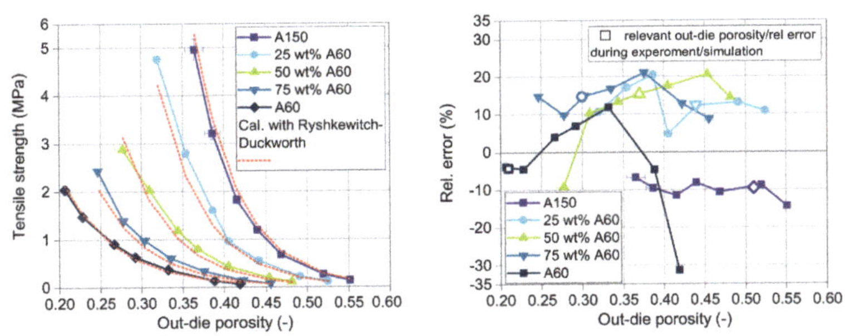

(**a**) (**b**)

Figure 12. (**a**) Measured (solid lines) and calculated (dotted lines) tensile strength of DI-CAFOS A60, A150, and mixtures, and (**b**) the relative error between the measured and the calculated data.

5.3. Comparison of the Simulative and Experimental Results

In order to determine the quality of the AB model, the resulting CQAs obtained by the simulation are compared with the experimental results achieved from the rotary press XL 100. Figure 13 shows the tablet weight as well as the mass fraction of the respective excipient as a function of time. The tablets weigh about 500 mg at the beginning of the experiment, containing 100 wt.% DCPA A150. After about twelve minutes, the tablet weight increases significantly as the DCPA A60 content starts to rise. This can be traced back to the increasing apparent bulk density with higher DCPA A60 content as its small particles fill pores and lead to a denser arrangement of the powder bed. Therefore, the weight filled into the die increases, resulting in a higher tablet weight. With increasing mass content of DCPA A60, the change in mass fraction as well as tablet weight flattens. Due to the residence time distribution of DCPA A150, it takes a certain time to fill remaining DCPA A150 particles into the dies, leading to change in content inside the feed frame.

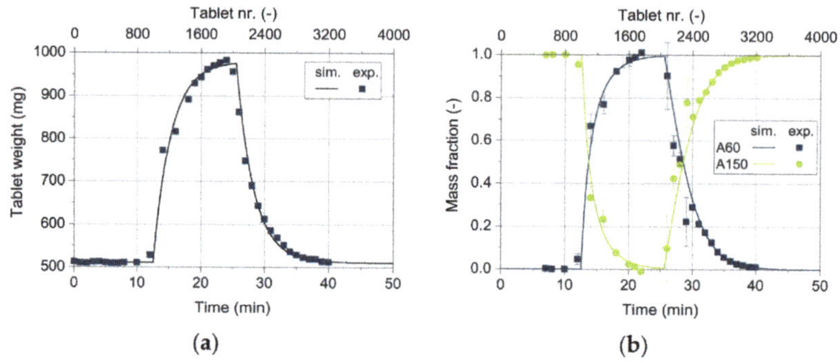

Figure 13. Comparison of the simulative data with experimental results for (**a**) the mass fraction of DCPA A150 and A60 and (**b**) the tablet weight.

The good agreement between the simulative data and the measured values for the tablet weight and mass fraction (Figure 13) are particularly worth to mention. Only for high DCPA A60 mass fractions, simulative data exhibit slightly lower tablet weights compared to the measured values. This underestimation might be linked to die filling, where apparent densities above the bulk density are possible due to particle rearrangements during forced feeding and especially during dosing [47]. To meet the tablet weight, obtained by the experiment, an apparent consolidated bulk density after filling of 0.73 g/cm^3 had to be used, which is considerably higher than the bulk density of 0.68 g/cm^3. As described in the literature, good flowing powder as dicalcium phosphate consolidates at high paddle speeds and low turret speeds inside the die. The process parameters selected here as well as the material properties support this hypothesis. The same observations were made for DCPA A60, so the apparent consolidated bulk density after filling was taken to be 1.41 g/cm^3 compared to the bulk density of 1.33 g/cm^3.

Although no specific mixing model is introduced in the AB simulation, mixing and the distribution of the newly entering DCPA A60 can be represented by the set-up of the simulation itself. As described in Section 3, the midstream with a diameter of 33 mm simulates a flow profile inside the filling pipe according to its geometry, by having double the velocity as the surrounding powder flow (compare Section 3). Thus, a powder flow similar to the actually expected is aspired. As soon as the powder, represented by the *blend in filling pipe*, enters the feed frame, the mass fraction of the respective compartment is recalculated. Therefore, the twelve compartments work as independent continuous stirred tank reactors (CSTRs), as the mixing ratio is constant within the entire compartment in each time step. Comparing this setting to the visual observations on the tablet press using a transparent feed frame, a completely homogenous concentration of the excipients is not realistic, especially not directly after new powder with a different composition entered the feed frame. Mixing in the feed frame over time can be observed, while the powder is not only mixed inside an interspace but is also able to change the compartment if the particles are close to the bottom or flow over the paddles in real experiments. Due to the high paddle speed of 60 rpm and the simultaneously low turret speed of 20 rpm, leading to a high residence time, the powder can mix quickly inside the feed frame. Puckhaber et al.

investigated the residence time distribution on the XL 100 for a pure dicalcium phosphate with similar powder properties as DCPA A150, showing strong intermixing for the process parameters used in this study [50]. This improves the model quality in so far that the simulated mass fraction, resulting from the compartments, modeled as CSTRs, is in good agreement with the experimental data. In future work different combinations of paddle and turret speed shall be looked at to investigate whether the concept of the CSTRs is still valid or must be refined for combinations of free flowing with poorly flowing materials, for example.

For the experimental data of the mass fraction, a very good correlation between experimental and simulative values was found (Figure 13b). However, higher standard deviations are observed for the experimentally determined mass fraction than for the tablet weight, which shows standard deviations below 1.5%. The mass fraction is determined by using dyed DCPA A60, which intrinsically shows higher standard deviations of the loading as ten samples à 100 mg presented a relative standard deviation of 11.7%. Therefore, the tablet weight ($n = 6$) is a more reliable and more facile to determine parameter to compare the simulative and experimental data.

Besides the tablet weight, the out-die porosity and the tensile strength are of particular interest regarding their possible correlation to disintegration time and the handling stability of the tablets, respectively. The out-die porosity exhibited by the simulation fits very well with the experimental data (Figure 14a). To achieve this degree of convergence, the models used to describe the maximum compression stress and the total elastic recovery as function of filling weight and blend density do present a very good applicability.

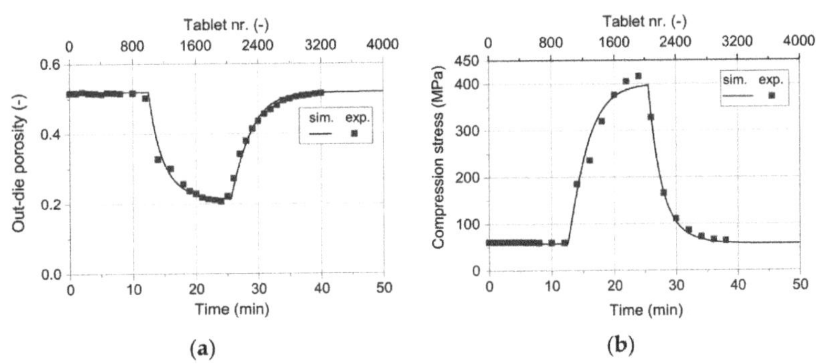

Figure 14. Comparison of the simulative data with experimental results for (**a**) the out-die porosity and (**b**) the compression stress.

As shown for the validation of the process models, the in-die tablet density and thus the in-die tablet porosity have a deviation between −2 and 4% above 50 MPa (Figure 10b). As the tablets contain only DCPA A150 at a compression stress of about 60 Mpa in this case, the apparent error due to the Kawakita model is close to 0%. As Figure 15a shows, the resulting simulated compression stress in this study for 75 wt.% DCPA A60 is around 200 Mpa. In this stress range, the deviation for the respective content is small (Figure 11b, hollow symbol). Although tablets with 50 wt.% and 25 wt.% DCPA A60 present high relative errors around 400 Mpa (Figure 11b), they do not affect the simulation results as the apparent compression stress is lower than 150 Mpa for these mass fractions. Therefore, the total mean error for the out-die porosity is low with 2.13% as listed in Table 8.

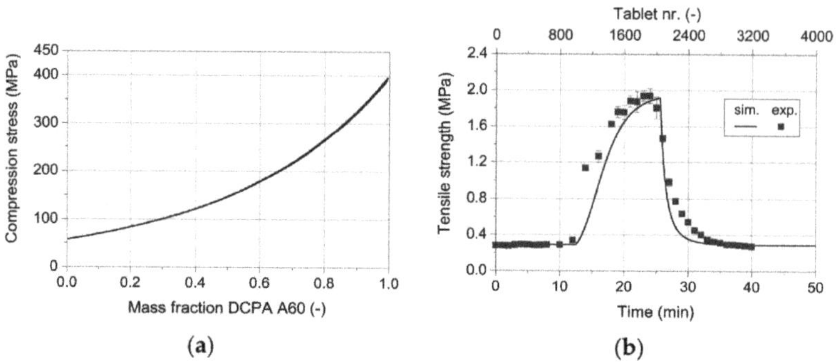

Figure 15. (**a**) Calculated compression stress over the mass fraction of DCPA A60. (**b**) Comparison of the simulative data with experimental results for the tensile strength.

Table 8. Final agent-based simulation result deviations of critical quality attributes.

Critical Quality Attributes		Tablet Weight	Mass Fraction	Out-Die Porosity	Tensile Strength
Mean absolute relative error (f) (%)		1.34	6.99	2.13	17.69
Deviation (%)	Max	7.60	43.82	8.93	112.74
	Upper quartile	2.15	9.34	3.25	27.06
	Median	0.83	2.41	1.55	5.21
	Lower quartile	0.20	0.00	0.63	2.38
	Min	0.00	0.00	0.02	0.00

For the comparison of the experimentally recorded compression stress on the rotary press and the simulated values, a very good match can be observed (Figure 14b). In the high stress range, the simulative data underestimates the experimental values. This might be due to the slightly lower simulated tablet weight at high DCPA A60 content. At such a high compression state, a low change in weight has a relatively strong effect on the corresponding compression stress (compare Figure 10a). For further research, the consolidation of the powder during and after filling has to be addressed in more detail to better calculate the tablet weight and thus predict the compression stress even more accurately.

Regarding the tensile strength, considerably higher deviations between the simulative and the experimental data are observed (Figure 15b). While the measured values match well for pure DCPA A150 and pure A60, the correct prediction of the tensile strength during the change in content is challenging (Figure 16). The tensile strength obtains the highest relative errors for the simulative data with over 100% with rising mass fraction of DCPA A60 and about 70% with decreasing content (Figure 16a). The lower quality of the predicted data of the tensile strength can be traced back to the model and especially the mixing rule used. For the model validation, already relative errors of -35 up to 20% are obtained, while no specific trend for the mass fraction is observable (Figure 12b). Nevertheless, the model of Ryshkewitch uses the out-die porosity as input parameter to calculate the corresponding tensile strength, which includes previous errors. Due to the strong increase in strength for low changes in porosity, small differences in the porosity have a high impact. A parabolic shape of the relative error over the mass content can be observed, which indicates a systematic error for the determination of the tensile strength (Figure 16b).

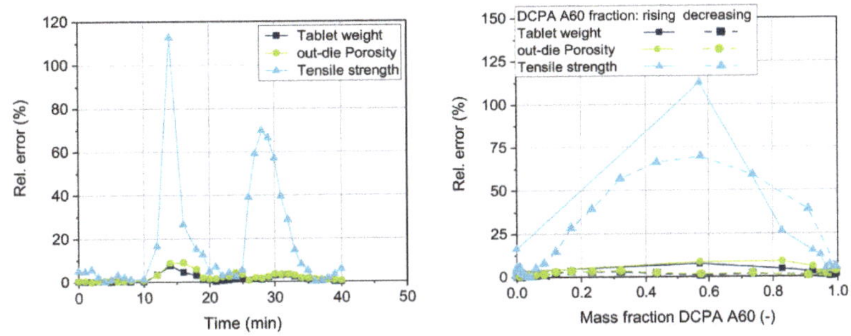

(**a**) (**b**)

Figure 16. Relative error of the simulative data compared to experimentally determined values for the (**a**) tablet weight, out-die porosity, and tensile strength over time and (**b**) the mass fraction of DCPA A60.

Interestingly, comparing the increasing and decreasing DCPA A60 fractions, the relative errors of all three parameters present higher values with increasing DCPA A60 fraction than vice versa (Figure 16b). This might be due to the change in the apparent bulk density as the actual volume of the powder is lower than the calculated one by adding the respective volumes of each powder. Additionally, the higher deviation with rising DCPA A60 content in comparison to the decreasing content can indicate different residence time distributions and thus, different distribution profiles of the two excipients. Both will be investigated in future research.

Considering the overall accuracy of the simulation results, the mean absolute relative error f (compare Equation (11)) of the CQAs can be seen in Table 8.

Besides the tensile strength of the tablets, all other CQAs have a mean relative error of less than or equal to 6.99%. However, the quality of the AB model can only be as good as the quality of the physical models used to calculate product properties. For the calculation of the tensile strength a better model and, moreover, a better bulk density determination of the component mixture is necessary.

From a simulation point of view, the AB modeling approach allows the observation of each agent over the complete simulation time. Furthermore, the input materials have been digitally marked and allow a tracking and tracing over the whole process. This fact allows to investigate the output in relation to the input on a specific matter. With this, several processing questions can be answered, e.g., the characterization of tablets that have fractions of a specific input material. This shows the potential of AB models to be capable of model particle-based processes chains at low computational cost, prospectively making their application in real time control of pharmaceutical production processes feasible.

6. Conclusions and Outlook

This study presents the AB simulation approach, which is, to the best knowledge of the authors, for the first time applied for the simulation of a pharmaceutical, particle-based process chain. The implementation of such an AB simulation framework necessitates a careful analysis of the individual process steps to derive simple but effective sub-models. Mindful assumptions are necessarily defined and calibrations of the systems must be performed to implement relations that are not directly assessable by experimental means. The results of the simulation compared with experimental data regarding several CQAs were predicted predominantly with only minor errors. It is shown that an AB modeling approach is an alternative to the classical simulation paradigm for solids simulation, combining the positive characteristics of the FSS and the DEMs, i.e. the ability to model process chains or sub-process chains and the representation of material-process interactions, respectively. As is known regarding FSSs, physical sub-models of high quality are required to describe the resulting product properties as a function of process time exactly.

Further research in this field should consider the effect of additional materials with different properties as well as predecessor and successor steps, e.g., blending, granulation and coating. The main research questions consider the interconnection between these models and the transfer of deep process knowledge into an AB simulation approach. This model development is by now rather complex and further modeling strategies should be explored. Moreover, the agent's behavior should be enhanced to foster better and more accurate simulation results. With this, several effects, as for example the segregation of components, can be modeled within an AB model. Furthermore, environmental influences should be integrated in AB models, e.g., the environmental relative humidity or temperature.

In order to contribute to the digitalization of manufacturing systems, the resulting AB process chain models need to be empowered for process control, e.g., within cyber-physical production systems. To this end, the development of an AB model can be based on or assisted by QbD approaches and use inline data by coupling PAT with the AB model to allow the forecast of product quality over complex process chains close to real time. Additionally, the design space for new products might be more easily defined. New formulations can already be evaluated in advance using the simulation before further experiments are conducted, which saves time and costs. By now, the simulation approach is able to perform descriptive as well as diagnostic analytics and in addition is much faster than the experimental process. In future research, this model needs to enable predictive or even prescriptive analytics. This will allow future manufacturing systems not to be mandatorily centralized and bound to the necessary process knowledge but will be able to work in decentralized locations with simulation models controlling the processes.

Author Contributions: Conceptualization, N.L.M., A.K.S., J.H.F., T.A., A.K. and C.H.; methodology, N.L.M. and A.K.S.; software, N.L.M.; validation, A.K.S., N.L.M. and J.H.F.; formal analysis, A.K.S. and N.L.M.; investigation, A.K.S. and N.L.M.; resources, N.L.M., A.K.S., J.H.F., T.A., A.K. and C.H.; data curation, A.K.S. and N.L.M.; writing—original draft preparation, N.L.M. and A.K.S.; writing—review and editing, J.H.F., T.A., A.K. and C.H.; visualization, N.L.M. and A.K.S.; supervision, A.K. and C.H.; project administration, A.K. and C.H.; funding acquisition, N.L.M., J.H.F., A.K. and C.H. All authors have read and agreed to the published version of the manuscript.

Funding: This paper evolved from the research project "Simulation of distributed product structures in combined discrete and continuous production processes for solid, particulate products", which is funded by the Deutsche Forschungsgemeinschaft (DFG, German Research Foundation)–413141366.

Institutional Review Board Statement: Not applicable.

Informed Consent Statement: Not applicable.

Data Availability Statement: Not applicable.

Acknowledgments: We acknowledge support by the German Research Foundation and the Open Access Publication Funds of Technische Universität Braunschweig. Furthermore, the authors acknowledge and thank KORSCH AG for the support and the provision of the tablet rotary press XL 100 and Chemische Fabrik Budenheim KG for the provision of testing materials (DI-CAFOS® A150 and DI-CAFOS® A60).

Conflicts of Interest: The authors declare no conflict of interest.

References

1. Lee, S.L.; O'Connor, T.F.; Yang, X.; Cruz, C.N.; Chatterjee, S.; Madurawe, R.D.; Moore, C.M.V.; Yu, L.X.; Woodcock, J. Modernizing Pharmaceutical Manufacturing: From Batch to Continuous Production. *J. Pharm. Innov.* **2015**, *10*, 191–199. [CrossRef]
2. Boukouvala, F.; Niotis, V.; Ramachandran, R.; Muzzio, F.J.; Ierapetritou, M.G. An integrated approach for dynamic flowsheet modeling and sensitivity analysis of a continuous tablet manufacturing process. *Comput. Chem. Eng.* **2012**, *42*, 30–47. [CrossRef]
3. Thiede, S. Environmental Sustainability of Cyber Physical Production Systems. *Procedia CIRP* **2018**, *69*, 644–649. [CrossRef]
4. Borshchev, A.; Filippov, A. From System Dynamics and Discrete Event to Practical Agent Based Modeling: Reasons, Techniques, Tools. In Proceedings of the 22nd International Conference of the System Dynamics Society, Oxford, UK, 25–29 July 2004; Kennedy, M., Ed.; System Dynamics Society: Albany, NY, USA, 2004. ISBN 0974532908.
5. Forrester, J.W. System dynamics, systems thinking, and soft OR. *Syst. Dyn. Rev.* **1994**, *10*, 245–256. [CrossRef]

6. Hildebrandt, C. Improving Die Filling in Pharmaceutical Tableting by Experimental and Numerical Means. Ph.D. Thesis, Kiel University, Kiel, Germany, 2018.
7. Hartge, E.-U.; Pogodda, M.; Reimers, C.; Schwier, D.; Gruhn, G.; Werther, J. Flowsheet Simulation of Solids Processes. *KONA Powder Part. J.* **2006**, *24*, 146–158. [CrossRef]
8. Dosta, M.; Litster, J.D.; Heinrich, S. Flowsheet simulation of solids processes: Current status and future trends. *Adv. Powder Technol.* **2020**, *31*, 947–953. [CrossRef]
9. Singh, R.; Ierapetritou, M.; Ramachandran, R. Hybrid Advanced Control of Flexible Multipurpose Continuous Tablet Manufacturing Process via Direct Compaction. In *23rd European Symposium on Computer Aided Process Engineering*; Elsevier: Amsterdam, The Netherlands, 2013; pp. 757–762. ISBN 9780444632340.
10. Ketterhagen, W.R.; Ende, M.T.A.; Hancock, B.C. Process Modeling in the Pharmaceutical Industry using the Discrete Element Method. *J. Pharm. Sci.* **2009**, *98*, 442–470. [CrossRef]
11. Skorych, V.; Dosta, M.; Heinrich, S. Dyssol—An open-source flowsheet simulation framework for particulate materials. *SoftwareX* **2020**, *12*, 100572. [CrossRef]
12. Rogers, A.J.; Hashemi, A.; Ierapetritou, M.G. Modeling of Particulate Processes for the Continuous Manufacture of Solid-Based Pharmaceutical Dosage Forms. *Processes* **2013**, *1*, 67–127. [CrossRef]
13. Macal, C.M.; North, M.J. Tutorial on agent-based modelling and simulation. *J. Simul.* **2010**, *4*, 151–162. [CrossRef]
14. Belič, A.; Škranjc, D.Z.-B.I.; Vrečer, F.; Karba, R. Artificial Neural Networks for Optimisation of Tablet Production. In Proceedings of the 6th EUROSIM Congress on Modelling and Simulation, EUROSIM 2007, Ljubljana, Slovenia, 9–13 September 2007; ISBN 978-3-901608-32-2.
15. Aksu, B.; Paradkar, A.; De Matas, M.; Özer, Ö.; Güneri, T.; York, P. Quality by Design Approach: Application of Artificial Intelligence Techniques of Tablets Manufactured by Direct Compression. *AAPS PharmSciTech* **2012**, *13*, 1138–1146. [CrossRef]
16. Jia, R.; Mao, Z.; Wang, F.; He, D. Self-tuning final product quality control of batch processes using kernel latent variable model. *Chem. Eng. Res. Des.* **2015**, *94*, 119–130. [CrossRef]
17. Westerhuis, J.A.; Coenegracht, P.M.J. Multivariate modelling of the pharmaceutical two-step process of wet granulation and tableting with multiblock partial least squares. *J. Chemom.* **1997**, *11*, 379–392. [CrossRef]
18. Wang, Z.; Pan, Z.; He, D.; Shi, J.; Sun, S.; Hou, Y. Simulation Modeling of a Pharmaceutical Tablet Manufacturing Process via Wet Granulation. *Complexity* **2019**, *2019*, 1–16. [CrossRef]
19. Kuentz, M.; Leuenberger, H. A new model for the hardness of a compacted particle system, applied to tablets of pharmaceutical polymers. *Powder Technol.* **2000**, *111*, 145–153. [CrossRef]
20. Boukouvala, F.; Chaudhury, A.; Sen, M.; Zhou, R.; Mioduszewski, L.; Ierapetritou, M.G.; Ramachandran, R. Computer-Aided Flowsheet Simulation of a Pharmaceutical Tablet Manufacturing Process Incorporating Wet Granulation. *J. Pharm. Innov.* **2013**, *8*, 11–27. [CrossRef]
21. Rogers, A.J.; Inamdar, C.; Ierapetritou, M.G. An Integrated Approach to Simulation of Pharmaceutical Processes for Solid Drug Manufacture. *Ind. Eng. Chem. Res.* **2013**, *53*, 5128–5147. [CrossRef]
22. Baroutaji, A.; Bryan, K.; Sajjia, M.; Lenihan, S. Mechanics and Computational Modeling of Pharmaceutical Tabletting Process. In *Reference Module in Materials Science and Materials Engineering*; Elsevier: Amsterdam, The Netherlands, 2017.
23. Lewis, R.W.; Gethin, D.T.; Yang, X.-S.; Rowe, R.C. A combined finite-discrete element method for simulating pharmaceutical powder tableting. *Int. J. Numer. Methods Eng.* **2005**, *62*, 853–869. [CrossRef]
24. Cabiscol, R.; Finke, J.H.; Kwade, A. Calibration and interpretation of DEM parameters for simulations of cylindrical tablets with multi-sphere approach. *Powder Technol.* **2018**, *327*, 232–245. [CrossRef]
25. Rantanen, J.; Khinast, J. The Future of Pharmaceutical Manufacturing Sciences. *J. Pharm. Sci.* **2015**, *104*, 3612–3638. [CrossRef]
26. Wu, C. DEM simulations of die filling during pharmaceutical tabletting. *Particuology* **2008**, *6*, 412–418. [CrossRef]
27. Guo, Y.; Kafui, K.D.; Wu, C.-Y.; Thornton, C.; Seville, J.P.K. A coupled DEM/CFD analysis of the effect of air on powder flow during die filling. *AIChE J.* **2008**, *55*, 49–62. [CrossRef]
28. Bierwisch, C.; Kraft, T.; Riedel, H.; Moseler, M. Die filling optimization using three-dimensional discrete element modeling. *Powder Technol.* **2009**, *196*, 169–179. [CrossRef]
29. Gopireddy, S.R.; Hildebrandt, C.; Urbanetz, N.A. Numerical simulation of powder flow in a pharmaceutical tablet press lab-scale gravity feeder. *Powder Technol.* **2016**, *302*, 309–327. [CrossRef]
30. Ketterhagen, W.R. Simulation of powder flow in a lab-scale tablet press feed frame: Effects of design and operating parameters on measures of tablet quality. *Powder Technol.* **2015**, *275*, 361–374. [CrossRef]
31. Siegmann, E.; Forgber, T.; Toson, P.; Martinetz, M.C.; Kureck, H.; Brinz, T.; Manz, S.; Grass, T.; Khinast, J. Powder flow and mixing in different tablet press feed frames. *Adv. Powder Technol.* **2020**, *31*, 770–781. [CrossRef]
32. Mateo-Ortiz, D.; Méndez, R. Microdynamic analysis of particle flow in a confined space using DEM: The feed frame case. *Adv. Powder Technol.* **2016**, *27*, 1597–1606. [CrossRef]
33. Cabiscol, R.; Finke, J.H.; Kwade, A. Assessment of particle rearrangement and anisotropy in high-load tableting with a DEM-based elasto-plastic cohesive model. *Granul. Matter* **2019**, *21*, 98. [CrossRef]
34. Giannis, K.; Schilde, C.; Finke, J.H.; Kwade, A.; Celigueta, M.A.; Taghizadeh, K.; Luding, S. Stress based multi-contact model for discrete-element simulations. *Granul. Matter* **2021**, *23*, 1–14. [CrossRef]

35. Macal, C.M.; North, M.J. Agent-Based Modeling and Simulation. In Proceedings of the 2009 Winter Simulation Conference (WSC), Austin, TX, USA, 13–16 December 2009; pp. 86–98.
36. Macal, C.M. Everything you need to know about agent-based modelling and simulation. *J. Simul.* **2016**, *10*, 144–156. [CrossRef]
37. Schönemann, M.; Bockholt, H.; Thiede, S.; Kwade, A.; Herrmann, C. Multiscale simulation approach for production systems. *Int. J. Adv. Manuf. Technol.* **2019**, *102*, 1373–1390. [CrossRef]
38. Celik, M. Overview of Compaction Data Analysis Techniques. *Drug Dev. Ind. Pharm.* **1992**, *18*, 767–810. [CrossRef]
39. Wünsch, I.; Finke, J.H.; John, E.; Juhnke, M.; Kwade, A. A Mathematical Approach to Consider Solid Compressibility in the Compression of Pharmaceutical Powders. *Pharmaceuticals* **2019**, *11*, 121. [CrossRef]
40. Heckel, R.W. Density-Pressure Relationships in Powder Compaction. *Trans. Metall. Soc. AIME* **1961**, *221*, 671–675.
41. Kawakita, K.; Lüdde, K.-H. Some considerations on powder compression equations. *Powder Technol.* **1971**, *4*, 61–68. [CrossRef]
42. Gurnham, C.F.; Masson, H.J. Expression of Liquids from Fibrous Materials. *Ind. Eng. Chem.* **1946**, *38*, 1309–1315. [CrossRef]
43. Cooper, A.R.; Eaton, L.E. Compaction Behavior of Several Ceramic Powders. *J. Am. Ceram. Soc.* **1962**, *45*, 97–101. [CrossRef]
44. Busignies, V.; Mazel, V.; Diarra, H.; Tchoreloff, P. Prediction of the compressibility of complex mixtures of pharmaceutical powders. *Int. J. Pharm.* **2012**, *436*, 862–868. [CrossRef]
45. Hirschberg, C.; Paul, S.; Rantanen, J.; Sun, C.C. A material-saving and robust approach for obtaining accurate out-of-die powder compressibility. *Powder Technol.* **2020**, *361*, 903–909. [CrossRef]
46. Ryshkewitch, E. Compression Strength of Porous Sintered Alumina and Zirconia. *J. Am. Ceram. Soc.* **1953**, *36*, 65–68. [CrossRef]
47. Schomberg, A.K.; Kwade, A.; Finke, J.H. The Challenge of Die Filling in Rotary Presses—A Systematic Study of Material Properties and Process Parameters. *Pharmaceutics* **2020**, *12*, 248. [CrossRef] [PubMed]
48. Fell, J.T.; Newton, J.M. Determination of Tablet Strength by the Diametral-Compression Test. *J. Pharm. Sci.* **1970**, *59*, 688–691. [CrossRef]
49. Nicklasson, F.; Alderborn, G. Analysis of the Compression Mechanics of Pharmaceutical Agglomerates of Different Porosity and Composition Using the Adams and Kawakita Equations. *Pharm. Res.* **2000**, *17*, 949–954. [CrossRef]
50. Puckhaber, D.; Eichler, S.; Kwade, A.; Finke, J.H. Impact of Particle and Equipment Properties on Residence Time Distribution of Pharmaceutical Excipients in Rotary Tablet Presses. *Pharmaceuticals* **2020**, *12*, 283. [CrossRef] [PubMed]

pharmaceutics

Review

Reviewing the Impact of Powder Cohesion on Continuous Direct Compression (CDC) Performance

Owen Jones-Salkey [1,2,†], Zoe Chu [1,2,*,†], Andrew Ingram [2] and Christopher R. K. Windows-Yule [2]

1 Oral Product Development, Pharmaceutical Technology & Development, Operations, AstraZeneca, Macclesfield SK10 2NA, UK; oxj807@student.bham.ac.uk
2 School of Chemical Engineering, University of Birmingham, Birmingham B15 2TT, UK
* Correspondence: zhc906@student.bham.ac.uk
† These authors contributed equally to this work.

Abstract: The pharmaceutical industry is undergoing a paradigm shift towards continuous processing from batch, where continuous direct compression (CDC) is considered to offer the most straightforward implementation amongst powder processes due to the relatively low number of unit operations or handling steps. Due to the nature of continuous processing, the bulk properties of the formulation will require sufficient flowability and tabletability in order to be processed and transported effectively to and from each unit operation. Powder cohesion presents one of the greatest obstacles to the CDC process as it inhibits powder flow. As a result, there have been many studies investigating potential manners in which to overcome the effects of cohesion with, to date, little consideration of how these controls may affect downstream unit operations. The aim of this literature review is to explore and consolidate this literature, considering the impact of powder cohesion and cohesion control measures on the three-unit operations of the CDC process (feeding, mixing, and tabletting). This review will also cover the consequences of implementing such control measures whilst highlighting subject matter which could be of value for future research to better understand how to manage cohesive powders for CDC manufacture.

Keywords: continuous direct compression (CDC); powder cohesion; loss in weight (LIW) feeding; blending; tabletting; continuous manufacture

Citation: Jones-Salkey, O.; Chu, Z.; Ingram, A.; Windows-Yule, C.R.K. Reviewing the Impact of Powder Cohesion on Continuous Direct Compression (CDC) Performance. *Pharmaceutics* **2023**, *15*, 1587. https://doi.org/10.3390/pharmaceutics15061587

Academic Editor: Colin Hare

Received: 3 April 2023
Revised: 8 May 2023
Accepted: 10 May 2023
Published: 24 May 2023

1. Introduction

The pharmaceutical industry is currently undergoing a paradigm shift from batch to continuous processing in order to improve manufacturing efficiency [1]. Continuous direct compression (CDC) (Figure 1) is considered highly efficient because of the reduced number of unit operations involved [2,3]; however, there currently exists only a limited range of formulations for which it is viable. Suitable formulations will generally have low drug load, a need for adequate flowability (as the material has to be transported through the system), and have sufficient tabletability, for the CDC process to work [4].

There are several critical material attributes (CMAs) which impact the flow of pharmaceutical powders and hence the successful formulation of oral solid dosage forms, of which cohesion is thought to be the most significant [5]. Powder cohesion refers to the affinity of particles to adhere to each other. It manifests in the bulk as resistance to powder flow often accompanied by adhesive behaviour, which refers to the tendency of a material to bind to surfaces and other materials [6]. Factors which contribute to cohesion include:

- Liquid bridges (liquid bridges will not be discussed in this literature review as there already exist comprehensive reviews of this topic [7–9]). Our focus in the present work will also be on the *effect* of cohesion in CDC processes rather than the *causes* of cohesion) [10];
- Van der Waals forces [11];

- Electrostatic forces [12];
- Frictional charging during handling [13];
- Particle shape and size [3].

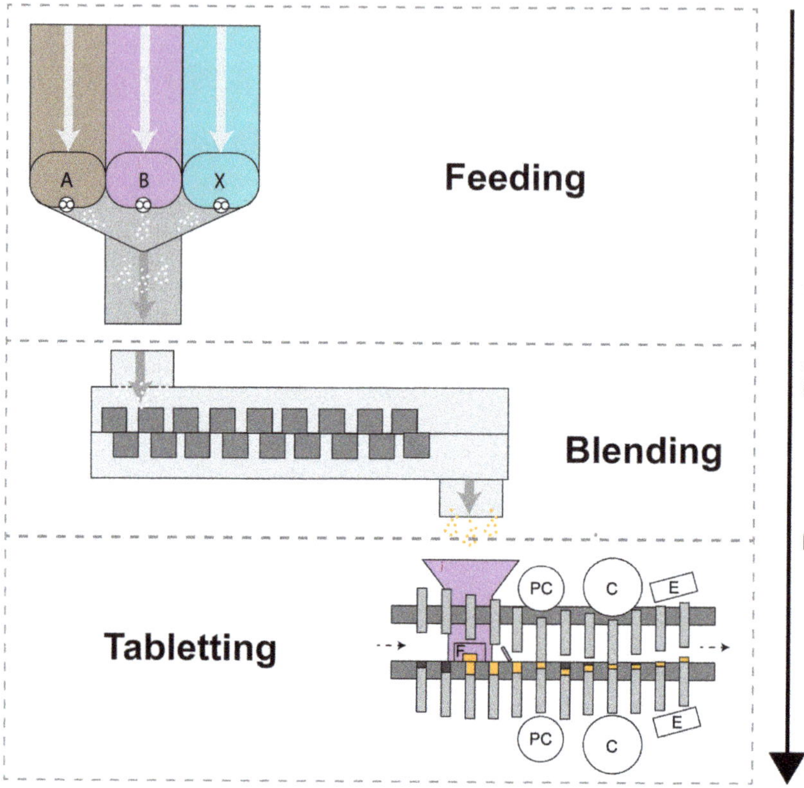

Figure 1. Schematic of the CDC process. Feeding: (where **A**, **B**, & **X**; are example labels of the formulation's constituents) powders are fed into the blending stage. Blending: Shows the constituents entering and being blended by the agitator, before entering the hopper of the rotary tablet press. Tabletting: (where **F**, **PC**, **C**, & **E**; corresponds to the Die Filling, Pre-Compression, Compression and Ejection, respectively) the final formulation fills the die cavity, undergoes compression forces—forming the tablet—before being ejected from the die.

Although humidity is controlled in a GMP environment, moisture is still present in the air and can be adsorbed and or absorbed into the powders, increasing cohesivity [14–16]. This increase in local moisture content in powders results in transient, complex interactions between powder surfaces and water, resulting in higher cohesion and decreased flowability [14–16]. Therefore, for every instance where good flowability is desired, an increase in humidity will have a negative effect. On the other hand, with compaction, there is evidence that some moisture can produce higher-strength tablets. However, too much moisture impacts the tablet strength negatively [17]. Thus, the positive effect of humidity, during tabletting, is largely outweighed by its negative effect across all the other unit operations in the CDC process. Therefore, generally or where suitable, moisture content should be kept to a minimum. In the interest of the review, relative humidity will be discussed when it is the only variable suited to solving an issue, for instance mitigating the triboelectric effect during powder feeding (Section 2). Furthermore, there are additional reviews [8,9] which discuss the influence of humidity on powder cohesion.

Active pharmaceutical ingredients (APIs) are typically the components of tablet formulations that give rise to handling issues, due to their often very high cohesivity, low bulk density, and highly aspherical geometries, in turn resulting in undesirable operating performance for CDC [6,18]. Many APIs are needle-like and very fine (<10 μm) resulting in a high contact area and cohesion (and thus poor flowability), whereas excipients are typically coarser and have a more favourable (spherical) particle shape, leading to better flowability. This difference in flow properties is perhaps unsurprising, as excipients are inactive substances whose purpose is to aid the formulation process by supporting or enhancing the stability and tabletability [19] of a pharmaceutical dosage form; as such, cohesive excipients would be omitted during the experimental design stages [20] as they are much easier to replace than APIs. Based on the above, it is clear that a high concentration of API may heavily affect the processability of powders, rendering the CDC manufacturing route non-viable [21].

In spite of the above, cohesion is not always a negative influence on CDC operations. In some instances, moderate cohesion has been found to aid the proper functioning of a given unit operation. For example, during powder feeding the addition of nano-sized silica was shown to result in reduced adhesion to feeder surfaces and improved powder flow as will be discussed later in this review [6,22]. However, in most cases of high cohesivity, flow is inhibited [23,24], which is problematic for feeding [25], mixing [26] and die filling [27,28]. Conversely, during the tabletting stage, powder cohesivity (specifically interparticle adhesion) is generally desirable, inviting smaller particle sizes and higher surface energy powders [29–36].

There are three main ways in which cohesion can be managed in tablet production:

1. Formulation modification, through the introduction of glidants or lubricants [37].
2. Operational process changes which are documented in many different studies for each unit operation [38].
3. Granulation, which also manages cohesion as poor flowing powder fines are made into coarser more uniform agglomerates. (This is beyond the scope of this literature review as this unit operation is not a part of CDC processes [39]).

However, most literature does not consider how cohesion controls impact downstream unit operations. As far as the authors are aware there are no literature reviews that encompass the mitigation of the effects of cohesion and how those controls may affect certain unit operations. Therefore, the primary aim of this literature review is to explore and consolidate current knowledge concerning the impact of powder cohesion and cohesion control measures on all unit operations in the CDC process (feeding, mixing and compression), discuss the consequences of cohesion control measures implemented, and establish what further studies could be valuably conducted in the future.

2. Powder Feeding

2.1. Feeding Introduction

Powder feeding for CDC processes needs to be accurate and consistent for CDC performance success. Feeding modes are commonly defined by the delivery and maintenance of material, in a given time frame. As such, there are two main categories for feeders: volumetric and gravimetric, which supply material according to volume per unit time, and mass per unit time, respectively [2,22,25,40,41]. Volumetric feeding is often used to understand the characteristics of powder behaviour within the hopper, whereas gravimetric feeding seeks to smooth out any volumetric perturbation through feedback control, arguably resulting in gravimetric being the superior of the two modes for mass delivery [2,22,25,40–42]. Powder feeding is the first unit of operation in the CDC process; thus disturbance or variation from the set point can propagate downstream, fundamentally impacting the tablet quality [25,43].

One of the tasks of the blending stage in ensuring that constituents are well-mixed is to smooth out the perturbations from the feeder. The lower the intensity and frequency of these perturbations, the lower the chance that content uniformity or manufacturability issues will arise downstream [22,44,45]. Moreover, it should be noted that not all disturbances result in a detrimental impact. It is both dependant on the magnitude and duration of such disturbances—Gyürkés et al. [46] aptly discuss this concept with the use of funnel plots.

Powder cohesion impacts feeding performance by introducing variation through complex bulk behaviour, which results in the variation of flow behaviour [22,25]. This contributes to both the restriction of power into the converting screws, and the resistance to flow within the swept volume of the screws. In addition, these materials are prone to electrostatic charging, causing attraction-repellent behaviour, which can lead to deposits of material residing and adhering to surfaces post-feeding [15,47]. This buildup can migrate into the bulk feeder, dosing it with a spike of material, therefore resulting in a compositional imbalance [15,47]. The following section aims to discuss the work of several publications, exploring the complexity surrounding powder feeding.

2.2. Feeding Discussion

The difficulty associated with powder feeding directly derives from the handling of larger volumes of powder (in the hopper) and converting this into smaller, consistent streams (at the outlet). Essentially, this refers to the driving of powder down a pathway of lowering power-to-powder contact, to instead increase powder-to-wall contact. When there is resistance to this gradient of increasing powder-to-wall interactions, it suggests there is a lack of flowability, which commonly manifests in the form of cohesivity. Though cohesion is ultimately a microscopic process (i.e., it occurs at the scale of particle-particle contacts), at present its effects are predominantly characterised via macroscopic (bulk) measurements [48].

The Flow Function Coefficient (FFC) is a measurement used to quantify the amount of stress required to cause a powder to yield, and is one of several manners in which a powder's 'flowability' may be characterised [49,50]. Gathered from a ring shear tester, for example, the measurement applies a normal consolidation force to a powder bed and measures the tangential force required to cause failure (incipient flow). A yield locus of shear stress vs normal stress is plotted and, applying Mohr–Coulomb failure theory, the major principal yield stresses of consolidated and unconfined material are obtained. FFC is the ratio of these values. The theory and methodology is discussed in various degrees of detail in the following papers: [5,49–51]. Leung et al. [5] performed a comprehensive analysis on 1130 ring shear measurements and effectively surmised that particle friction coefficients are negligible on FFC measurements, suggesting that cohesion is the primary characteristic which influences the measurement. In addition, the authors state that mitigating the effects of cohesion is far more influential than mitigating the frictional aspect(s) of the powder when it comes to improving powder flow, and also suggest that altering interparticle forces is the best approach to take when combating cohesive powders.

Leung et al. [5] demonstrated the effect of introducing 1% w/w of colloidal silicon dioxide into two excipients on FFC, cohesion and angle of internal friction. The first excipient—hypromellose—showed a significant drop in cohesion and an increase in FFC (Figure 2). Meanwhile, the second excipient—dicalcium phosphate anhydrous—showed the inverse behaviour, but to a much lower degree. No significant changes were seen in the angle of internal friction for both materials. Additional work regarding the use of colloidal silicon dioxide nanoparticles for formulation adjustment is seen in [52,53]. The mechanism at work involves the surface coating of SiO_2 nanoparticles, which introduce additional distance, therefore lowering the strength of the van der Waals attraction between the two surfaces of the more cohesive species [5,6,52,53]. Furthermore, Leung et al. state that particles with uneven shapes and rough surfaces may receive little beneficial effect from the SiO_2 surface coating, and therefore minimal improvement in flowability. The SiO_2 will fail to act as a barrier to the contact of the carrier species as it is shown that using SiO_2 as an additive can be detrimental to FFC.

Tran et al. [54], investigated these underpinning mechanisms which govern flow modification through the use of colloidal silica—their findings agree with those of Leung et al. [5].

Tran et al. studied the differences in silica loading for two different excipients using flow characterisation, static image analysis, and particle size distribution. It was found that materials with higher surface area (described as rugged) were able to accommodate higher silica loads, and that the optimal amount of silica loading is dependent on both the surface area and the size of the carrier particle but, most notably, their results discussed the possibility of taking a 'quality by design' approach by determining silicate (glidant) loading by carefully assessing the carrier powder's physical properties.

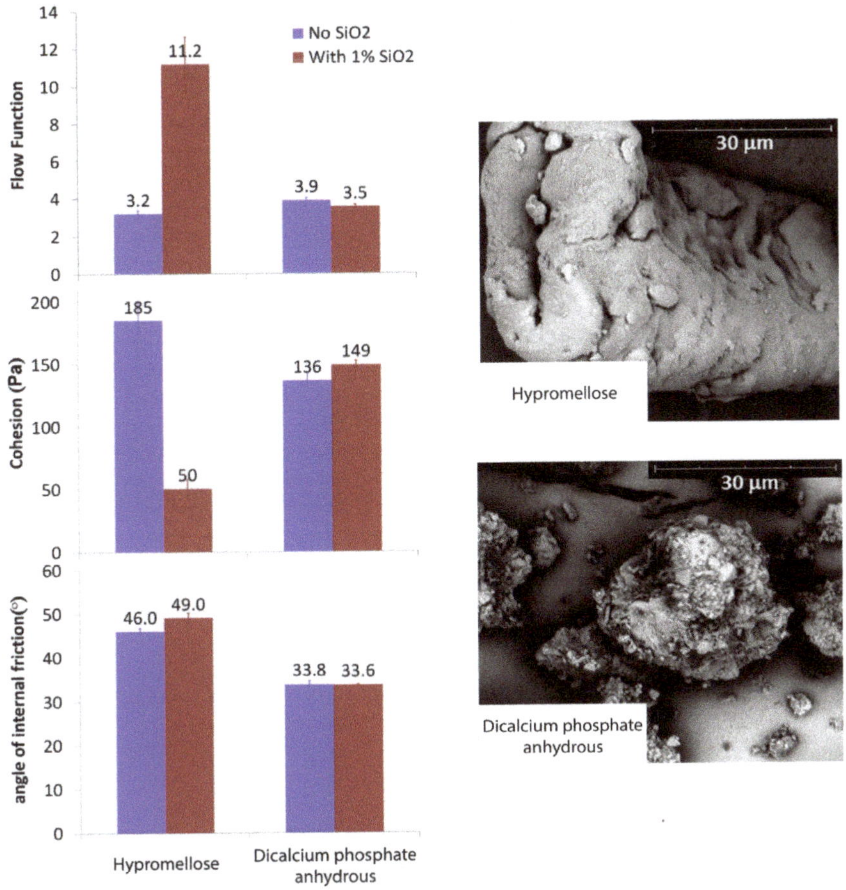

Figure 2. (**Left**): Graphs highlighting the difference in characteristics (Flow Function (FFC), Cohesion, and angle of internal friction) of the two excipients: Hypromellose and Dicalcium Phosphate Anhydrous with and without colloidal silica. (**Right**): Scanning Electron Microscope (SEM) of the two excipients mixed with 1% *w*/*w* colloidal silica. Adapted from Leung et al. [5].

Lopez et al. [55] published a DEM study on the effect of increasing cohesion on feeding performance. The researchers began by calibrating the powder used in the system, first identifying and assigning the physical properties and particle shape of paracetamol, before performing a sensitivity analysis across different adhesive stiffness values (K_{adh}) and conveying impeller rotation rates. When the conveying impeller was kept at a constant rotation rate (10 rad/s), the particles saw an increase in translational velocity in conveying barrel with increasing K_{adh} (ranging from 0 to 0.5). This effect was lesser but present on the interface between the conveying barrel and hopper. On the other hand, keeping K_{adh} at a constant 0.2, across increasing conveying speeds (10, 20, and 50 rad/s) saw an increase in the overall translational velocity of the particles. Perhaps unsurprisingly, for the range of parameters explored, increasing conveying speed was observed to have a greater influence on particle velocity in the hopper than altering cohesion.

The DEM simulation of Lopez et al. [55] also provides valuable insight into the expected power draw; given the CDC system is continuous, it would mean the process would be running for days, if not weeks, at a time, making this an important consideration. The authors demonstrated the average mass discharged per Watt as a function of the K_{adh} (between 0.1 to 0.5 K_{adh}), which showed an exponential decay with increasing K_{adh} Figure 3. Accordingly, K_{adh} values higher than 0.5 showed arching: a phenomenon which occurs when the cohesive forces between particles are stronger than forces due to gravity. Arching (also referred to as bridging) is a semi-stable instance of particles forming particle–particle structures suspended in a hopper or opening, which are resistant to the displacement of powder beneath them [25]. For additional information on the influence of particle properties on arching, see the following papers: [25,49,56]. Lopez et al. described this behaviour with absolute translational velocity where they revealed instances which lead to inconsistent feeding of the conveying screw, which in turn lead to flow irregularities Figure 3. The authors link this artefact of transient macro behaviour to the micro by stating that inconsistencies in flow result in the improper filling of the screw pitch, giving rise to an inconsistent mass flow rate. This is best demonstrated by the plot of mass per screw pitch vs. K_{adh} shown in Figure 3). Finally, the researchers indicated what future work would be of value, notably suggesting the mapping of the transient cohesive behaviour by altering the particle properties, feeder's geometry and rotational speed.

To build upon the above suggestions, it would be interesting to see the implementation of a DEM study on a more commercially representative feeder, which possesses both a hopper agitator and twin screw conveying elements. Allowing the exploration of concepts such as agitator design and rotation speed to increase barrel/screw filling consistency or the breakage/mitigation of arching. What is more, the research conducted by Leung et al. [5], which discussed the minimal effect of friction in powder flow consistency, could utilise simulation tools such as DEM to more directly support their claims.

Escotet-Espinoza et al. [6] present a publication showcasing the use of silication to improve feeder performance. The researchers began by pre-blending the three different APIs explored with 1% silica, before characterising the bulk behaviour of the pre-and post-silicated API. The APIs (and their silicated counterparts) were then fed through the feeder, whilst a catch scale continuously measured the mass exiting the feeder. The response was then measured over time, with attention being paid to the consistency and accuracy of the feeding process.

Like Leung et al. [5], Escotet-Espinoza et al. [6] also found that the introduction of silica improved flowability by reducing the interparticle cohesion. Feeding improved with the addition of silica, with each of the API case studies showing improved screw speed consistency, a reduction of mass flow RSD (relative standard deviation), a reduction in powder adhesion to the hopper surfaces, and a reduction of remaining mass. This suggests that the addition of silica is hugely advantageous for feeding. The improvement can potentially be attributed to an increase in flowability, allowing the powder to better fill the swept volume of the screws, thus leading to an increase in mass per revolution with the addition of silica. This observation is supported by the work of Lopez et al. [55] which shows that, with a reduction of K_{adh}, there follows an increase in mass per screw pitch. Furthermore, with FFC (as previously discussed) being a strong indicator for powder cohesion, it would suggest that it would be useful to understand this screw-filling behaviour. This conveniently leads to another suggestion to further Lopez et al's work [55], whereby similar DEM studies could be undertaken to understand the effect of additives on feeding and powder conveying. Escotet-Espinoza et al. [6] also describe the material used generating electrostatic charge, which enabled the API to stick to the agitator. However, in the silicated API case study, there was very low powder adhesion, suggesting that adding silica also suppressed the effects of electrostatic adhesion.

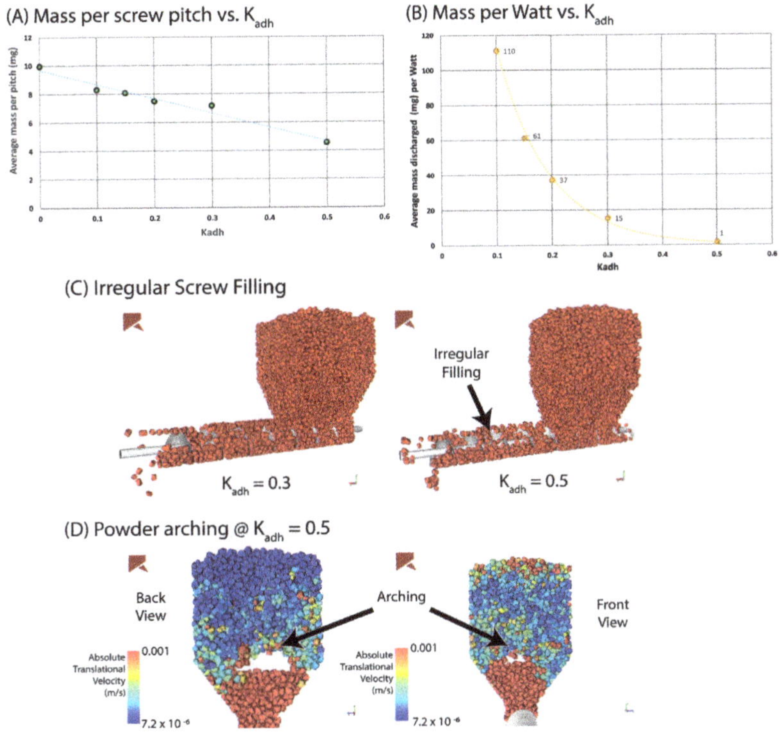

Figure 3. (**A**): Mass per screw pitch as a function of cohesive stiffness (K_{adh}). (**B**): Mass per Watt as a function of cohesive stiffness (K_{adh}). (**D**): Visualisation of the powder arching/bridging within the hopper volume. The images show both the back and front of the hopper from two different simulations using the same K_{adh}. (**C**): Irregular filling of the conveying volume due to cohesive stiffness (K_{adh}). The images compare K_{adh} = 0.3 and K_{adh} = 0.5. Adapted from Lopez et al. [55].

A study by Lumay et al. [57] explicitly investigated the influence and mechanisms of mesoporous silica (MPS) on the electrostatic charge, again finding that the addition of silica species improves powder flow. The experiments differed from those discussed previously in several manners. Firstly, the study differed in the materials and methods used—the authors tested three different grades of silica, varying in particle and pore size, in three common excipients (microcrystalline cellulose, lactose, and maize starch) using a rotating drum. Secondly, the authors applied different experimental analyses: the dynamic effect of silication was evaluated in this case by measuring the cohesive index [58] of materials (with and without MPS) across a range of rotation speeds. Finally, through the use of a standardised charge density measurement technique [59], the charge density of the material blends is shown in the presence of different types and amounts of MPS.

It was found that the MPS with the smallest particle size provided the greatest improvement in flowability. Furthermore, the addition of just 0.5% w/w made a considerable impact on the cohesive index—see Figure 4A. The authors attribute this improvement to the particle size of the MPS, explaining that there would be an increase in effective surface area, which would suggest greater coverage, and thus intervention between, cohesive species.

An additional striking finding of the study was that the addition of silica was also found to change the shear-dependent rheological properties of certain tested materials, with typically rheopectic materials becoming thixotropic when mixed with a small volume of silica. Specifically, the study evaluated the cohesive index of the powder dynamically by testing the excipients against their 2% *w*/*w* silica counterparts across a range of rotation rates (see Figure 4B). Both materials, maize and lactose, were shown to be more cohesive (thickening) with increasing rotating speed (ranging from 2 rpm to 10 rpm) and in the presence of silica, less cohesive (thinning) with increasing rotating speed. The silicated maize showed a greater decrease in cohesion than the silicated lactose with increasing rotation speed.

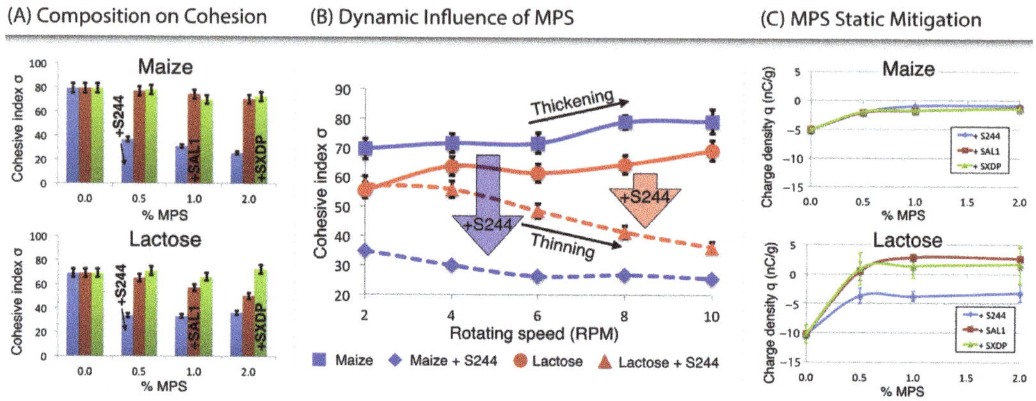

Figure 4. (**A**) Comparing the differences in the dynamic cohesive index (σ) with the addition of different types and weight percentages of Mesoporous Silicates (MPS). (**B**) Dynamic influence of MPS on the cohesive index, demonstrated through plotting the cohesive index over different rotating speeds. (**C**) Comparison of Charge Density (nC/g) in two excipients and resultant performance with the addition of 2% weight silicates. Where: +S244 is Syloid® 244FP, +SAL1 is Syloid® AL–1FP, and +SXDP is Syloid® XDP3050. Adapted from Lumay et al. [57].

Finally, the inclusion of MPS was shown to decrease the static charge density of the powder—see Figure 4C. Across all materials and MPS grades, the addition of 1% (of MPS) saw similar results to the addition of 2—suggesting 1% was sufficient for charge mitigation, with any greater amount being surplus. MPS incorporated into maize saw a decrease in charge after just 0.5% *w*/*w*, with the grade of MPS used making a minimal difference. Conversely, MPS in lactose still produced a decrease in the magnitude of charge density, though different grades responded differently. This suggests there may be some complexity associated with the use of MPS with lactose. It is potentially important to note that Lumay et al. [57] stated that the lactose varied in positive or negative charge on the day, which was dependent on the air's relative humidity. Given the magnitude is roughly the same but the charge differs, if the S244 (Syloid®FP244) was positive it would suggest that the grade of MPS used would have little effect on the charge density and vice versa for SXDP (Syloid®XDP3050). Research by Ramires-Dorronsoro et al. [60] supports this: when completing a similar study—with a different measurement technique—they also witnessed spray-dried lactose exhibiting a positive charge density. Ultimately, this suggests that the use of silica in poorly flowing hygroscopic powders greatly improves the flowability, and decreases the sensitivity to triboelectric charging.

The previously discussed papers showcase the underpinning behaviour of cohesive powders during feeding, and the discussion of silication demonstrates the advantage of understanding the mechanism underlying a powder's cohesive properties. Despite this, when considering CDC in its entirety, the predictive capability is a large part of reducing experimentation and understanding the process. Therefore it is desirable to be able to predict the feeding performance from a few key powder characteristics [42,43,61], one of which is cohesion.

Van Snick et al. [43] and Garcia-Munoz et al. [45] provided some of the first in-depth analysis of the end-to-end CDC process, thus by definition including aspects of feeding. The authors discuss the correlation of bulk density, tapped bulk density and FFC to feeding consistency (Figure 5). The experiments involved the comparison of different materials running volumetrically (at a constant screw speed) from full to empty, whilst measuring the feed factor response over time. The maximum (ff_{max}) and minimum (ff_{min}) feed factors (g/revolution) were then taken and used for further calculation. Feed factors change throughout the hopper fill as powder above the hopper bears down upon the powder below, due to gravity. When the hopper is at low fill, there is not only a reduction of gravitational bulk powder compaction but a potentially inconsistent feed into the conveying volume.

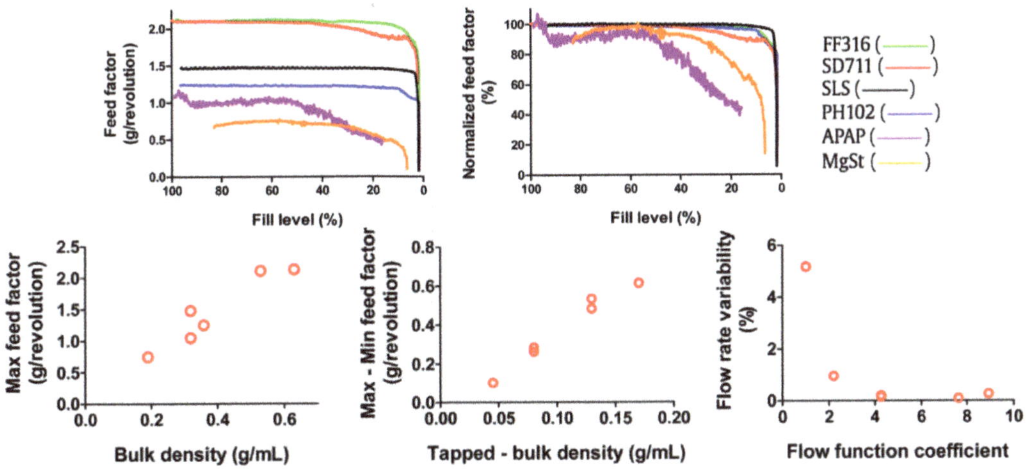

Figure 5. (**Top**): Two graphs detailing the feed factor throughout the volumetric empty, with absolute values (**left**) and normalised values (**right**), the key indicates the material used. Where: FF316 is Lactose Fast Flo® 316, SD711 is Sodium Croscarmellose Ac-Di-Sol® SLS is Sodium Lauryl Sulphate, PH102 is MCC Avicel® PH-102, APAP is Acetaminophen, and MgSt is Magnesium Sterate. (**Bottom**): Three graphs compare feeder performance to bulk characteristics. Adapted from Van Snick et al. [43].

As a result, bulk density increased linearly with max feed factor, tapped bulk density increased linearly with delta feed factor (where delta feed factor is: $ff_{\Delta} = ff_{max} - ff_{min}$), and finally flow rate variability exponentially decayed with increasing FFC. Since both bulk density and FFC are common descriptors for powder cohesion, it then can be used as a proxy to describe behaviour. The study highlighted FFC values greater than 3 would give sub 1% flow rate variability, and that powders with low bulk density and low tapped bulk density are prone to large ff_{Δ}—which suggests instability when processing. Escotet-Espinosa [22,62] also looked into both the influence of powder properties on feed factor and modelling such behaviour using the following (simplified) equation:

$$ff_w = ff_{max} * (ff_{max} - ff_{min}) exp[-\beta * w] \quad (1)$$

where ff_w is the feed factor as a function of $'w'$ the weight, while ff_{max}, ff_{min} and β are the maximum feed factor, minimum feed factor, and a fitting constant based on feeder geometry, respectively.

The ff_{max} and ff_{min} were then plotted in a correlation heat map (Figure 6), relating the responses to a plethora of recorded bulk characteristics. Interestingly, the highest correlation was seen for cBD, conditioned bulk density (g/mL). cBD is akin to tapped bulk density, obtained from the initial standardised conditioning step of a powder rheometer. It involves a blade passing through a volume of powder with a set torque before measuring the weight of the known volume. Paying attention to the units, it is a given that there is a correlation between the two density measurements, but the interest is with 'compressibility at 15 kPa' which also scored high for ff_{min}, which would be a method for attaining higher bulk density. This suggests that powder that is capable of being compressed, and has a high conditioned bulk density, will deliver a high ff_{min}. Similar studies by Shier et al. [63] and Bekaert et al. [42] similarly look to predict feeder performance using bulk characteristic properties.

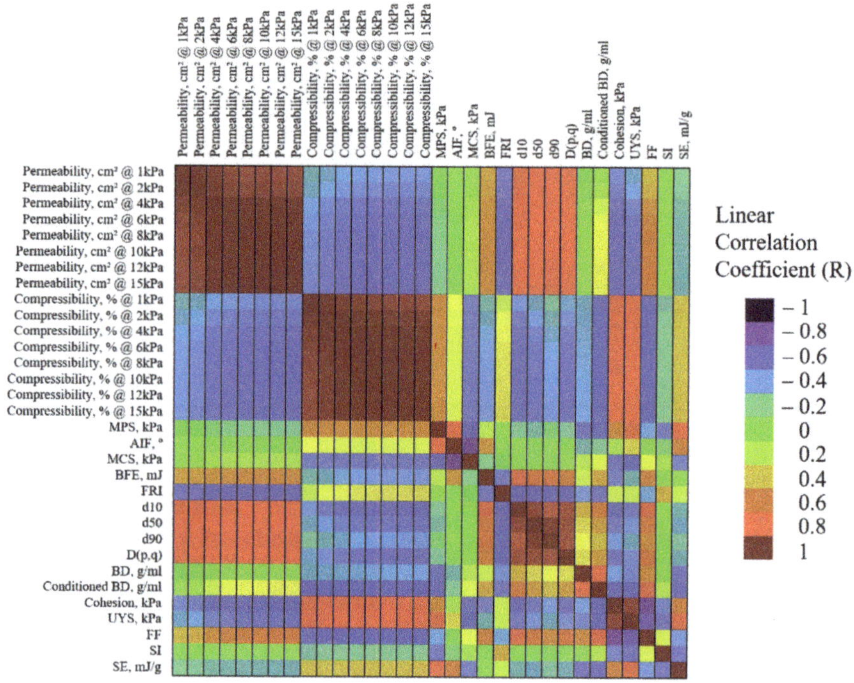

Figure 6. A correlation heat map acquired, with permission, from Escotet-Espinosa [62], detailing the magnitude of linear correlation (R) between bulk powder characteristics and regressed feed factor parameters.

Engisch & Muzzio [41] performed a study on varying hopper refilling conditions on powders with varying cohesiveness and measured the resultant effect on feeding accuracy and consistency. Typical Loss in Weight (LIW) feeder operation is gravimetric, until topping up (or refilling), where the system operates volumetrically to ensure a consistent feed is maintained whilst an influx of mass is registered on the feeder's load cell. This method results in a trade-off between the number of times the system has to enter the 'less accurate' volumetric mode and ensuring the hopper is sufficiently full, such that the feed factor is stable. For an example of the effect of decreasing fill on feed factor, see Figure 5. The conclusion from Engisch & Muzzio is that refilling the feeders with less material, and therefore more frequently, results in better overall performance. This should be kept in mind when paying attention to the duration of time that the system enters volumetric mode and reducing the overall disturbance from the setpoint over the whole duration of running the feeder. Irrespective of the frequency and volume of fill, Engisch & Muzzio state refilling should be gentle as this causes the least disruption to feeding. Furthermore, it is shown that refilling sensitivity is much lower at higher refill levels, Figure 7. Refilling at higher fill levels reduces the maximum set point deviation, time of deviation, and the total amount of mass during deviation.

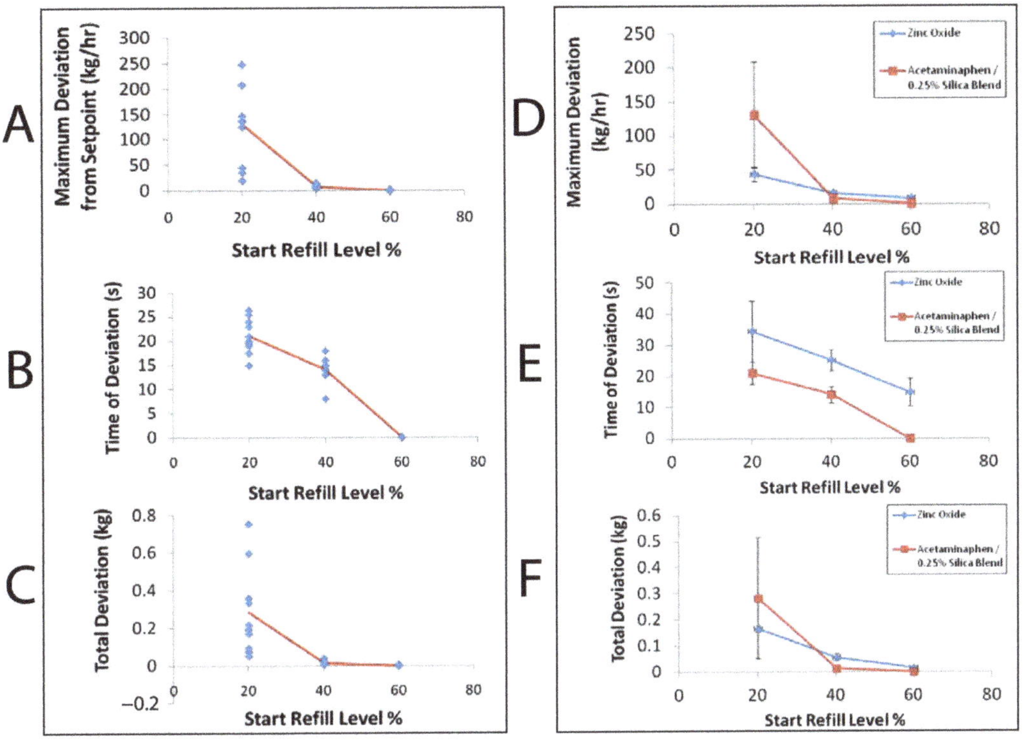

Figure 7. (**A–C**) Shows the feeder performance for just semi-fine APAP & 0.25% silica over 3 refill levels, each with 10 manual refills. (**A**) Maximum deviation from setpoint (kg/h), (**B**) Time of deviation (s), and (**C**) Total deviation (kg). (**D–F**) Similarly, shows the comparison between zinc oxide powder and semi-fine APAP & 0.25% silica blend. (**D**) Maximum deviation from setpoint (kg/h), (**E**) Time of deviation (s), and (**F**) Total deviation (kg). The deviation is defined as the sum of the surplus powder delivered during refill. Adapted from Engisch & Muzzio [41].

A similar methodology could also be applied to combat cohesive powders through compaction during refilling, by attempting to keep the hopper volume as constant as possible, during their most stable ff_w. To build a theoretical case study, using the APAP being fed volumetrically in Van Snick et al. [43] (Figure 5) and the supporting evidence of refilling stability from Engisch & Muzzio [41], see Figure 7. A top-up of 25–30% of the volume of the hopper could be used and triggered at the 50–60% fill level, refilling to 75–80%. This would occur during the most stable feed factor window, whilst the hopper is already half full. During the window in which powder is dumped into the system, there is a transient influx of powder onto the existing powder bed. This impulse of vertical pressure on the powder would increase powder compression within the bed, increasing the bulk density of the powder local to the conveying screws. Corroboration is seen by adding Escotet-Espinosa's [22] correlation measures, as there is an improvement in the conditioned bulk density, which increases ff_{min} reducing the ff_{Δ}, essentially reducing the feeder inconsistency related to lower fill levels. Albeit this theoretical case study is generalising by not considering phenomena such as disturbances due to air flows during discharging mass, and an increased volume of powder that may be prone to triboelectric charging due to the agitator. DEM studies, such as Lopez et al. [55], open a stage for exploring some of the fundamental behaviours seen in these systems, in particular when looking to evaluate the micro and macro behaviour attributed to conveying inconsistency.

Finally, Yadav et al. conducted a comprehensive study with a loss in weight feeder [64] utilising a range of screw speeds, gear ratios, and screw types, across a range of materials. The result was a PCA model which identified the influence of these processing parameters and how they interacted with the materials' bulk characteristics. The principal components consisted of one comprising the powder's bulk characteristics, a second representing the feeder's processing parameters, and a third representing an interaction component. Most notably, the interaction component related screw free volume (cm^3) to density, particle size (d_{10}, d_{50}, d_{90}), wall friction angle, and Feed Factor. The authors note that a different screw type will have a different screw-free volume, meaning that the comparison between a smaller or larger volume to dispense per revolution must be taken into consideration when evaluating the feed factor. Similar findings are seen with Engisch et al. [2], where their suggestion is the use of both self-cleaning concave screws and the removal of any outlet screens for very cohesive materials.

2.3. Feeding Summary

Cohesion has been shown, across multiple studies, to worsen the flowability of powders, in turn reducing the consistency with which powder flows into the conveying volume [5,43,55]. The transient inconsistency of material flowing into this volume persists and has been shown to result in turn in an inconsistency in the delivered mass powder flow—and therefore a deviation from the desired set point [2,41,55,65]. Thus, it is very important to ensure that powder, despite its cohesive nature, is consistently motivated to enter and be transported through the conveying section of the feeder.

This review outlines just a few examples that have been used so far to reduce the negative influence of cohesion on feeding: reducing interparticle contact between cohesive species [5,57]; reducing the effect of triboelectric charging [57]; refilling the hopper at moderate fill levels [41,65]; and adjusting screw speed and screw type [2,64].

When applying the knowledge gained from Lumay et al. [57] to the work of Escotet-Espinoza et al. [6] and Lopez et al. [55] it would perhaps be sensible to suggest that smaller sized silica nanoparticles would be ideal for improving flowability at this stage of the CDC process. Silica nanoparticles have also been shown to change the blend from shear-thickening to shear-thinning, implying that, when the powder contacts the hopper's agitator or the conveying screws, one might expect a further transient, localised increase in flowability.

Despite silicates holding the much of spotlight for discussion, surrounding improving powder flowability, it should be noted that several alternative materials could be selected. However, every material in question, for flow modification, would need to undergo consideration for its end-to-end impact on the process. Silicates, for instance, have been shown to negatively impact compatibility [66,67] but also improve flowability and therefore improve feeder processing consistency [5,43,55], whereas magnesium stearate has also been shown to increase flowability, but due to its hydrophobicity—coating the surface of the API—would hinder the dissolution of the API within the formulation [68]. This would result in the material being greatly undesirable for any scenario which may involve the intimate blending, or over-mixing, of magnesium stearate with API. However, pre-blending API with other materials such as tricalcium phosphates or microcrystalline cellulose [69,70] propose the ability to improve flow behaviour at a minimised trade-off, and as such, more research should be carried out into both the use (for improving feeding performance) and the end-to-end implications of selecting such materials.

Our review of the literature has also highlighted the benefit of future DEM studies which could be undertaken to better understand the fundamental powder movement that governs powder feeding, in particular using commercially-relevant geometries. Notable areas of study could include modelling the effect of silication; altering, and potentially optimising, agitator and screw design; further investigation of the influence of hopper refill regimes. The development of dynamic DEM simulations could also be used to develop and test various control strategies, both existing and novel. Intuitively, it could be possible to develop a control proposition for shear thickening powders involving a sudden reduction in RPM followed by a slow build-up of screw speed back to setpoint.

3. Blending

3.1. Blending Introduction

Despite diverse designs available commercially and in the academic literature, continuous blenders can, in general, be reduced to four main components: an inlet, an outlet, a mixing volume, and a mechanism for promoting blending whilst in the volume of the blender. Typically, this last component will be some form of rotating agitator, the geometry of which may differ from mixer to mixer, or even formulation to formulation. Recognising the different types and geometries of mixers that exist within this space is fundamental when evaluating the role of powder mixing in CDC performance, as the features that differentiate one mixer from another can have significant impacts on the behaviour of the powders being blended, and thus may require different measures to handle the blending of cohesive powders.

Achieving good mixing is integral to the performance of any given pharmaceutical tablet; the therapeutic efficacy and mechanical properties of the tablet will suffer without the effective distribution of its constituent materials, ultimately compromising the safety of the product [43,71–75]. Thus, mixing inherently defines the final quality of the formulation (albeit that the final product cannot be delivered without the consistency and accuracy of the upstream and downstream processes [43,45]).

Practically defining whether a formulation is well-mixed is a complex endeavour with issues such as frequency of sampling [71,76,77], location of sampling [71,76,77], and scrutiny (size) of the sample [71,76,77]. However, describing the mechanisms of mixing may be ever-so-slightly easier, and is of more direct relevance to the present work.

The two commonly-defined types of mixing—macro- and micro-mixing—are both essential when looking to create ideal mixtures [71,72,75,78,79]. Macro-mixing is responsible for large, global movements, of material—or bulk transport—allowing the material to flow and circulate around the vessel [71,75,79–82]. Micro-mixing, on the other hand, involves the movement and displacement of material at the particle scale, i.e., the local interactions between particle species [71,79]. Accordingly, micro-mixing achieves the detail of the mixing action due to the mixing happening at the smallest possible scale: the particle–particle level [71,79], whereas macro-mixing serves to smooth out the upstream mass flow rate perturbations, due to the flow and bulk motion of the powder [71,79].

Fan et al. [83], succinctly discuss the evaluation of mixing in terms of cohesive mixtures. Figure 8 and Equation (2) encapsulate the majority of this discussion. It is shown that by increasing the surface force—which is also a function of particle size, and therefore surface area—it is possible to reduce the influence factor defined in Equation (2). A low influence factor means surface forces dominate, and the particle species adhere to the surface of another particle species, creating a perfectly-ordered blend. Small, less dense particles would be the ones to adhere to the more dense 'carrier' particles- due to gravitational forces. Then, if the surface forces are minimal, gravity dominates, and thus the mixture would involve non-cohesive species, following more typical solids mixing criteria.

$$Influence\ Factor = \frac{Gravitational\ Force}{Surface\ Force} \tag{2}$$

Conversely, de-mixing occurs through several mechanisms, such as: segregation, the separation of a formulation into constituent particles of similar properties, for example, size and density [71,76,84]; agglomeration, the formation of particle–particle dense macro-particles, due to interparticle attractive forces [85,86]; and electrostatic repulsion, or attraction, due to inter-particle charge imbalances, which predominantly result from contact-induced electron transfer, as particles move relative to each other and the equipment surfaces [26,87]. The part cohesion plays differs significantly in each of these mechanisms, and can either subdue or exacerbate de-mixing [84]. For instance, electrostatic charges could be worked into powder, through the triboelectric effect, creating attractive or repulsive forces between different species [26,77]. These interactions underpin much of the following discussion of how cohesion affects the process of mixing and the competition between mixing and de-mixing.

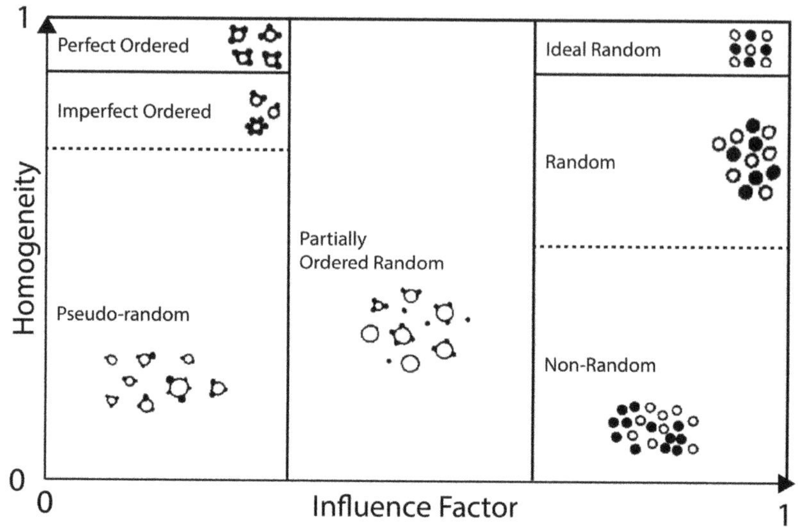

Figure 8. Cohesive mixtures described by homogeneity vs influence factor (Equation (2)). Adapted from Fan et al. [83].

3.2. Blending Discussion

Processing parameters such as agitator RPM, feed rate (throughput), and agitator design, are the primary methods of manipulating powder through a given blending volume [43,75,88]. For example, trends have been identified for improving mixing quality through the use of processing parameters: macro mixing as a function of mean residence time; micro mixing as a function of RPM (agitation); residence time being a function of throughput [43,75,82,89,90]. These examples signify the importance of process parameters on mixing performance. However, the intricacies of powder behaviour in response to these systems are seldom discussed in the wider literature.

As discussed in the preceding section, cohesion can be overcome through the use of formulation additives and/or changes to the process [5,37,57,78]. Additives, such as glidants (e.g., SiO_2), are typically used to make the bulk powder properties easier to deal with, by increasing flowability [5,37,57]. However, additives change the overall formulation composition, which can have detrimental effects on downstream processing-namely tabletting [66,67]. In contrast, process changes (processing parameters) can potentially overcome issues relating to cohesion without altering the formulation's composition [43,75,88]. Utilising these favourable processing parameters is thus the more practical of the two solutions for CDC processes, but the development and transfer of knowledge between different systems appears to be a substantial hurdle. This is largely due to this approach requiring a much deeper understanding of the specific unit operation being used in the study. Therefore, the knowledge gained is often less transferable; each particular set of process parameters may be unique to system geometry, orientation, or method of operation. With this in mind, it is useful to note the use of dimensionless numbers/groups within the discussed research, as it allows greater understanding and application of relative powder behaviour between publications.

Portillo et al. [91] performed trials on an axial continuous blender (see Figure 9)), the axis of which was adjustable between ±30° to the horizontal. The researchers found that at (+30°) incline there was an increase in mean residence time which led to a decrease in the relative standard deviation (RSD) of Acetaminophen (APAP) concentration in two grades of lactose. This demonstrates the effect of longer mean residence time on macro mixing and a reduction in APAP RSD. The grades of lactose differed in terms of average particle size and size range: Lactose 100 (70–250 μm, average 130 μm) and Lactose 125 (55 μm), whereas APAP (36 μm) remained the same in both tests, and made up 3% *w*/*w* of the mass flow rate. They also discussed that, despite typical behaviour, the smaller (and more cohesive) grade of lactose did not affect mixing performance at either the high or the low rotation speeds. It was observed that while agglomerates were readily formed by the cohesive material, these were relatively weak and agitation from the impeller was sufficient in breaking the clusters up, dispersing the powder.

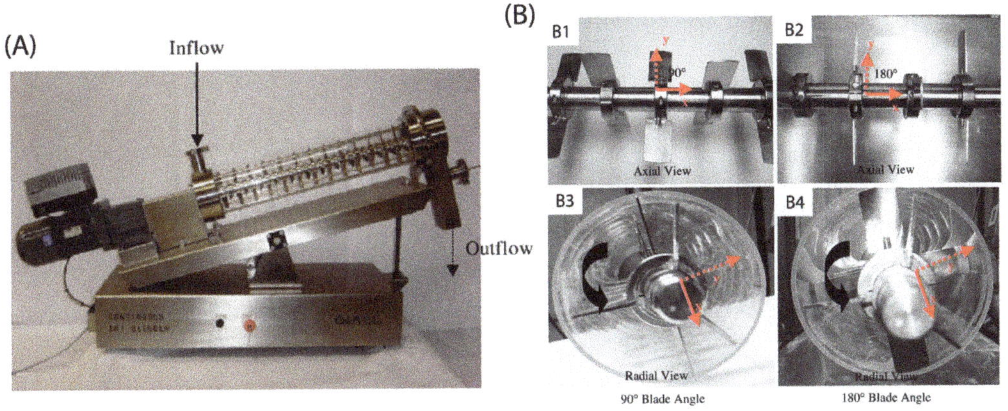

Figure 9. Example linear blender used in Portillo et al. [91]. Profile view of the blender (**A**) view of the agitator shaft and blade angle orientation (**B**).

Later work by the same group [78] discussed the influence of cohesivity on powder behaviour in a different continuous mixer (albeit similar to the blender shown in Figure 9), inclined at +17°. The study used two different grades of lactose, Fast Flo and Edible, which displayed flow indices (measured using a Gravitational Displacement Rheometer) of 24.9 and 34.8 respectively, suggesting them both to be highly flowable powders. In addition, the bulk density tapped density of the Fast Flo Lactose were 0.626 (g/mL) and 0.704 (g/mL), respectively, while the Edible Lactose measured 0.629 (g/mL) and 0.981 (g/mL), respectively. Portillo et al. [78] found that cohesivity negatively impacted axial mixing, becoming a statistically significant factor at higher levels of agitation; thus demonstrating cohesion's effect in resisting the separation or relative movement of particles. In addition, the authors observed that particles with high cohesion experience longer residence times, with similar path lengths, suggesting the particles move at a slower velocity through the blender. This postulates that methods which increase the shear rate, and therefore the relative particle velocity, may combat the effects of cohesion by increasing dispersion [92], and consequently improve micro-mixing. This could also be described as increasing the dynamic granular temperature as a function of the increased shear rate [93,94]. This finding may, at face value, seem to contradict the findings discussed in the previously mentioned study by Lopez et al. [55], in which the translational velocity of the powder going through the conveying portion of the system was observed to be higher with increasing cohesivity. However, despite the *net* velocity being higher, the *relative* velocity (which drives micro-mixing) may nonetheless be lower.

It should be noted that any residence time distribution (RTD) data will be a function of both the volume % of the tracer and the interaction of the tracer within the bulk that it aims to measure. For more detail on RTD behaviour in continuous blenders see

Escotet-Espinoza et al. [61,80], where the authors provide a comprehensive two-part study, analysing the behaviour of tracers with different properties.

Vanarase et al. [89] investigated the blending performance of a horizontal linear blender, both with and without the addition of a co-mill. The authors stated that under optimal macro-mixing conditions, micro-mixing was identified as the limiting factor, leading to poor mixing performance. The inverse was also found for optimum micro-mixing conditions. This highlights the independence of these two types of mixing, and the consequent requirement, to balance a blending system's capability to deliver both sufficient micro and macro mixing.

Interestingly, research by Portillo et al. [91] contains contradictory findings, when compared to Vanarase et al.'s study [89]. They found that using an incline blender increases the overall residence time when compared to a horizontal system—see Figure 10. Thus, for the same residence time as was demonstrated by Vanarase et al. and therefore similar macro behaviour, Portillo et al. showed that the agitator speed could be increased, which in turn provides greater shear, achieving improved micro-mixing conditions with cohesive powders.

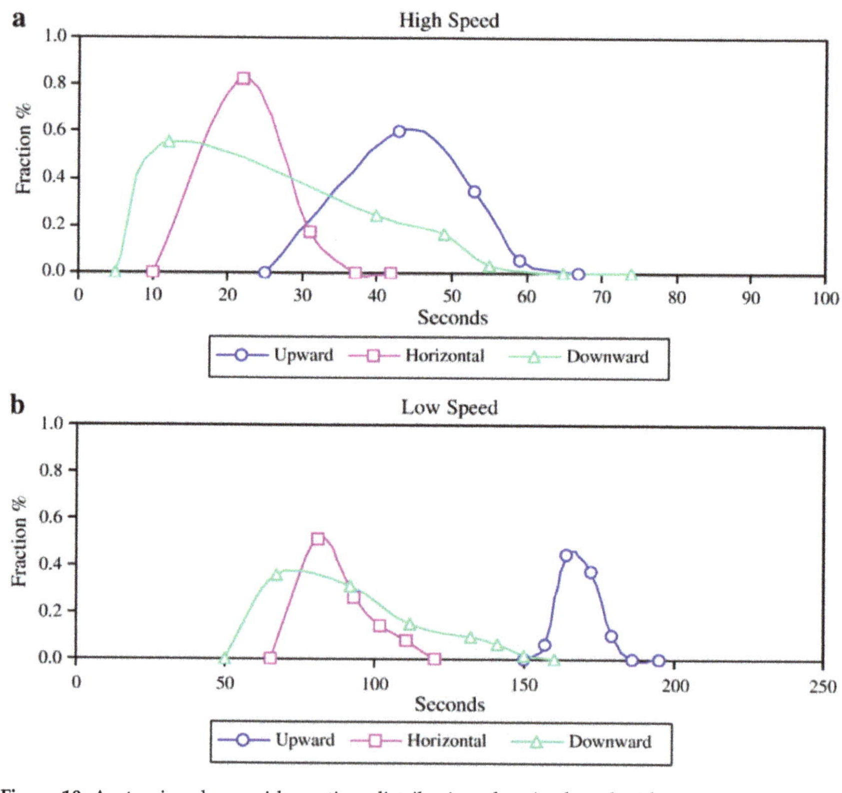

Figure 10. Acetaminophen residence time distribution plots (gathered with permission) from Portillo et al. [91], demonstrating the effect of blender volume inclination and RPM on RTD; (**a**) 78 RPM, (**b**) 16 RPM.

Van Snick et al. [43] investigated the performance of different mixing blade configurations (Figure 11, graphically demonstrates the orientation of blades along the agitator's axis) and found that despite the 'P16' blade configuration not delivering the highest dispersion, it offered the best consistency (see Figure 12). The 'P16' blade config' utilised a series of two sets of transport blades (45°), followed by a set of radial mixing blades (90°). Whereas the other example—'D8'—delivered the highest powder dispersion, with the downside of higher variability. This is supported by Vanarase et al.'s [89] findings, with both surmising that a balance is required between micro and macro mixing. Based on 'P16's performance, in Van Snick et al. [43], it seems as if the 'D8' configuration provides a surplus of dispersion-lending to poor macro conditions. The authors also attribute the good performance of 'P16' to the increased mass hold-up within the blender's volume, as the mixing blades are shown to increase fill level, and therefore total mass hold-up [43]. Ultimately, the study suggests that varying position, number, and angle of the blades could aid in controlling API within the blending phase, and that different blade configurations may be selected according to the formulations' properties, such that the desired throughput is attained, whilst output variation is minimised.

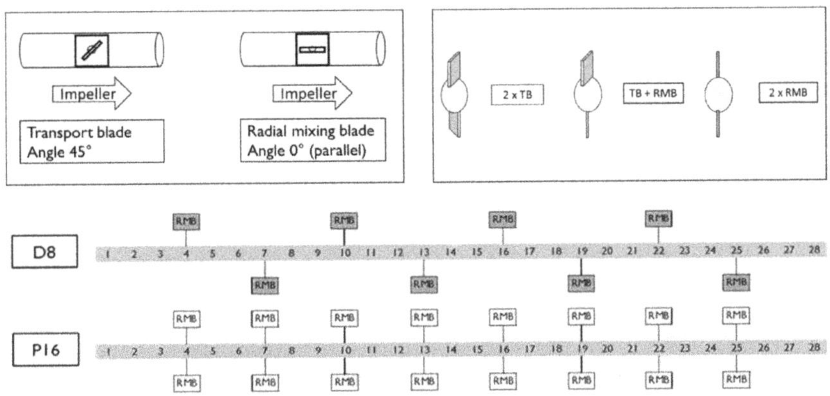

Figure 11. Blade configurations refer to a series of different blade orientations along the agitator's axis. (**Top left**): side-by-side comparison of transport blade and radial mixing blade (RMB). (**Top right**): example of a blade combination that would take a position along the agitator's axis, denoted by a number in the configurations D8 and P16 below. D8 and P16 depict blade configurations; all positions that are not labelled as RMBs are transport blades, and each collar position sees the blade combination rotate 60° clockwise to the direction of the impeller. Adapted from Van Snick et al. [43].

Gao et al. [95] developed a two-blade (periodic) section of a continuous horizontal mixer (see Figure 13), and evaluated mixing performance using a monodisperse DEM simulation of the blender. Gao et al. [95] subjected four different simulated powders (classified as: control, lower density, larger particle size, and cohesive) to the same tests, and mapped the axial conveying efficiency in fill level-rpm space. This conveying efficiency was quantified by normalising the mean axial velocity of particles with the rotor speed and effectively giving the mean distance conveyed per revolution. The powders were subjected to various fill levels (from 25–75%) and a range of rpms (50–250). Figure 14 shows that the relationship between normalised axial speed, and fill level/rotational speed was very different for the cohesive simulation: axial velocity is lower, when compared to the control and to the other powder types, especially at low speeds (50–125 rpm) with a moderate-to-high fill level (50–75%). In addition, the cohesive particles achieved a max velocity of only 38 mm/revolution, the lowest of the four samples [95]. These results are in agreement with the discussion gained from the work of Portillo et al. [78] work.

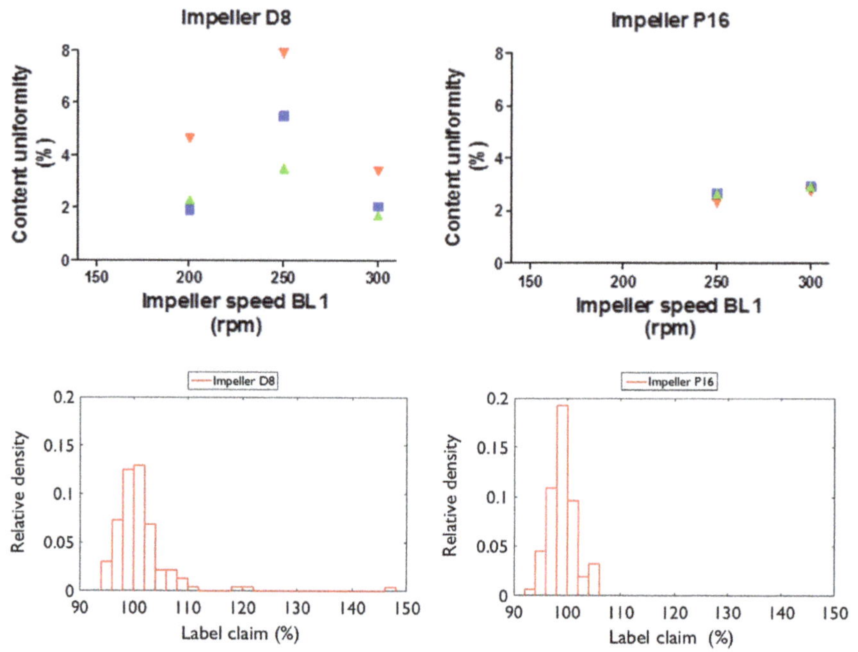

Figure 12. Tablet uniformity corresponding to the blade configurations seen in Figure 11 (D8 and P16); Upper graphs: content uniformity as a function of impeller speed and mass rate (green triangle = 24 kg/h, blue square = 30 kg/h, red triangle = 36 kg/h). Lower graphs: relative density distribution based on the target API dosage at the centre point of the experimental design (250 rpm and 30 kg/h). Gathered with permission from Van Snick et al. [43].

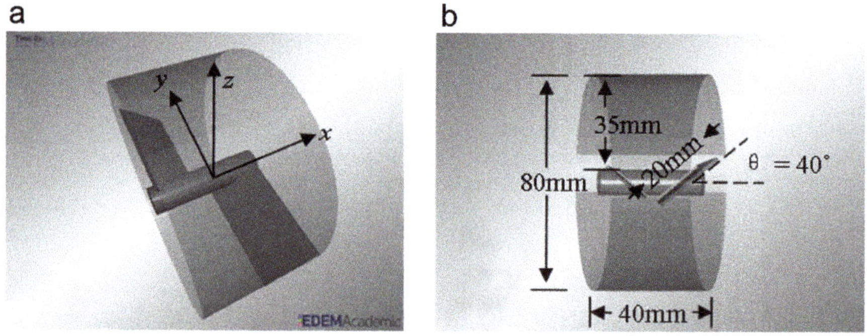

Figure 13. Periodic section of a continuous blender including two blades (**a**) Isometric View, (**b**) Side profile view with dimensions. Gathered with permission from Gao et al. [95].

It is worth noting that, for all non-cohesive powders used in Gao et al.'s work, when going from high to moderate fill levels (i.e., from 75% to 40%) at 100 rpm, normalised axial velocity increases. Cohesion is still shown to have an effect at higher rpms, demonstrated by the increased distance between the contours. This finding corroborates with the rpm-cohesion-significance result from Portillo et al. [78], clearly showing that the relative effect of cohesion will depend on operating conditions (speed and fill/mass rate), and not necessarily in a predictable way, reinforcing the need for more work in this area. This finding, surrounding the three non-cohesive powders, was further supported by the work of Sarkar and Wassgren [96], where the normalised axial velocity presented the same trend, despite different geometry, particle properties, and operating conditions.

Gao et al. [95] also mapped the continuous blending rate (k_c), which is the ratio of the normalised mean axial velocity to the time-dependent RSD decay rate, where k_c essentially details the ratio between the axial (macro) and radial (micro) behaviour of the system. The results obtained were a similar contour plot to Figure 14, but in terms of k_c. Across the different particle properties, it was shown that high RPM and low fill produced the greatest k_c. Moreover, there was little difference seen in k_c at <125 rpm irrespective of fill level. For rpm > 125: going from high-to-low fill showed a sluggish increase in k_c, as the space between contours and colour changes progressed slowly. This highlights that cohesion does not only affect the axial aspect of the mixing process, but also contributes to the hampering of radial mixing. Finally, Gao et al. [95] suggest further investigation on segregating materials, which would prompt several questions on a constituent's experience within the bi-or-polydisperse system. Particularly, it is worth considering what might be the best excipients to pair with different levels of cohesive materials, in order to best aid their macro- and micro-mixing. This has the potential to result in a CDC formulation, of the same drug, differing from their original batch formulation. Different excipient(s) may offer a significantly improved CDC performance, without impacting the tablet's therapeutic efficacy.

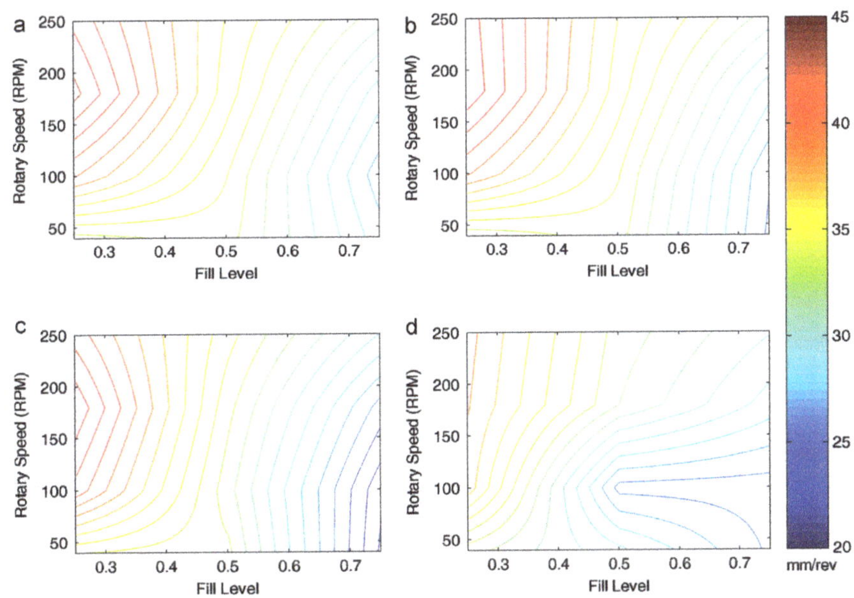

Figure 14. Contour plots of normalized mean axial velocity: (**a**) control, (**b**) lower density, (**c**) larger particle size, and (**d**) cohesive particles. Gathered with permission from Gao et al. [95].

Tomita et al. [97] utilised an atypical blending set-up, operating with both an impeller and a scraper in a horizontal mixer (see Figure 15). The inclusion of the scraper allowed for the control of the mean residence time in a manner that—unlike the previously-discussed systems—was quasi-independent of the magnitude of the impeller's agitation. In this system, the rpm of the axial impeller was driven separately to the ribbon-shaped scraper, which circumscribed the internal wall of the blender. Figure 15B shows the RTD curves of spiked acetaminophen using different scraper rpm speeds, whilst running at a feed rate of 10 kg/s and constant axial agitator speed of 3000 rpm. There is a considerable difference in the RTD performance between the slowest (5 rpm), and highest (50 rpm) scraper speed, which delivered a mean residence time of 84.3 s and 12.1 s, respectively. The mixer provides a method of achieving higher micro-mixing whilst managing the macro-mixing through either increasing or decreasing the mean residence time. A similar result could be achieved with both the horizontal, or incline, blender through the alteration of either the angle of inclination, and/or the quantity of blades, blade angle, and/or blade positions. These design choices are, however, less practical to implement within a feedback loop or control strategy—though the incline angle of a blender may conceivably be varied dynamically.

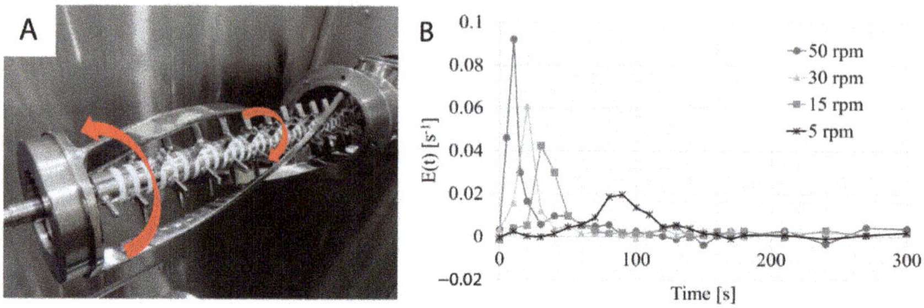

Figure 15. (**A**) Impeller and scraper continuous mixer with arrows detailing rotation direction, (**B**) APAP RTD spike response. Adapted from Tomita et al. [97].

Despite the advantages of an incline blender from a high-shear perspective, practical considerations must be made when attempting to maximise mixing capabilities. For example, the use of two units in series (horizontal blender and co-mill), as shown in Vanarase et al. [89], would allow independent, de-coupled, control in real-time. This reduces the risk that the process would leave a state of control, as each unit would have the capacity to compensate for the other. Conversely, if the incline blender is the only unit responsible for delivering both micro and macro-mixing, given it has the capacity to do so, it is likely the domain of control will be smaller when compared to the two units in series. This is because the process parameters controlling the micro-mixing also control the macro-mixing, and so there is a smaller window for advanced process control to operate within to ensure the mixing balance is maintained. This, therefore, showcases the strength of Tomita et al's [97] blending unit, as it possesses the ability to independently influence the macro and micro aspects of mixing, in real-time, using process control.

Palmer et al. [75] evaluated fill level and content uniformity across a range of processing parameters in an inclined (+15°) continuous blender. These processing parameters include: throughput, number of mixing blades, and agitator RPM. The study then explored the same parameters, using the same formulation, with different grades of API (APAP)—see Table 1. The result is the proposed exponential decay model (see Figure 16), and micro-mixing model (see Figure 17), which uses both the Peclet Number (Pe, see Equation (4)) and the strain (ϵ, see Equation (3)). In addition, Figure 18, showcases a series of contour plots describing the effect of processing parameters on the fill level, for each APAP formulation. The study explored a range of Froude numbers (Fr) between 1.5, and 13.4 (for 150 and 450 rpm, respectively).

Table 1. Material properties of the APAP grades used in Palmer et al. [75].

Material	FFC	Bulk Density ($g\ cm^{-3}$)	Tapped Density ($g\ cm^{-3}$)	Hausner Ratio	Carr's Index
Micronized APAP	1.4	0.19	0.30	1.58	36.67
Powdered APAP	1.9	0.31	0.53	1.71	41.51
Special Granular APAP	19.4	0.73	0.83	1.14	12.05

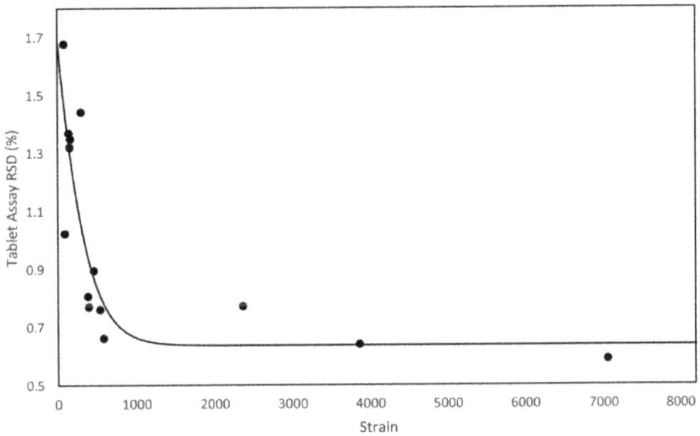

Figure 16. Tablet assay RSD plotted against Strain (see Equation (3)), an exponential decay model is fitted to the data. Gathered with permission from Palmer et al. [75].

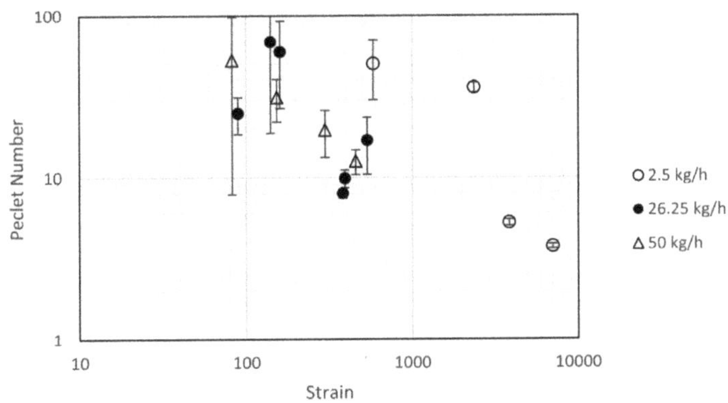

Figure 17. Peclet Number (Pe) and Strain relationship plotted on a log–log graph, grouped by throughput. Gathered with permission from Palmer et al. [75].

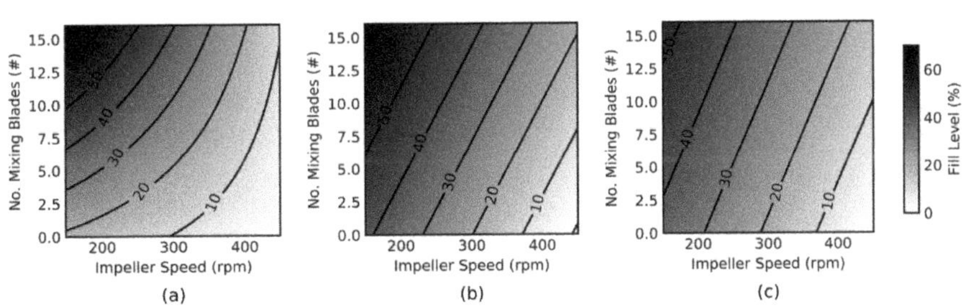

Figure 18. Contour plots mapping the impeller speed, number of mixing blades, and fill level at a throughput of 26.25 kg/h (**a**) Micronized APAP, (**b**) powder APAP, (**c**) special granular APAP. Gathered with permission from Palmer et al. [75].

Comparing the bulk characteristics in Table 1, Micronized APAP (mAPAP) has closer bulk characteristics to Powdered APAP (pAPAP), but shows a unique shape contour in Figure 18. Despite pAPAP and Special Granular APAP (sgAPAP) being opposed in flowability indices, they present similar contour shapes. This indicates a similar fill-level, and therefore bulk powder response, to the number of radial mixing blades (see Figure 11) across a range of RPMs. For the mAPAP, however, increasing the number of radial mixing blades has more impact on the fill level at lower RPMs, and this impact reduces with increasing RPM. Van Snick et al. [43], discussed the necessity of sufficient mass hold-up (fill level) to attain good levels of macro mixing. The combination of these two studies suggests (i) that it is critical to determine the fill level landscape for the desired formulation to achieve and maintain sufficient mixing, and (ii) that different highly cohesive powders would be expected to exhibit different complex behaviours. What is more, using the mAPAP contour from Palmer et al. [75], considerations have to be made regarding what variables can be controlled in real-time. Since the radial mixing blades cannot be changed in real-time, it could be inferred from Figure 18, that using the highest number of mixing blades would be more optimum, as it would allow the system to have the largest range of fill levels.

Furthermore, Figure 16, from Palmer et al. [75], models the exponential decay of content uniformity RSD with increasing Strain (Strain, Equation (3)). Strain is a confounded variable for this system (an incline (+15°) linear blender), as increasing the RPM decreases the mean residence time, meaning that the number of blade passes also affects the mean residence time [75,91]. Thus, strain is dependent on the throughput, RPM, and the powder's material properties, making it essential to understand how the powder responds to processing parameters. Understanding in this area can be gleaned from the work of Portillo et al. [78,91], Gao et al. [90], and Lopez et al. [55] regarding powder velocity, fill level, and resultant changes in residence time due to cohesion. More cohesive powders would be subject to longer residence times, but lower relative velocities. Therefore, more cohesive powders should be more resistant to increasing RPM in turn reducing cohesion's effectiveness; Figure 18 demonstrates the added complexity. Ultimately, strain is a great, yet elusive, measure of effective work done to the powder. The model shows there is an optimum amount of strain required (around 1500) to attain a high-quality blend.

Lastly, Palmer et al. [75] shows a linear log–log relationship between Peclet Number (Pe) and strain, see Figure 17. Pe is a dimensionless number describing the ratio between advective and diffusive mixing in the system (see Equation (4)). To give an example, a high Peclet number would be expected to produce a narrow RTD. The model, at higher throughputs (and therefore, higher residence masses), showed reduced scattering and more linear behaviour, implying that the model may be favoured in predicting systems with high residence mass, whilst also implicitly suggesting that systems with low residence mass may present unstable micro-mixing capability and therefore may be unfavourable. Despite this, across all throughputs, it was seen that with reducing strain, the Peclet number increased.

$$Strain = \epsilon = MRT * \omega \tag{3}$$

where MRT is the mean residence time and ω is the agitator rotation rate.

$$Peclet\ Number = Pe = \frac{advective\ transport}{diffusive\ transport} = \frac{uL}{D} \tag{4}$$

where u is the flow velocity, L is the characteristic length and D is the mass diffusion coefficient.

$$Froude\ Number = Fr = \frac{centripetal\ force}{gravitational\ force} = \frac{\omega^2 R}{g} \tag{5}$$

where ω is the blade rotation rate, and R is the geometrically relevant length-scale—i.e., the radial distance from the centre axis to the tip of a blade.

While, as outlined above, strain is considered to be beneficial to powder blending, it has also been shown that powders which experience high levels of strain (number of blade passes) are also more likely to be subject to triboelectric charging [13]. Naturally, the effect will present some interest in how electrostatics influence/interact with powder cohesion. Karner and Urbanetz [13] investigated the influence of pharmaceutical mixing in a batch mixer and found that particle size, the fraction of fine particles, and mixing volume were significantly impacting the charging behaviour of the powder, stating that small particles were more prone to generate higher charge density. How these findings would translate to a continuous blender represents a potentially interesting and valuable line of inquiry.

Beretta et al. [15,47] performed a two-part study on the triboelectric effect of charging powder via. powder transport in twin-screw feeding. This was followed by a study using relative humidity (RH) to subdue the electrostatic effect. Given that it is typical to maintain a RH between 30–60% in ISO-graded GMP environments [16] and that additional air moisture can increase the presence of liquid bridging, promoting cohesion [57], it becomes difficult to assess purely on the grounds of cohesion. The downside of controlling relative humidity is that it becomes difficult and expensive to maintain HVAC systems, especially as temperature and humidity fluctuate across the year. There are other options to consider, such as silication, outlined by Lumay et al's study [57], which was shown to be capable of electrostatic dissipation and localised relative humidity control. Therefore, the use of silicates becomes a much cheaper and simpler option, despite what relative humidity may offer. However, this assumption is limited by current knowledge and awareness of the author(s) at the time of composing this review—novel literature may render this assumption moot.

Beretta et al. [15,47] stated that the charging forces are primarily recruited through particle–particle friction, while particle–wall friction did not directly contribute to charging. In addition, the authors made considerations for impacts on powder mixing; interestingly, the discussion touched on utilising these static forces to aid the mixing process [47]. Huang et al. [70] provide evidence of this, as dry powder coating is seen in a high-shear co-mill. Huang et al. [70] investigated the effect of dry powder coating of different APIs and excipients and found that the coating of fine colloidal silica improved both bulk density and FFC performance. SEM images displaying the silica coating are shown in Figure 19. Therefore, charge- or cohesion-based coating could provide an avenue to attain what Fan et al. [83] (see Figure 8) would describe as perfect ordered mixtures.

Figure 19. SEM images showing the surface coatings of (**top**): excipient (lactose) and (**bottom**): API (Ascorbic Acid). (**Left**): Non-coated and (**right**): 1% w/w Aerosil® R972 colloidal silica coating. Adapted from Huang et al. [70].

Lastly, alternative silicates—such as magnesium aluminosilicates (MAS)—have been trialled in some flowability studies; demonstrating an increase in flowability, without a constraint on mixing time and intensities, which is a limitation of several glidants [98]. Such materials could be used in place of the more traditionally used MPS grades for processes which may operate with different materials or mixing regimes.

3.3. Blending Summary

The scope of this topic is multidimensional, and therefore increasingly difficult to surmise without reducing the complexity of the situation; despite this, several trends—and avenues for future work—have emerged throughout this review.

Firstly, and perhaps most obviously, the literature strongly suggests that, regardless of differences between continuous blenders, cohesion provides a barrier to dispersion, and thus a barrier to achieving micro-mixing [78,90], which therefore decreases the quality of the mix. Cohesion has also been shown to exhibit complex behaviour at higher fill levels and lower rpms [75,78,90], where the particle–particle contacts are expected to be at their highest.

There is a common theme across the literature emphasizing the need to devise solutions to mitigate the effects of cohesion through the optimisation of processing parameters [43,45,75,88,99]. The literature suggests multiple manners in which this may be achieved, for example through alterations in system or agitator geometry, agitator rotation rate, feed rate, and various other factors. The differences in geometry for each blending system, however, means that each system has its own approach to solving the cohesion problem, thus making it difficult to generalise findings, and methodology, when attempting to understand the mechanisms and behaviour which explicitly govern cohesive blending. This is then exacerbated when considering that different formulations respond differently under the same conditions. An important task for future research is to define common parameters—e.g., dimensionless numbers—which can successfully generalise previous findings.

In the literature to date, several variables have been determined to affect the influence of cohesion on mixing. These include the Froude number (Fr, Equation (5)) or rpm, the magnitude of powder cohesivity, and the fill level of the blender. It has also been widely suggested that, due to cohesive particle–particle contact, a given system will require either a higher shear or more frequent agitation to break apart the additional cohesive forces. Strain (ϵ, Equation (3)) is defined as the mean residence time multiplied by the rotation rate of the agitator (RPM), and is a measure used to quantify the amount of work the agitator does per unit of space-time (τ) [43,75,82,88]. This quantity has been suggested as a valuable indicator of the degree of shear in a system and thus of blending success [43,75,82,88].

It has also been suggested that cohesion loses its significance, and complex behaviour, when rpm increases, implying that by running a continuous blender at high RPM, one may improve (micro-)mixing. However, the work of Van Snick et al. [43] suggested the need for a sufficiently large mass hold-up in order to ensure good macro-mixing. However, these seemingly contradictory requirements can potentially be reconciled by ensuring that RPM is high enough to minimise cohesion's effect, thus improving micro-mixing, whilst altering the system's geometry (or other system-specific variables) to ensure a suitable amount of hold-up mass is retained, thus retaining good macro-mixing.

Regarding future work, it would be interesting to see where the lines of significance lie for each of the three cohesion-dependent variables and their interactions. Such a study would most likely have to be conducted via DEM and should follow the suggestions left by Gao et al. [90]. In general, there is potentially significant value to be found in using numerical modelling methods such as DEM to address a number of the experimental observations discussed in this section.

A Perspective on Continuously Blending Cohesive Species

A final thought on continuous blending; is that the tablet's powder formulation could be thought of as an emulsion, whereby work (in the form of strain) is given by the blender to shear through existing API clusters, reducing their size, and promoting the distribution and circulation of the new, finer API clusters around the bulk mass. The cycle then continues, by imparting even-more work and further distributing the API clusters. The remainder of the tablet's formulation (which is mostly excipient) should, therefore, promote and stabilise the small API clusters, ensuring that they are maintained before they receive more work. Similarly, emulsions require thickeners and surface active agents to develop and maintain immiscible fluids [100]. Formulation development for CDC processes could look to use the same methodology, namely, through the exploitation of dry powder coating [70] and electrostatic charging [13]. However, considerations have to be made to ensure that, if the API is coated, it does not affect the therapeutic efficacy of the tablet.

4. Tabletting

4.1. Tabletting Introduction

The tabletting unit operation consists of three main steps:

- Die filling;
- Compaction;
- Tablet ejection.

The tabletting unit operation starts with the die-filling stage which fills the die with a consistent amount of powder mixture to achieve an adequate fill depth ready for the compaction step while also keeping good drug content uniformity [101,102]. The upper punch then lowers into the die increasing the compressional pressure applied to loose powder in the tabletting die; the powder consolidates as particles rearrange, filling void spaces and finally, as the availability of void spaces reduces, particles deform elastically, and ultimately plastically, before beginning to fragment [21] once the target pressure has been reached the upper die lifts releasing the pressure off the compact. Finally, the lower punch lifts the tablet to the top of the die ejecting the tablet. However, die filling and compaction are strongly influenced by cohesion whereas tablet ejection is not [103]. Qu et al. [103] study showed that additives such as silica which lowered cohesion did not lower ejection stresses effectively. Therefore, as this literature review focuses on cohesion, this section will only discuss die filling and compaction.

4.2. Tabletting Discussion

4.2.1. Die Filling Introduction

As mentioned previously, die filling is a critical process step for tablet manufacture as mass and content uniformity are heavily dependent on the die-filling performance. If the die filling is not consistent, the mass and drug content uniformity is not accurate [104]. This will have an impact in the downstream processes as the quality attributes such as tensile strength and the dissolution profile will be affected [102]. The die-filling process is known to be affected by many bulk powder properties: flowability, cohesion, particle size, and morphology [101,102]. Many studies have shown a good correlation between these bulk powder properties and the die-filling performance [27,28,104]. Normally, dies are filled gravitationally using a shoe-die system where power is dispensed from a feed hopper which is connected to a box-shaped shoe which then moves over the opening of the die and allows powder to flow [38] (Figure 20). However, there are different mechanisms for die filling such as:

- Forced feeders [102] which use moving components (usually paddles) to feed the powder into the die. This is commonly used in turret presses [105] (Figure 20);
- Suction filling is where the lower punch also known as the 'suction punch' moves down creating a negative air pressure gradient promoting powder to move into the

die, promoting the movement of the powder into the die [106] which removes the effect of air entrapment [38,107] (Figure 20).

Similar to Sections 2 and 3, to have a good die-filling performance, the powder will need to have lower cohesive properties [3] These include larger particle size, spherical particle shape powder with lower surface energy. (Lower surface energy will mean that powder will find it more difficult to form strong interparticle bonds [103,108]), which all contribute to improved flowability [27,38,107]. To have content uniformity, the segregation must be kept to a minimum during the die-filling stage so as not to undo the work done during the mixing stage [107,109]. This section will focus on the die fill performance where the mass, segregation and fill depth are important to ensure the compression process will be successful.

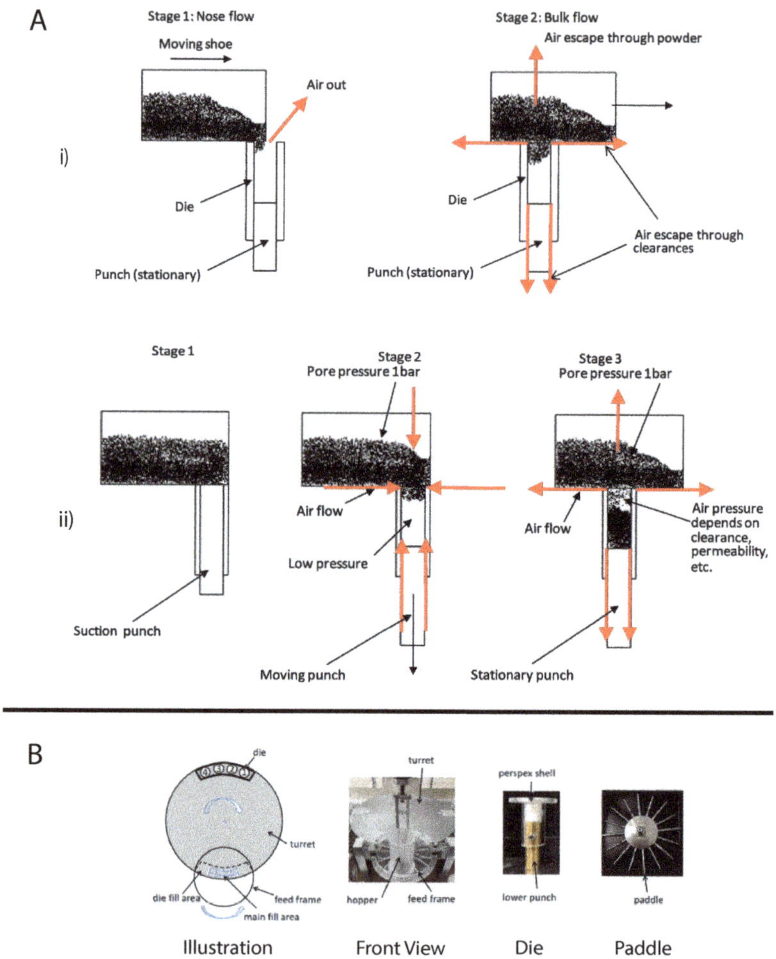

Figure 20. (**A**) Schematic diagram of air flow mechanism in (**i**) gravitational die-filling mechanism (**ii**) suction fill mechanism in a shoe and die system obtained from Mills and Sinka [38]. (**B**) Schematic of a rotary die-filling system, a type of force feeder, typically found in a turret tabletting machine obtained from Tang et al. [105].

4.2.2. Die Filling Discussion

Cohesion has been seen to affect the die-filling performance in many studies as the flowability of the powder is reduced with increasing cohesivity [27,110]. As mentioned previously, CDC processes have fewer unit operations as the granulation step is bypassed. Granulation is where powder fines which have poor flowability are made into agglomerates. The powder fines adhere to one another commonly using a binder fluid (wet granulation) to form a larger multiparticle entity which are also known as granules [111]. The granulation step helps control key attributes such as flowability and the processability of the powder and therefore controls cohesion [112]. However, as the CDC process does not have a granulation step, the tablet formulation will need to have adequate flowability and tabletability [3] which limits the amount of tablet formulation that can be used with the CDC process. However, there are die-filling mechanisms which can improve the die-filling performance of poorer-flowing powders without the granulation step.

Wu [110] established 'critical velocity', which is defined as the maximum velocity which the shoe passes over the die with the die being filled after one pass [27]. The critical velocity is a common way to measure the flowability of the powder. A strong correlation is seen between critical filling velocity and mean particle diameter; specifically, an increase in particle diameter and a decrease in surface energy increases the critical velocity (Figure 21) [101]. This indicates that when the powder is more cohesive, the critical velocity is lower as the flowability is poorer, which means that cohesivity can have a negative impact on die-filling performance and cohesion control measures will be needed.

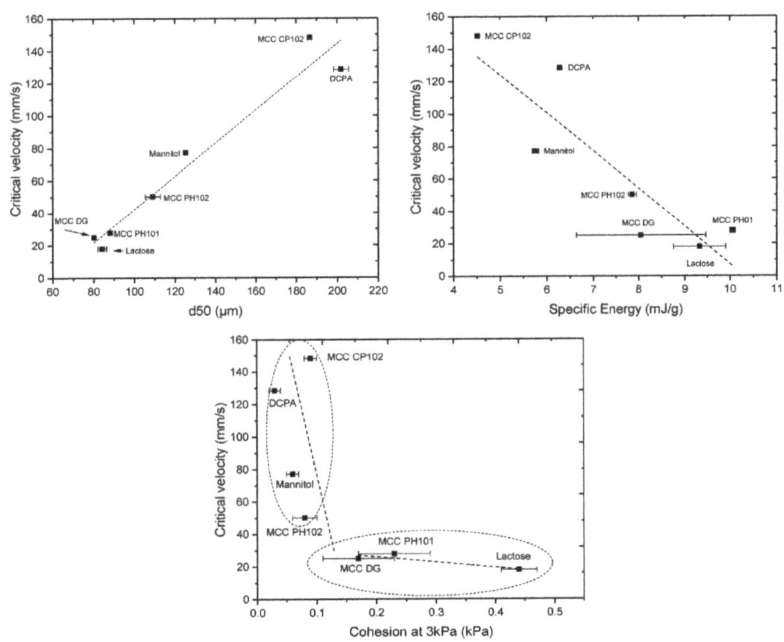

Figure 21. Graphs obtained from Zakhvatayeva et al. [101] showing the change of critical velocity as a function of: average particle size (**top left**), specific energy (**top right**) and cohesion at 3 kPa consolidation stress measured using the FT4 (**bottom**) of a variety of different materials.

The effect of paddle speed in forced feeders was investigated by Goh et al. [107] which saw with a higher paddle speed there was an increase in die fill weight and reduced die variation. However, it was seen in other papers that the paddle speed only had an effect on the weight variability, not the fill weight itself [84] and only affected fair flowing materials such as MCC [113]. Therefore, a higher paddle speed can allow for cohesive powders to 'flow better' as they can essentially move as a solid block. On the other hand, in the study of Peeters et al. [114], results showed that the paddle speed affected the tensile strength. However, for the formulation with just microcrystalline cellulose, the tensile strength outcome was not affected by paddle speed. The only formulations that were affected were with magnesium stearate that saw a decrease in tensile strength with a higher paddle speed. Lubricants can have a negative effect on the tabletability if excessive shear force and higher levels of mixing occur. This is often referred to as 'over-lubrication' in the literature [115]. Therefore, the decrease in tensile strength is most likely due to these lubrication effects which is promoted with a higher paddle speed [116]. To understand the effects of lubrication in a forced feeder with higher paddle speeds, work should be conducted comparing the tensile strength outcomes of formulations with and without magnesium stearate. However, it should be noted that brittle materials are reported to not be as susceptible to the effects of lubricants as much as plastic materials [117]. Therefore, formulations with majority plastic powders will most likely be affected more heavily than formulations with a majority of brittle powders.

Another method to increase the flowability is by introducing air into the system, as flowing air gives a lubrication effect allowing particles to flow more easily with less resistance, increasing critical velocity [27]. The air prevents the percolation of fine particles, which reduces vertical segregation [118]. However, the presence of air can also create an adverse pressure gradient which resists the motion of particles, and pressure built up can further oppose the flow of the powder [27]. Suction filling has also seen positive effects with overcoming poorly flowable powders [38,106,109,119] and helps with the air entrapment which hinders powder flow as it creates a negative pressure gradient [38,107]. Furthermore, suction fill is generally known to lower the risk of segregation and improve the packing density [38]. This is seen in Figure 22 obtained from the study of Zakhvatayeva et al. [109]. The degree of segregation was measured using a sampling unit which separates the die into five sections; the process was repeated twice. The degree of segregation was calculated using Equation (6) where the segregation index indicates the level of uniformity in the powder blend in the die. Although the conclusion in Zakhvatayeva et al.'s [109] study is that suction filling improves die-filling efficiency and reduces segregation, in Figure 22 it is seen that gravitation filling with a fast die-filling velocity had a lower segregation index compared to suction filling. However, the slow die-filling velocity may not be representative of a real-life scenario in industry. Another interesting observation from Figure 22 is that with acetylsalicylic acid concentration of 10 percent, this had the worst overall segregation index across all cases which was explained as generally lower API concentrations are known to have a higher risk of segregation [120].

$$Segregation\ Index = \sum_{i}^{n} \frac{c_i - c_t}{c_t} \tag{6}$$

where c_i is the starting concentration of the blend and c_t is the concentration of the analysed sample.

All these results show that there are promising ways to overcome the effect of particle cohesion on die-filling processes; however, it is often concluded that these results are highly dependent on the particles' material properties [102,107]. Active pharmaceutical ingredients (APIs) are new and complex molecules that will often have unknown material properties [18]. Further studies should measure bulk properties and be able to understand what process parameters need to be utilised during the die-filling stage. APIs are known to be very cohesive and poorly flowing which gives the limitation that formulations for direct compression must have a maximum of 30% drug content [21].

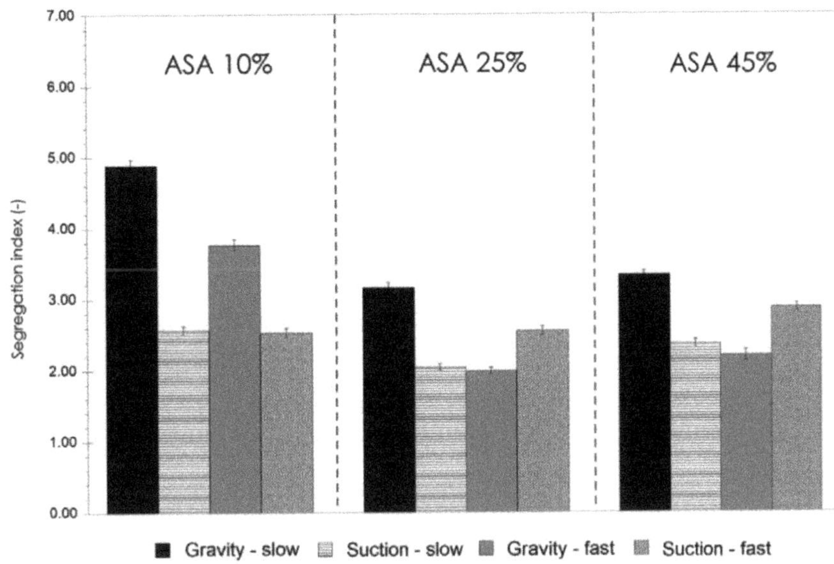

Figure 22. Segregation index of three different blends of Acetylsalicylic acid (ASA) with mannitol using gravity and suction filling where suction filling mostly has a lower segregation index compared to gravity from Zakhvatayeva et al. [109] the fast scenario is where the die-filling velocity is 260 mm/s and slow has a die-filling velocity of 70 mm/s.

A possible method to allow formulations with more than 30% drug content to undergo direct compression is spherical agglomeration/crystallisation, which will improve both processability and tabletability of the API. This is completed by changing the way the API is crystallised by either modifying the solvent addition rate and/or cooling rate [3]. Spherical agglomeration allows the API to crystallise spherically (Figure 23), which increases flowability and therefore lowers cohesion [3]. There are many promising studies using spherical agglomeration/crystallisation such as a study by Chen et al. [18] which saw an increase in flowability and therefore die-filling performance. Furthermore, the tabletability also improved (Figure 23) for the spherical agglomerated/crystallised produced using the quasi-emulsion solvent diffusion (QESD) method. The tensile strength was considerably higher compared to the 'as received' ferulic acid exhibits a needle-like elongated shape (Figure 23f). While spherical agglomeration/crystallisation is beneficial this should not be confused with the negative kind of agglomeration that occurs during secondary manufacture that can have a poor effect on blending, feeding and mixing. It is important to distinguish the two.

On the other hand, there are some disadvantages to spherical agglomeration/crystallisation; for example, solvent selection is a difficult process and may be limiting due to the growing concerns of using more green solvents in industry [121]. This will need to be further researched and optimised in order to be used as a commercial technique [121,122]. Furthermore, the optimisation of process parameters required for the spherical agglomeration/crystallisation method is difficult and will need considerable work to scale up the processes [123]. Therefore, although this technique will help produce many more formulations suitable for direct compression for most drugs to take place, the preparation times at this current state will probably outweigh the benefits of just granulating. However, this technique is very promising and something future research should focus on.

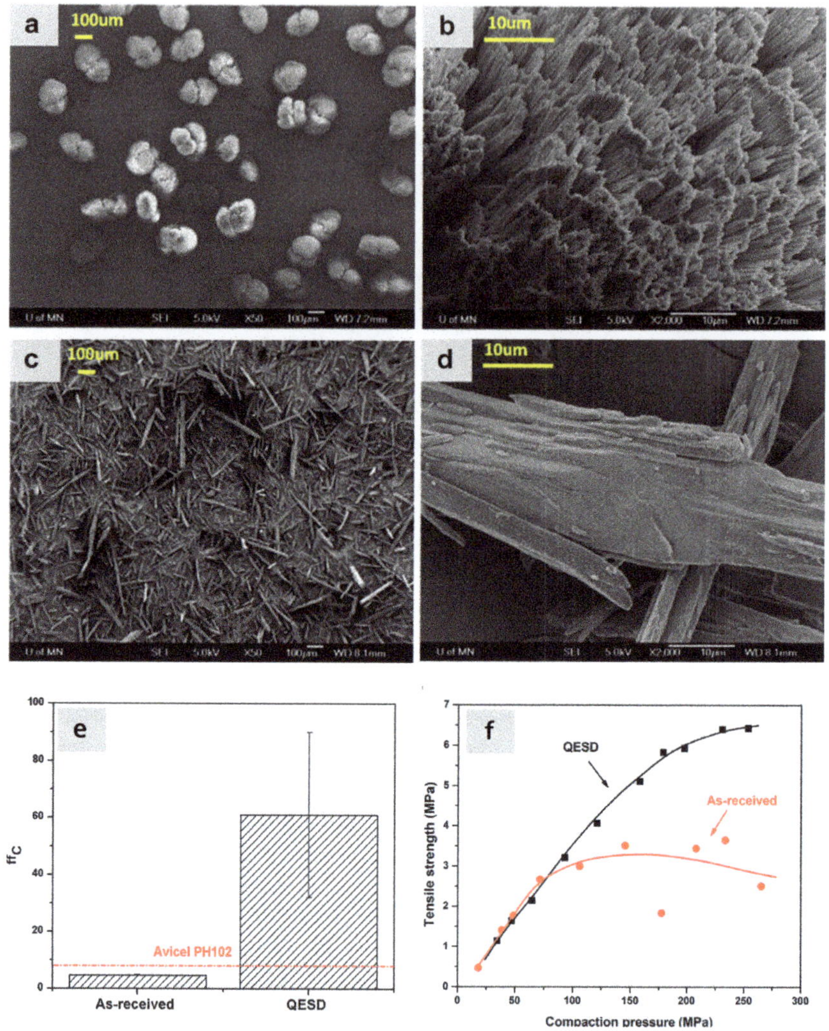

Figure 23. (**a**–**d**): Scanning electron microscope images of ferulic acid (FA) particles (**a**) Quasi-emulsion solvent diffusion (QESD) at 50× magnification, (**b**) QESD powder at 2000× magnification, (**c**) as-received FA powder at 50× magnification, and (**d**) as-received FA powder at 2000× magnification. (**e**,**f**): Figures to show improved tabletability and flowability with spherical agglomerated/crystallised (QESD) ferulic acid obtained from Chen et al. [18].

4.2.3. Compression Introduction

The aim of the compression process is to consolidate the powder mixture consisting of excipients and API into solid and cohesive compacts that can withstand the pressures of coating, packing, and shipping without fracturing. Therefore, the tablets will need to have a high enough tensile strength; however, if the tensile strength is too high this will typically lead to low porosity [30]. If the porosity is too low this could negatively affect the dissolution rate required to deliver the drug to the patient [124]. There are many aspects that contribute to the tablet's strength, such as moisture content [125], compression speed [126], and granulation [127]. However, as this literature review is focusing on the effect of cohesion and assumptions can be made that the tabletting process is completed under strict humidity controls [100], this section will just be focussing on surface energy and particle size and shape. These three particle properties contribute heavily to the magnitude of cohesion [128]. Although cohesion effects for other process operations in the tabletting manufacturing process are known to have a mostly detrimental effect (see Sections 2 and 3), in contrast, cohesion properties are important to the compression as cohesivity helps produce a strong solid compact. Therefore, it is important to understand how the cohesion properties contribute to the tabletting stage so potential compromises can be taken to help the formulation perform well throughout the manufacturing process.

4.2.4. Compression Discussion

The effect of particle size on powder compaction has been the focus of many papers where it is generally understood that smaller particle sizes form a stronger tablet [29–32,129]. Smaller particles have been found to have a larger specific surface area where interparticle bonding can take place resulting in stronger tablets [130,131]. However, smaller particles are typically more cohesive and have poorer flowability which will affect die filling (Section 4.1).

Particle shape is commonly linked with particle size effects in tabletting and has similar effects on the flowability and cohesivity of the powder [132]. A more irregular shape will exhibit poorer flow properties and be more cohesive compared to regular or spherical-shaped powders [133]. Additionally, irregular-shaped powders have been shown to form a stronger tablet, which was demonstrated by Johansson and Alderborn's [30] study. The authors found that tablets containing irregular shaped MCC granules had a higher tensile strength than spherical-shaped pellets (Figure 24). Both granules exhibited plastic deformation as the dominant deformation mechanism, but at higher pressure, there was some fragmentation/attrition which resulted in a closer pore structure. Šimek et al. [133] investigated the effects of different particle shapes (Figure 24) of a brittle material, paracetamol. Their findings were similar to Johansson and Alderborn's [30], where more irregular shaped paracetamol had increased compressibility shown with the Heckel plot. The Heckel plot is a commonly used equation and plot to help understand the densification of the powder during compaction and allows interpretation of how plastic the powder is which is taken from the yield pressure which is derived from the reciprocal of the linear region gradient of the curve. The lower the yield pressure, the more plastic the material is [134]). Figure 25 compares this to spherical-shaped paracetamol.

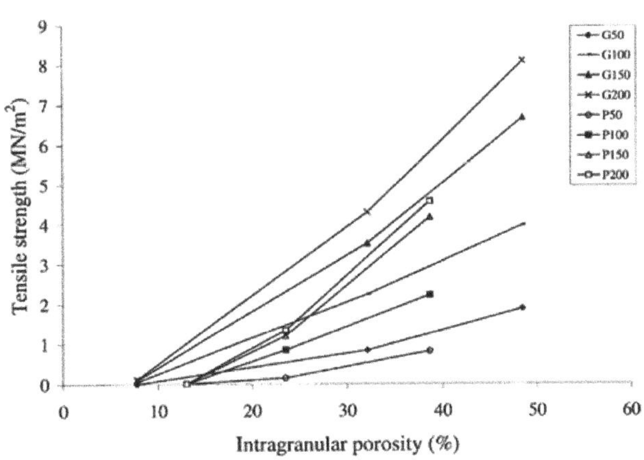

Figure 24. Johansson and Alderborn [30] G = granules (irregular), P = pellets (nearly spherical) the number is the compressional pressure it was compressed to e.g., G200 = granules compressed to 200 MPa.

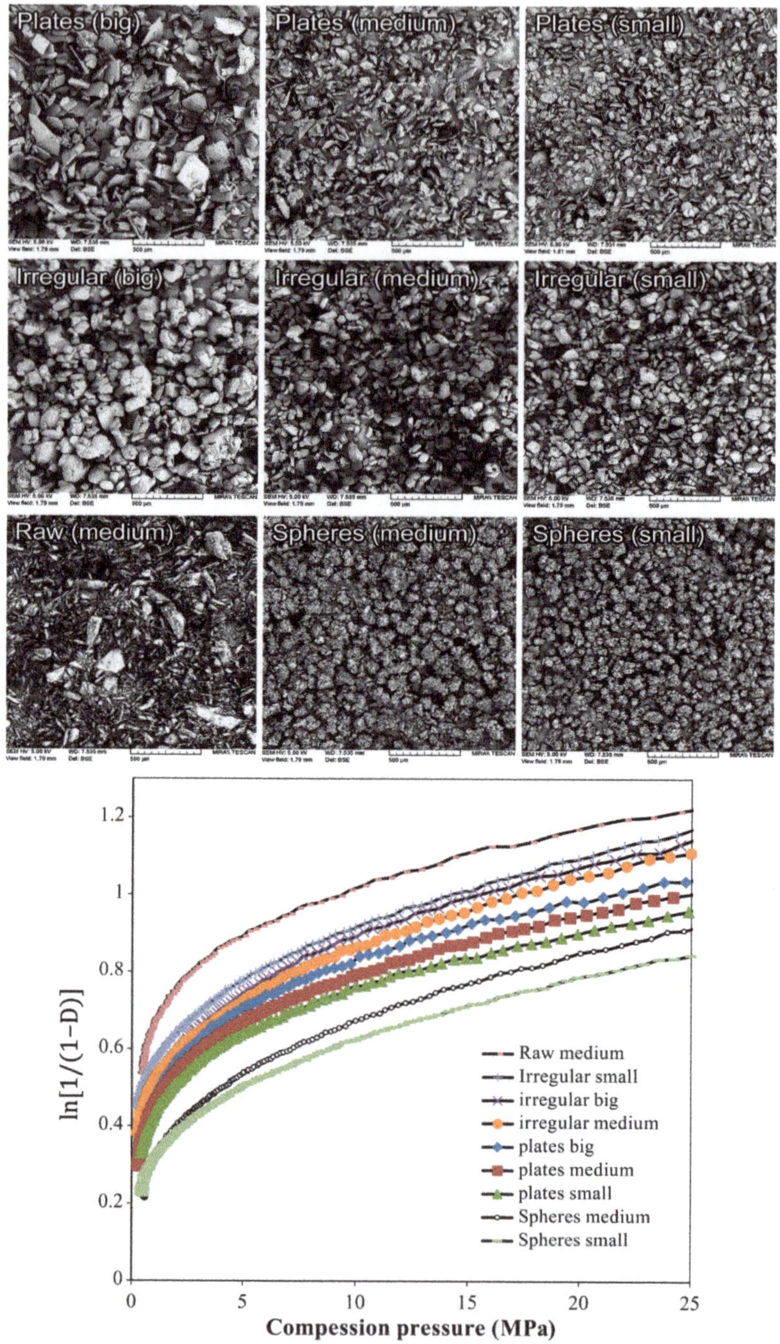

Figure 25. (**Top**): SEM micrographs of the paracetamol samples made (prepared and raw). The prepared samples were dissolved in ethanol and mixed at a set RPM to achieve the modified shapes. (**Bottom**): Heckel plot of different sizes and shapes of paracetamol. Both figures obtained from Šimek et al. [133].

Although the general trends of particle size and shape' effects on tabletting is mentioned in the previous section, there are some conflicting results regarding how they affect the compressibility and therefore tablet strength. This can be differentiated by material properties or the dominating deformation mechanism of the powder. Brittle materials such as lactose and paracetamol deform by fragmentation [135]. In studies, the strength of tablet consisting of brittle materials have been seen to be affected by particle size [29,128,136,137] and shape [138]. Skelbæk-Pedersen et al. [137] found that larger brittle powder particles such as lactose and calcium hydrogen phosphate dehydrate (DCP) fragmented more. In Figure 26, with increasing particle size in DCP and lactose, the peaks are more flatlines for the lines corresponding to the higher compressional pressure which means the particles fragmented into smaller particle sizes more readily. This agrees with De Boer et al. [29] Sun and Grant, [130] and Skelbæk-Pedersen et al. [137]. The more extensive fragmentation allows the void spaces to be filled with the smaller particles which further densifies the powder bed to reduce its negative influence of tensile strength. Nonetheless, smaller particles have higher tensile strength outcomes after compression [130]. Surprisingly, in contrast, Almaya and Aburub [117] found that dibasic calcium phosphate dihydrate did not have a difference with different particle sizes. Suggestions into why this may be occurring will be discussed below.

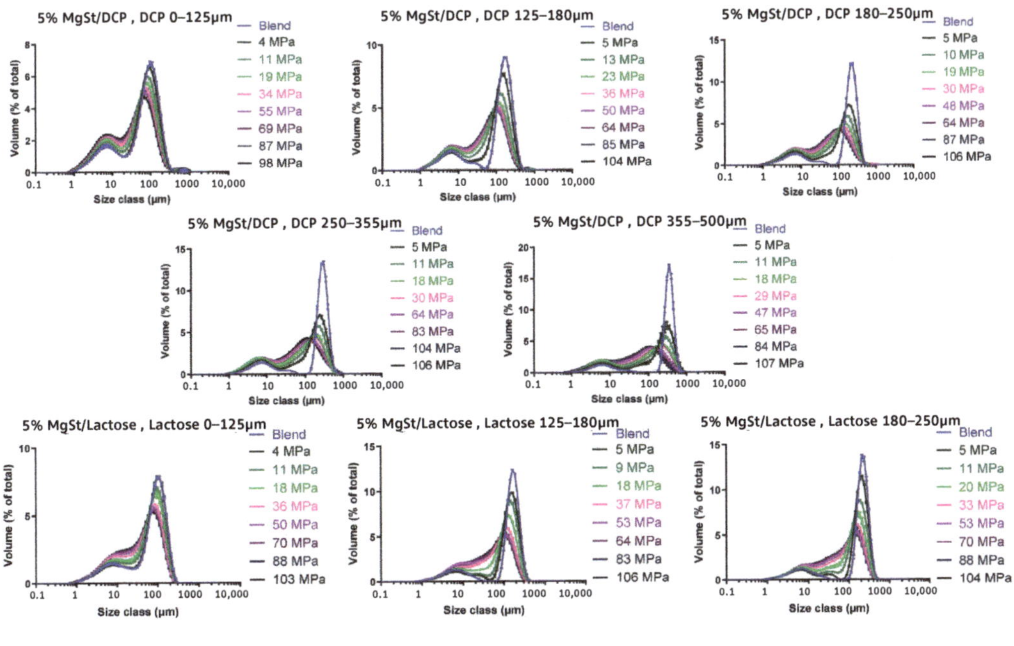

Figure 26. Skelbaek-Pedersen [137] PSD of calcium hydrogen phosphate dihydrate (**top**) and lactose (**bottom**) with different initial particle size distribution in the graph titles. The plots show the change in PSD with increasing levels of compressional pressure.

Plastically deforming powders are known not to be affected by particle size [117,128,136] due to the way they deform. McKenna and McCafferty [128] found that different sizes of microcrystalline cellulose (MCC) did not affect the tensile strength. However, Herting and Kleinebudde [116] found contradicting results that decreasing particle size of MCC and theophylline resulted in higher tensile strength. Wunsch et al. [132] found that there was an increase in tensile strength with decreasing particle size for all materials investigated (MCC and vildagliptin) (Figure 27) However, with vildagliptin, internal defects such as cracks during elastic recovery may have occurred resulting in large error bars. No fundamental understanding of these differing results is given in the papers, particularly regarding the contradictions with past literature.

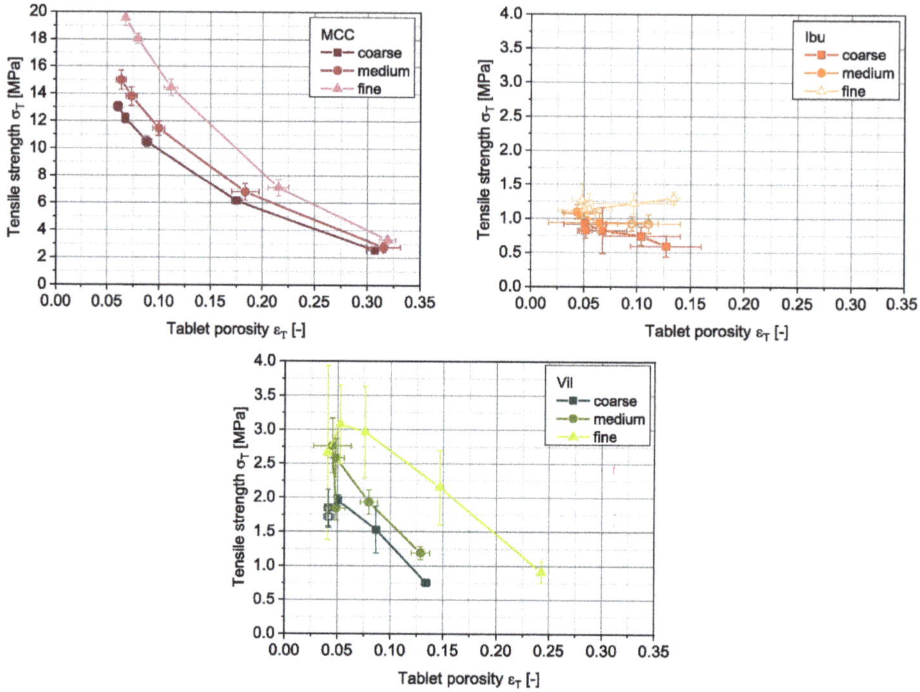

Figure 27. Compactability graphs of coarse, medium and fine grades of MCC, Ibuprofen and vildagliptin showing the change of compactability with particle size, obtained from Wunsch [132].

Despite these contradictions, which need to be further understood, a common conclusion can be made. Many papers agreed that, when the change in particle shape and size saw an increase in tensile strength, this was due to the specific shape or size which resulted in an increase in area of contact between particles under a compressional pressure [29,30,67,130,132]. Wünsch et al. [132] found that the reduction of initial particle size changes the deformation behaviour and therefore the effect of particle size is dependent on the specific material. With increasing initial particle sizes, the specific plastic energy rises but the specific elastic energy and elastic recovery is not affected as much. Therefore, when reducing sizes of materials, this could affect their mechanical response to the exerted compressional pressure. Furthermore, crystal structures may have important information about mechanical properties of the powder which is also seen in Sun and Grant's [130] study which saw two different polymorphs of sulfamerazine exhibit different compaction properties (Figure 28) Similar results were seen in Upadhyay et al.'s [139] study, where it was found that the different polymorphs of ranitidine hydrochloride exhibited different compressibility results. In a later study using the same polymorphs, polymorph I had better tabletability compared to polymorph II at all size fractions [140].

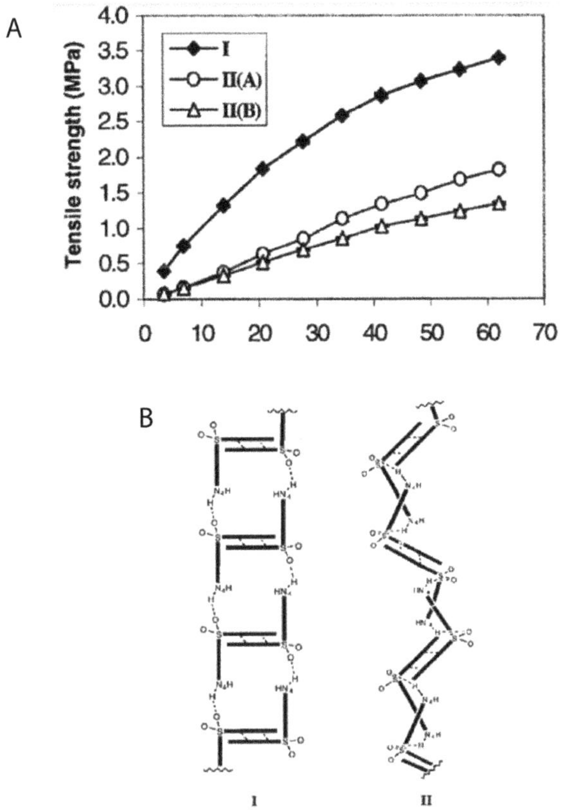

Figure 28. (**A**) Tabletability curves showing the difference of tabletability of three polymorphs of sulfamerazine I, II(A) and II(B). (**B**) Schematic of crystal structure of polymorphs: I and II. Hydrogen bonding are the broken lines. These crystal structure diagrams show that, even though the hydrogen bonding connectivity is the same for both polymorphs, the secondary structures are different. All figures obtained from Sun and Grant [130].

Future studies should pay more attention to crystal structure and understanding the corresponding mechanical properties of the materials being investigated before and after they are processed to a certain particle size which could affect the mechanical properties of the powder. For example, Simek et al. [133] modified their paracetamol samples in ethanol and changed the stirring speed to achieve the desired particle shape (Figure 25). Understanding these discrepancies could allow potential compromises to be made. For example, if a certain material is very cohesive but the particle size is independent of the tabletability properties, a larger particle size could be used to reduce cohesivity while not affecting the tabletability of the tablet formulation.

Surface energy is the excess energy on the surface compared to the bulk [141] which is affected by a number of intermolecular forces such as van der Waals [142]. A more cohesive powder will have a higher surface energy with interparticle bonding that will hinder the powder flow [143]. However, it is well documented that surface energy has a profound effect on producing successful tablets after powder compaction. Many results have suggested higher surface energy leads to stronger bonds forming between powder particles, which form higher tensile strength tablets [33–36]. For example, Wünsch et al. [132] suggested that formation of hydrogen bonds between MCC particles is why there was a considerable increase of tensile strength compared to ibuprofen and vildagliptin, as seen in Figure 29. However, the way that MCC, ibuprofen and vildagliptin deforms was not considered in this discussion point.

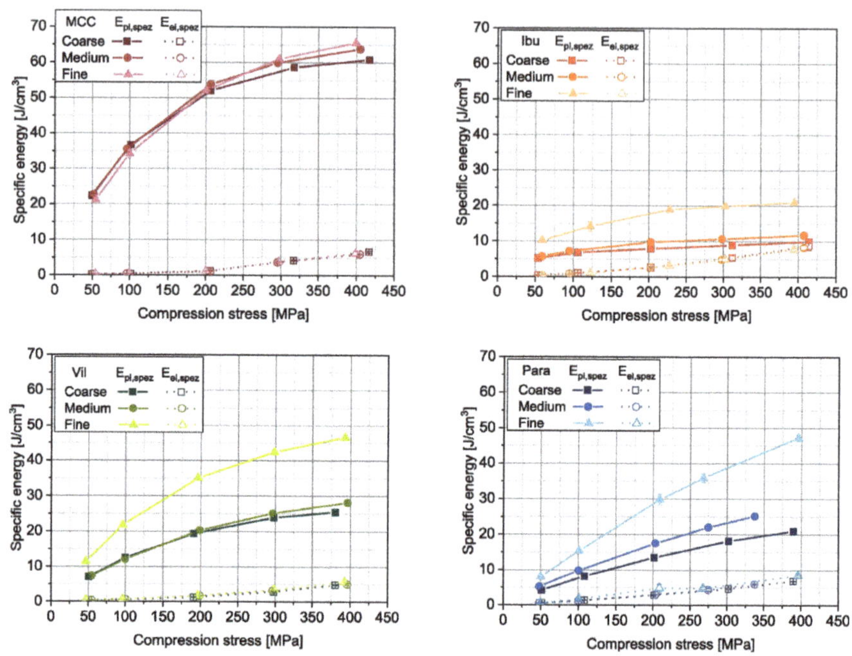

Figure 29. Graphs to show the specific plastic (solid lines) and elastic energies (dotted lines) of fine, medium and coarse grades of MCC, ibuprofen (Ibu) and vildagliptin (Vil) and paracetamol (para) obtained from Wunsch et al. [132] These graphs show that coarse grades have lower specific surface energy across all cases.

To intrinsically measure the effect of surface energy of the powder on the tensile strength is immensely difficult. Fichtner et al. [144] and Chen et al. [145] have attempted to change the surface energy of the powder by dry coating with polysorbate and Aerosil 200 respectively. This showed a decrease in tensile strength with the decrease of surface energy in both studies. However, the effect of changing particle size was not well documented and as previously stated in Section 4.2, depending on the specific material, particle size is another factor that changes the tablet outcome after compression. Ho et al. [146] counteracted this issue by changing the surface energy without changing any other powder particle property by modifying the functional groups on the surface of the powder without disrupting the surface morphology and particle size by silanisation. This treatment profoundly decreased the tensile strength of the tablet as the surface energy was reduced [147].

Another method, which was also mentioned in Section 2.1 to decrease the surface energy of a powder and improve flowability is to add a glidant, which is seen to increase the flowability by reducing cohesion [6]. How these additives affect the tabletability of the powder formulation was looked at in many different studies. Magnesium stearate (MgSt) is known to improve flowability [148]. However, it has a detrimental effect on the tablet strength especially when magnesium stearate is mixed for too long and vigorously. This is an effect which is known as 'over-lubrication' [116,149]. The MgSt is theorised to create a thin hydrophobic layer around powder particles preventing bonding between the powder particles to take place and therefore decreasing the tensile strength [149]. Furthermore, as MgSt is hydrophobic, the powder being 'over-lubcricated' will effect the dissolution [150].

On the other hand, MgSt has excellent lubrication properties and is commonly used in tablet formulation as a lubricant [144] rather than a glidant.

A glidant that is discussed frequently in literature is silicon dioxide. Apeji and Olowosulu [67] used talc and colloidal silica which appears to have had a negative influence on the tensile strength of the tablet. This paper did not have any comparison to a control data without any glidant to compare with. What the previous papers mentioned do not include is whether they are using hydrophobic or hydrophilic silicon dioxide. Kunnath et al. [151] looked at both hydrophilic and hydrophobic nano-silica and their impact on tablet tensile strength and dissolution (Figure 30). The nano-silica was dry coated onto the API with similar methods to those implemented by Chen et al. [145], mentioned previously. This saw an increase in tensile strength for both hydrophilic and hydrophobic nano-silica and did not see an effect on the dissolution profile. The possible explanation of this is that the dry coating of API led to deagglomeration of the powder and increased the total contact area where interparticle bonding can take place. On the other hand, in Mužíková et al.'s [66] study, when colloidal silica (Aerosil 200 and 255) was added, this saw a decrease in tensile strength, therefore the disintegration time also decreased. The max compressional pressure used to produce the tablets was very low (26.37 MPa) compared to what is normally used (200–250 MPa). However, Wunsch [132] found that it is the number of bonds which determine the strength of the tablet, not the bonding strength. Similar to the particle size conclusion, the fundamentals of the mechanical properties and therefore deformation properties of the powder are vital to understand as enhancing the bonding area may be more important than increasing the surface energy. Furthermore, this could mean a compromise can be taken on surface energy which will lower the cohesivity of the powder but would not affect the tablet outcomes if the contact area between particles is sufficient enough to gain the number of bonds needed. This further points to the importance of understanding how the powder deforms and how that effects and governs the contact area between particles [19].

There have been promising results for silicon dioxide alternatives such as magnesium aluminosilicate (MAS) and tricalcium phosphate (TCP). Both of these novel glidants have a positive effect on flowability [98]. However, there is just a single study investigating the tabletability of MAS, and none with TCP; in which, Hentzchel et al. [152] compared the tabletability of MAS with other silicates, when mixed with MCC of different concentrations. MAS was the only silicate that did not decrease in tablet strength, with increasing concentration, which is a positive result. Although the tablet strength is not affected by MAS, a study on how these novel glidants affect the dissolution rate at different concentrations will need to be conducted, as this is a critical quality attribute of the tablet. A study by Khunawattanakul et al. [153] saw the MAS in a tablet film coat increased the dissolution time, suggesting a balance will be required to ensure the right concentration to aid flowability is achieved, without negatively affecting dissolution. Although Tran et al. [98] shows that MAS aids flowability, this will need to be further investigated into whether MAS does have an impact on the feeding and blending parts of the CDC process. As far as the authors are aware there are no extensive studies that analyse the effect of these alternative glidants on these unit operations, essentially showing that additional studies are required to understand the complex interactions between not only CDC processing but the resultant in vivo performance of the tablet. The aforementioned publication Tran et al. [54] takes a perspective of using a range of measurement techniques to inform quantities or types of flow modifiers, practically taking a quality by design approach. A similar methodology could be applied in terms of glidants (namely, MAS) and their subsequent impact on tablet quality CQAs such as tensile strength of binary mixtures with excipients other than MCC.

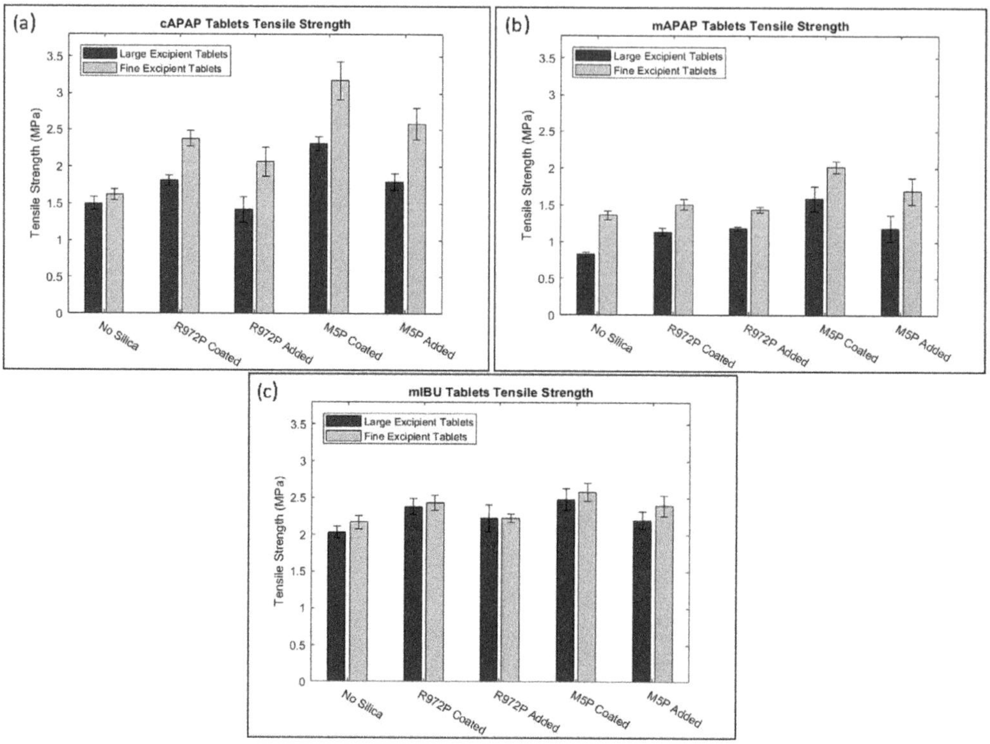

Figure 30. Tensile strength of the tablets made from (**a**) coarse acetaminophen (cAPAP), (**b**) micronised acetaminophen (mAPAP), and (**c**) micronised ibuprofen (mIBU) where Aerosil R972P nano-silica (hydrophobic) and CAB-O-SIL M5P nano silica (hydrophilic) obtained from Kunnath et al. [151].

4.3. Tabletting Summary

In this section, cohesion was split between two main contributing factors, particle size and shape, and surface energy, where smaller particle size and higher surface energy generally resulted in a higher magnitude of cohesion [24]. Although there are some contradictions seen in the particle size section, generally the tabletting process needs a good level of cohesion to be successful. However, spherical agglomeration is a promising technique that can lower the effects of cohesion seemingly without negatively affecting the tabletability [3,122]. More research will need to be conducted to optimise the solvent selection process [123].

Furthermore, both particle size and shape and surface energy research pointed towards the fundamentals of the mechanical properties of the powder. Particle size and shape discrepancies may be due to how the particle size and shapes were manipulated and how the crystal structure and therefore the material properties have changed. Additionally, it was found in Wunsch's et al. [132] paper that it is more the number of bonds rather than the strength of individual bonds that influences the overall tablet strength. Therefore, a higher contact area between particles to allow for a higher number of bonds could be more important than having higher surface energy powders to obtain stronger tablets.

The addition of glidants which increased flowability are seen to have contradicting results from commonly used silicon dioxide discussed in the previous section. On the other hand, promising initial results from modern alternative glidants, such as magnesium aluminosilicate (MAS), showed an increase in flowability and no change in tensile strength of the tablet regardless of concentration [98]. However, more studies need to be completed to scope what kind of an effect MAS has on dissolution and the tensile strength of binary mixtures with different commonly used excipients with different deformation properties than MCC.

5. Summary Table

Table 2 summarises the effect cohesion has on the unit operation and what control measures are outlined in the discussion. How the control measures from one-unit operation can affect another downstream has yet to be discussed. This section will outline how some cohesion controls that have positive effects on unit operations can promote issues in downstream unit operations in the continuous direct compression line.

The consistency and accuracy of which powders are dispensed from a feeder will directly influence the downstream unit operations [43,45]. Therefore, it is of utmost importance that understanding is gained on how to maintain a consistent feed rate for very cohesive materials [2,41,55]. The use of silica shows great promise, as it seemingly both improves flowability and reduces electrostatic charging, leading to better screw filling and thus resulting in improved feeding consistency [5,47,57]. Lumay et al. [57] also showed silica's ability to influence the thixotropic behaviour; in future work, it would be interesting to see a development on this in the context of LIW feeding. However, the addition of silica has potentially negative effects on tabletting [66,67], while its effect on blending has not been explicitly described. Bridging/arching within the hopper volume has been shown to suffocate the flow of powder into the conveying volume. As a result, both additives and refilling strategy show ideas for reducing the processing variability [2,41,55,65]. Furthermore, using DEM for complex agitator design would further help the development of arching minimisation.

Despite the benefits of silication, attention should be paid to the end-to-end impact of formulation additives. For instance, nano-silicas were seen to have a lesser impact, compared to MPS, on the tablet strength, but greatly improved the flowability [57]. Alternatively, novel glidants, such as magnesium aluminosilicates, show promise in minimising the respective trade-off of improving flowability and affecting the tablet's tensile strength [152].

Blending seeks to intimately mix all of a given formulation's constituents and deliver them to the final, tabletting stage of the process with minimal variation [45,154]. Cohesion, in most cases, serves as a direct barrier to blending, as it provides resistance to intimate mixing [75,78,90]. Thus, higher levels of agitation are required to break apart these cohesive clusters, but systems must devise their own method to balance both the micro and the macro aspects of mixing [43,88]. In addition, complex bulk behaviour occurs when there is a moderate amount of cohesive particle–particle contacts [90]. Thus, it is of interest to develop a greater understanding of the mechanisms which govern this complex bulk behaviour. Moreover, despite the viewpoint to increase RPM to deal with cohesion, it is not understood what these new shear environments look like, and whether they have an impact on tabletability.

Spherical Agglomeration, which was discussed in Section 4.2, is a very promising technique to control the amount of cohesivity APIs add to tablet formulations without compromising tabletability to the same extent as the addition of glidants and lubricants which is beneficial to the whole CDC process. Further work will, however, need to be carried out to improve the efficiency of selecting the right solvent and process parameters [123]. Process parameters especially will need to be focussed on as some parameters are highlighted in studies that may be difficult to scale up. However, there have been good strides looking into the scale-up process and making spherical agglomeration/crystallisation continuous [155,156]. Ensuring the API crystals are not too large will be important to ensure content uniformity within tablets [123].

Table 2. Summary table with all the unit operations discussed in this literature review and their respective cohesion effects and controls. Note: Cohesion compromise is denoted with a ⋄, future work with ∘, considerations △.

Unit Operation		Cohesion Effects Results in …	Cohesion Control or Compromise ⋄	Future Work ∘/Considerations △
Feeding		Inconsistent screw filling	Modifying bulk density and cohesive particle-particle contact through additives, such as Silica [5,47,57]	Understand and be able to modify thixotropic powder ∘ behaviour
			Increase/Decrease conveying screw speeds [57]	
		Bridging/Arching	Refill Strategy [41,65]	DEM simulations exploring optimum hopper agitator design ∘
			Agitator Design	
Blending		Reduced micro-mixing	Increase in rpm (Fr) balanced by changes to system geometry to maintain macro [75,78,90]	
		Complex bulk behaviour, which results in poor mixing	Modify the three contributing factors: Fill level, Magnitude of Cohesion, and Froude Number [43,75,88,90]	
			Mechanistic understanding of how these factors interact [90]	
Tabletting				Find a more efficient way to find the right solvents ∘
	Die filling	Incomplete die fill and weight [102,110]	Spherical agglomeration [3]	Understand how spherical agglomeration will affect other CDC unit operations (feeding and blending) ∘
				Investigate the possibility of a scale-up method as the will be hard as the process is difficult and time consuming on small scale ∘
			Paddle speed [107]	Overlubrication can occur if the paddle speed is too high [115]. A compromise will need to be taken to ensure an increase of paddle speed will not effect tablet strength △
			Introduction of air in the system acting like a lubricant [27]	As this can create an adverse pressure gradient which will then resist the motion of particles. Find the ranges when air introduction is beneficial by Quality by Design experiments △
	Compaction	Better tabletability; generally lower particle size and higher surface energy results in stronger tablets but can have bad effects on dissolution if the tablet strength is too high [157]	Nano-silica and other novel glidants (magnesium aluminosilicate) can help promote better flowability and has little to no effect on tablet strength ⋄ [151]	Understand the impacts of different types of nano-silica. Magnesium aluminosilicate will need to be tabletted with other common excipients and dissolution testing will to be conducted to assess the limitations∘
			Number of bonds is more important than the strength of the bond could be able to compromise lower surface energy/cohesivity if there is sufficient contact area gained during compaction ⋄	Understanding crystal structure fundamentally and how that changes material properties. This may be what is causing discrepancies regarding particle size and the tablet strength outcome ∘ [130]

6. Conclusions

This literature review consolidates the impact of powder cohesion and mitigation for each of the unit processes in CDC, whilst considering the consequences of upstream/downstream processing. For a tablet formulation to be viable for the CDC process, the formulation must possess both suitable flowability *and* tabletability [19,25]. Cohesion is known to cause powders to resist flow [23,24] which has a negative impact on the performance of the feeding [25], mixing [85,87], and die filling [3] unit operations involved in the CDC process whereas, for the compression stage, cohesivity is desired [29,30,33,35].

Understanding the cohesive nature of APIs and excipients and then manipulating them, where possible, is useful for improving the overall performance of the CDC process. The use of additives such as glidants and lubricants may aid the flow of powders, thus improving the feeding [6], mixing [37] and die filling [110] performance; however, these additives can have a negative impact on compaction [29]. Flow variability should be managed (specifically during the feeding stage) as any perturbations will be propagated into successive unit operations [2,158].

Despite the volume of literature in this area, there remain many important open questions and thus valuable future research avenues:

- Silication has benefits similar to glidants and lubricants as it can lower the cohesive behaviour of the powder and therefore increase flowability [6]. However, there are contradictions on whether silication negatively affects tabletability; this will need to be further investigate [67,145]. Alternative glidants, e.g., magnesium aluminosilicates, are a promising option however, more work needs to be carried out to understand the impact on tablet performance.
- Discrete Element Method (DEM) modelling, allowing researchers to create a digital twin(s) of an existing experiment and conduct statistical analysis on the simulated powder behaviour would allow researchers to gather metrics to quantify mixing/feeding performance (as discussed in Escotet-Espinoza et al. [80]) that were not accessible experimentally.
- Utilising triboelectric charging for the blending stage: static surface charge is developed on the surface of particles due to the strain given to the powder during mixing/transport [47], potentially allowing a formulation to be altered to promote attraction/repulsion between constituent species in order to gain a more-ordered well-mixed system.
- Spherical agglomeration/crystallisation can improve the flowability and tabletability of APIs, which is normally the most cohesive component of the tablet formulation [3,18]. However, further work needs to be performed to improve the solvent selection and process parameters to allow for scale-up [123].
- The many discrepancies in the literature regarding the manner in which particle size affects the tabletability of the powder should be comprehensively addressed, as this represents an important gap in the fundamental understanding of powder compression [39,128]. Understanding which types of powders are affected by particle size and which are not could lead to compromises to have larger particle sizes to lower cohesivity and therefore improve flow while not affecting the tabletability.

Author Contributions: Conceptualization, O.J.-S. and Z.C.; formal analysis, O.J.-S. and Z.C.; investigation, O.J.-S. and Z.C.; writing—original draft preparation, O.J.-S. and Z.C.; writing—review and editing, C.R.K.W.-Y. and A.I.; visualization, O.J.-S. and Z.C.; supervision, C.R.K.W.-Y. and A.I.; project administration, C.R.K.W.-Y. and A.I.; funding acquisition, C.R.K.W.-Y. and A.I. All authors have read and agreed to the published version of the manuscript.

Funding: The authors would like to thank the Engineering Doctorate programme funded by EPSRC through the Centre for Doctoral Training in Formulation Engineering (grant no. EP/L015153/1), and from AstraZeneca plc.

Acknowledgments: The authors would like to also thank Adedoyin Yahyi for the initial insight and discussion surrounding powder feeding.

Conflicts of Interest: The authors declare no conflict of interest. The company had no role in the design of the study; in the collection, analyses, or interpretation of data; in the writing of the manuscript, and in the decision to publish the results.

Abbreviations

The following abbreviations are used in this manuscript:

CDC	Continuous Direct Compression
CQAs	Critical Quality Attributes
CMAs	Critical Material Attributes
API	Active Pharmaceutical Ingredient
FFC	Flow Function Coefficient
SEM	Scanning Electron Microscope
DEM	Discrete Element Method
MPS	Mesoporous Sillica
rpm	Revolutions Per Minuite
cBD	Conditioned Bulk Density
LIW	Loss in Weight
APAP	Acetaminophen
RSD	Relative Standard Deviation
PCA	Principle Component Analysis
RTD	Residence Time Distribution
MCC	Microcrystalline Cellulose
MgSt	Magnesium Stearate
MAS	Magnesium Aluminosilicate
TCP	Tricalcium Phosphate
mAPAP	Micronised Acetaminophen
pAPAP	Powdered Acetaminophen
sgAPAP	Special Granular Acetaminophen
RH	Relative Humidity
GMP	Good Manufacturing Practice
HVAC	Heating Ventilation & Air Conditioning
FT4	Freeman Powder Rheometer
QESD	Quasi-emulsion solvent diffusion
DCP	Calcium hydrogen phosphate dehydrate

References

1. Vercruysse, J.; Delaet, U.; Van Assche, I.; Cappuyns, P.; Arata, F.; Caporicci, G.; De Beer, T.; Remon, J.P.; Vervaet, C. Stability and repeatability of a continuous twin screw granulation and drying system. *Eur. J. Pharm. Biopharm.* **2013**, *85*, 1031–1038. [CrossRef] [PubMed]
2. Engisch, W.E.; Muzzio, F.J. Loss-in-Weight Feeding Trials Case Study: Pharmaceutical Formulation. *J. Pharm. Innov.* **2015**, *10*, 56–75. [CrossRef]
3. Chattoraj, S.; Sun, C.C. Crystal and Particle Engineering Strategies for Improving Powder Compression and Flow Properties to Enable Continuous Tablet Manufacturing by Direct Compression. *J. Pharm. Sci.* **2018**, *107*, 968–974. [CrossRef] [PubMed]
4. Liu, L.X.; Marziano, I.; Bentham, A.C.; Litster, J.D.; White, E.T.; Howes, T. Influence of particle size on the direct compression of ibuprofen and its binary mixtures. *Powder Technol.* **2013**, *240*, 66–73. [CrossRef]
5. Leung, L.Y.; Mao, C.; Srivastava, I.; Du, P.; Yang, C.Y. Flow Function of Pharmaceutical Powders Is Predominantly Governed by Cohesion, Not by Friction Coefficients. *J. Pharm. Sci.* **2017**, *106*, 1865–1873. [CrossRef] [PubMed]
6. Escotet-Espinoza, M.S.; Scicolone, J.V.; Moghtadernejad, S.; Sanchez, E.; Cappuyns, P.; Van Assche, I.; Di Pretoro, G.; Ierapetritou, M.; Muzzio, F.J. Improving Feedability of Highly Adhesive Active Pharmaceutical Ingredients by Silication. *J. Pharm. Innov.* **2021**, *16*, 279–292. [CrossRef]
7. Visser, J. Van der Waals and other cohesive forces affecting powder fluidization. *Powder Technol.* **1989**, *58*, 1–10. [CrossRef]
8. Juarez-Enriquez, E.; Olivas, G.I.; Zamudio-Flores, P.B.; Perez-Vega, S.; Salmeron, I.; Ortega-Rivas, E.; Sepulveda, D.R. A review on the influence of water on food powder flowability. *J. Food Process Eng.* **2022**, *45*, e14031. [CrossRef]
9. Zafar, U.; Vivacqua, V.; Calvert, G.; Ghadiri, M.; Cleaver, J.S. A review of bulk powder caking. *Powder Technol.* **2017**, *313*, 389–401. [CrossRef]

10. Wei, Z.; Zhao, Y.P. Growth of liquid bridge in AFM. *J. Phys. D Appl. Phys.* **2007**, *40*, 4368. [CrossRef]
11. Hou, Q.; Dong, K.; Yu, A. DEM study of the flow of cohesive particles in a screw feeder. *Powder Technol.* **2014**, *256*, 529–539. [CrossRef]
12. Allenspach, C.; Timmins, P.; Lumay, G.; Holman, J.; Minko, T. Loss-in-weight feeding, powder flow and electrostatic evaluation for direct compression hydroxypropyl methylcellulose (HPMC) to support continuous manufacturing. *Int. J. Pharm.* **2021**, *596*, 120259. [CrossRef] [PubMed]
13. Karner, S.; Urbanetz, N.A. Arising of electrostatic charge in the mixing process and its influencing factors. *Powder Technol.* **2012**, *226*, 261–268. [CrossRef]
14. Abouzeid, A.Z.M.; Fuerstenau, D.W. Effect of Humidity on Mixing of Particulate Solids. *Ind. Eng. Chem. Process Des. Dev.* **1972**, *11*, 296–301.
15. Beretta, M.; Hörmann, T.R.; Hainz, P.; Hsiao, W.K.; Paudel, A. Investigation into powder tribo-charging of pharmaceuticals. Part II: Sensitivity to relative humidity. *Int. J. Pharm.* **2020**, *591*, 120015. [CrossRef] [PubMed]
16. Haycocks, N.; Goldschmidt, N.A.; Thomsen, U. Temperature & Humidity Requirements in Pharmaceutical Facilities. 2023. Available online: https://ispe.org/pharmaceutical-engineering/september-october-2021/temperature-humidity-requirements-pharmaceutical (accessed on 31 March 2023).
17. Sun, C.C. Mechanism of moisture induced variations in true density and compaction properties of microcrystalline cellulose. *Int. J. Pharm.* **2008**, *346*, 93–101. [CrossRef]
18. Chen, H.; Aburub, A.; Sun, C.C. Direct Compression Tablet Containing 99% Active Ingredient—A Tale of Spherical Crystallization. *J. Pharm. Sci.* **2019**, *108*, 1396–1400. [CrossRef]
19. Sun, C.C. Decoding powder tabletability: Roles of particle adhesion and plasticity. *J. Adhes. Sci. Technol.* **2011**, *25*, 483–499. [CrossRef]
20. Haywood, A.; Glass, B.D. Pharmaceutical excipients—Where do we begin? *Aust. Prescr.* **2011**, *34*, 112–114. [CrossRef]
21. Jivraj, M.; Martini, L.G.; Thomson, C.M. An overview of the different excipients useful for the direct compression of tablets. *Pharm. Sci. Technol. Today* **2000**, *3*, 59–63. [CrossRef]
22. Escotet-Espinoza, M.S. Phenomenological and Residence Time Distribution Models for Unit Operations in a Continuous Pharmaceutical Manufacturing Process. Ph.D. Thesis, Rutgers University-School of Graduate Studies, New Brunswick, NJ, USA, 2018; p. 312.
23. Shah, U.V.; Olusanmi, D.; Narang, A.S.; Hussain, M.A.; Tobyn, M.J.; Hinder, S.J.; Heng, J.Y. Decoupling the contribution of surface energy and surface area on the cohesion of pharmaceutical powders. *Pharm. Res.* **2015**, *32*, 248–259. [CrossRef]
24. Shah, R.B.; Tawakkul, M.A.; Khan, M.A. Comparative evaluation of flow for pharmaceutical powders and granules. *AAPS PharmSciTech* **2008**, *9*, 250–258. [CrossRef] [PubMed]
25. Engisch, W.E.; Muzzio, F.J. Method for characterization of loss-in-weight feeder equipment. *Powder Technol.* **2012**, *228*, 395–403. [CrossRef]
26. Mazumder, M.K.; Sims, R.A.; Biris, A.S.; Srirama, P.K.; Saini, D.; Yurteri, C.U.; Trigwell, S.; De, S.; Sharma, R. Twenty-first century research needs in electrostatic processes applied to industry and medicine. *Chem. Eng. Sci.* **2006**, *61*, 2192–2211. [CrossRef]
27. Wu, C.Y.; Dihoru, L.; Cocks, A.C. The flow of powder into simple and stepped dies. *Powder Technol.* **2003**, *134*, 24–39. [CrossRef]
28. Freeman, R.; Fu, X. Characterisation of power bulk, dynamic flow and shear properties in relation to die filling. *Powder Metall.* **2008**, *51*, 196–201. [CrossRef]
29. De Boer, A.H.; Vromans, H.; Leur, C.F.; Bolhuis, G.K.; Kussendrager, K.D.; Bosch, H. Studies on tableting properties of lactose—Part III. The consolidation behaviour of sieve fractions of crystalline α-lactose monohydrate. *Pharm. Weekbl. Sci. Ed.* **1986**, *8*, 145–150. [CrossRef]
30. Johansson, B.; Alderborn, G. The effect of shape and porosity on the compression behaviour and tablet forming ability of granular materials formed from microcrystalline cellulose. *Eur. J. Pharm. Biopharm.* **2001**, *52*, 347–357. [CrossRef] [PubMed]
31. Chang, S.Y.; Sun, C.C. Insights into the effect of compaction pressure and material properties on interfacial bonding strength of bilayer tablets. *Powder Technol.* **2019**, *354*, 867–876. [CrossRef]
32. Cabiscol, R.; Shi, H.; Wünsch, I.; Magnanimo, V.; Finke, J.H.; Luding, S.; Kwade, A. Effect of particle size on powder compaction and tablet strength using limestone. *Adv. Powder Technol.* **2020**, *31*, 1280–1289. [CrossRef]
33. El Gindy, N.A.; Samaha, M.W. Tensile strength of some pharmaceutical compacts and their relation to surface free energy. *Int. J. Pharm.* **1982**, *13*, 35–46. [CrossRef]
34. Fichtner, F.; Mahlin, D.; Welch, K.; Gaisford, S.; Alderborn, G. Effect of surface energy on powder compactibility. *Pharm. Res.* **2008**, *25*, 2750–2759. [CrossRef] [PubMed]
35. Luangtana-Anan, M.; Fell, J.T. Bonding mechanisms in tabletting. *Int. J. Pharm.* **1990**, *60*, 197–202. [CrossRef]
36. Etzler, F.M.; Pisano, S. Tablet tensile strength: Role of surface free energy. *Adv. Contact Angle Wettability Adhes.* **2015**, *2*, 397–418.
37. Pingali, K.; Mendez, R.; Lewis, D.; Michniak-Kohn, B.; Cuitino, A.; Muzzio, F. Mixing order of glidant and lubricant—Influence on powder and tablet properties. *Int. J. Pharm.* **2011**, *409*, 269–277. [CrossRef]
38. Mills, L.A.; Sinka, I.C. Effect of particle size and density on the die fill of powders. *Eur. J. Pharm. Biopharm.* **2013**, *84*, 642–652. [CrossRef]
39. Herting, M.G.; Kleinebudde, P. Roll compaction/dry granulation: Effect of raw material particle size on granule and tablet properties. *Int. J. Pharm.* **2007**, *338*, 110–118. [CrossRef]

40. Kehlenbeck, V.; Sommer, K. Possibilities to improve the short-term dosing constancy of volumetric feeders. *Powder Technol.* **2003**, *138*, 51–56. [CrossRef]
41. Engisch, W.E.; Muzzio, F.J. Feedrate deviations caused by hopper refill of loss-in-weight feeders. *Powder Technol.* **2015**, *283*, 389–400. [CrossRef]
42. Bekaert, B.; Van Snick, B.; Pandelaere, K.; Dhondt, J.; Di Pretoro, G.; De Beer, T.; Vervaet, C.; Vanhoorne, V. In-depth analysis of the long-term processability of materials during continuous feeding. *Int. J. Pharm.* **2022**, *614*, 121454. [CrossRef]
43. Van Snick, B.; Holman, J.; Vanhoorne, V.; Kumar, A.; De Beer, T.; Remon, J.P.; Vervaet, C. Development of a continuous direct compression platform for low-dose drug products. *Int. J. Pharm.* **2017**, *529*, 329–346. [CrossRef]
44. Simonaho, S.P.; Ketolainen, J.; Ervasti, T.; Toiviainen, M.; Korhonen, O. Continuous manufacturing of tablets with PROMIS-line—Introduction and case studies from continuous feeding, blending and tableting. *Eur. J. Pharm. Sci.* **2016**, *90*, 38–46. [CrossRef]
45. García-Muñoz, S.; Butterbaugh, A.; Leavesley, I.; Manley, L.F.; Slade, D.; Bermingham, S. A flowsheet model for the development of a continuous process for pharmaceutical tablets: An industrial perspective. *AIChE J.* **2018**, *64*, 511–525. [CrossRef]
46. Gyürkés, M.; Madarász, L.; Köte, Á.; Domokos, A.; Mészáros, D.; Beke, K.; Nagy, B.; Marosi, G.; Pataki, H.; Nagy, Z.K.; et al. Process design of continuous powder blending using residence time distribution and feeding models. *Pharmaceutics* **2020**, *12*, 1119. [CrossRef] [PubMed]
47. Beretta, M.; Hörmann, T.R.; Hainz, P.; Hsiao, W.K.; Paudel, A. Investigation into powder tribo-charging of pharmaceuticals. Part I: Process-induced charge via twin-screw feeding. *Int. J. Pharm.* **2020**, *591*, 120014. [CrossRef]
48. Windows-Yule, C.R.; Neveu, A. Calibration of DEM simulations for dynamic particulate systems. *Pap. Phys.* **2022**, *14*, 140010. [CrossRef]
49. Rhodes, M.J. *Introduction to Particle Technology*; John Wiley & Sons: Hoboken, NJ, USA, 2008.
50. Windows-Yule, K.; Nicuşan, L.; Herald, M.T.; Manger, S.; Parker, D. Comparison with other techniques. In *Positron Emission Particle Tracking: A Comprehensive Guide*; IOP Publishing: Bristol, UK, 2022.
51. Thornton, C. *Granular Dynamics, Contact Mechanics and Particle System Simulations: A DEM Study*; Springer's International Publishing: Berlin/Heidelberg, Germany, 2015; Volume 24, pp. 1–195. [CrossRef]
52. Kojima, T.; Elliott, J.A. Effect of silica nanoparticles on the bulk flow properties of fine cohesive powders. *Chem. Eng. Sci.* **2013**, *101*, 315–328. [CrossRef]
53. Meyer, K.; Zimmermann, I. Effect of glidants in binary powder mixtures. *Powder Technol.* **2004**, *139*, 40–54. [CrossRef]
54. Tran, D.T.; Majerová, D.; Veselý, M.; Kulaviak, L.; Ruzicka, M.C.; Zámostný, P. On the mechanism of colloidal silica action to improve flow properties of pharmaceutical excipients. *Int. J. Pharm.* **2019**, *556*, 383–394. . [CrossRef]
55. López, A.; Vivacqua, V.; Hammond, R.; Ghadiri, M. Analysis of screw feeding of faceted particles by discrete element method. *Powder Technol.* **2020**, *367*, 474–486. [CrossRef]
56. Krantz, M.; Zhang, H.; Zhu, J. Characterization of powder flow: Static and dynamic testing. *Powder Technol.* **2009**, *194*, 239–245. [CrossRef]
57. Lumay, G.; Pillitteri, S.; Marck, M.; Monsuur, F.; Pauly, T.; Ribeyre, Q.; Francqui, F.; Vandewalle, N. Influence of mesoporous silica on powder flow and electrostatic properties on short and long term. *J. Drug Deliv. Sci. Technol.* **2019**, *53*, 101192. [CrossRef]
58. Neveu, A.; Francqui, F.; Lumay, G. Measuring powder flow properties in a rotating drum. *Measurement* **2022**, *200*, 111548. [CrossRef]
59. Patel, D.; Francqui, F.; Defrêne, R.J.L.; Neveu, A. Influence of additives on the electrostatic charge build-up of excipients. *Powder Technol.* **2003**, *132*, 192–198. [CrossRef]
60. Ramirez-Dorronsoro, J.C.; Jacko, R.B.; Kildsig, D.O. Chargeability measurements of selected pharmaceutical dry powders to assess their electrostatic charge control capabilities. *AAPS PharmSciTech* **2006**, *7*, 103. [CrossRef] [PubMed]
61. Escotet-Espinoza, M.S.; Moghtadernejad, S.; Oka, S.; Wang, Z.; Wang, Y.; Roman-Ospino, A.; Schäfer, E.; Cappuyns, P.; Van Assche, I.; Futran, M.; et al. Effect of material properties on the residence time distribution (RTD) characterization of powder blending unit operations. Part II of II: Application of models. *Powder Technol.* **2019**, *344*, 525–544. [CrossRef]
62. Escotet-Espinoza, M.S.; Moghtadernejad, S.; Scicolone, J.; Wang, Y.; Pereira, G.; Schäfer, E.; Vigh, T.; Klingeleers, D.; Ierapetritou, M.; Muzzio, F.J. Using a material property library to find surrogate materials for pharmaceutical process development. *Powder Technol.* **2018**, *339*, 659–676. [CrossRef]
63. Shier, A.P.; Kumar, A.; Mercer, A.; Majeed, N.; Doshi, P.; Blackwood, D.O.; Verrier, H.M. Development of a predictive model for gravimetric powder feeding from an API-rich materials properties library. *Int. J. Pharm.* **2022**, *625*, 122071. [CrossRef]
64. Yadav, I.K.; Holman, J.; Meehan, E.; Tahir, F.; Khoo, J.; Taylor, J.; Benedetti, A.; Aderinto, O.; Bajwa, G. Influence of material properties and equipment configuration on loss-in- weight feeder performance for drug product continuous manufacture. *Powder Technol.* **2019**, *348*, 126–137. [CrossRef]
65. De Souter, L.; Waeytens, R.; Van Hauwermeiren, D.; Grymonpré, W.; Bekaert, B.; Nopens, I.; De Beer, T. Elucidation of the powder flow pattern in a twin-screw LIW-feeder for various refill regimes. *Int. J. Pharm.* **2023**, *631*, 122534. [CrossRef]
66. Mužíková, J.; Louženská, M.; Pekárek, T. A study of compression process and properties of tablets with microcrystalline cellulose and colloidal silicon dioxide. *Acta Pol. Pharm.—Drug Res.* **2016**, *73*, 1259–1265.
67. Apeji, Y.E.; Olowosulu, A.K. Quantifying the effect of glidant on the compaction and tableting properties of paracetamol granules. *J. Res. Pharm.* **2020**, *24*, 44–55. [CrossRef]

68. Uzunović, A.; Vranić, E. Effect of magnesium stearate concentration on dissolution properties of ranitidine hydrochloride coated tablets. *Biomol. Biomed.* **2007**, *7*, 279–283. [CrossRef] [PubMed]
69. Bolhuis, G.K.; Anthony Armstrong, N. Excipients for Direct Compaction—An Update. *Pharm. Dev. Technol.* **2006**, *11*, 111–124. [CrossRef]
70. Huang, Z.; Scicolone, J.V.; Gurumuthy, L.; Davé, R.N. Flow and bulk density enhancements of pharmaceutical powders using a conical screen mill: A continuous dry coating device. *Chem. Eng. Sci.* **2015**, *125*, 209–224. [CrossRef]
71. Bridgwater, J. Fundamental powder mixing mechanisms. *Powder Technol.* **1976**, *15*, 215–236. [CrossRef]
72. Cooke, M.H.; Stephens, D.J.; Bridgwater, J. Powder mixing—A literature survey. *Powder Technol.* **1976**, *15*, 1–20. [CrossRef]
73. Laurent, B.; Bridgwater, J. Influence of agitator design on powder flow. *Chem. Eng. Sci.* **2002**, *57*, 3781–3793. [CrossRef]
74. Van Snick, B.; Grymonpré, W.; Dhondt, J.; Pandelaere, K.; Di Pretoro, G.; Remon, J.P.; De Beer, T.; Vervaet, C.; Vanhoorne, V. Impact of blend properties on die filling during tableting. *Int. J. Pharm.* **2018**, *549*, 476–488. [CrossRef]
75. Palmer, J.; Reynolds, G.K.; Tahir, F.; Yadav, I.K.; Meehan, E.; Holman, J.; Bajwa, G. Mapping key process parameters to the performance of a continuous dry powder blender in a continuous direct compression system. *Powder Technol.* **2020**, *362*, 659–670. [CrossRef]
76. Venables, H.J.; Wells, J.I. Powder mixing. *Drug Dev. Ind. Pharm.* **2001**, *27*, 599–612. [CrossRef]
77. Asachi, M.; Nourafkan, E.; Hassanpour, A. A review of current techniques for the evaluation of powder mixing. *Adv. Powder Technol.* **2018**, *29*, 1525–1549. [CrossRef]
78. Portillo, P.M.; Vanarase, A.U.; Ingram, A.; Seville, J.K.; Ierapetritou, M.G.; Muzzio, F.J. Investigation of the effect of impeller rotation rate, powder flow rate, and cohesion on powder flow behavior in a continuous blender using PEPT. *Chem. Eng. Sci.* **2010**, *65*, 5658–5668. [CrossRef]
79. Poux, M.; Fayolle, P.; Bertrand, J.; Bridoux, D.; Bousquet, J. Powder mixing: Some practical rules applied to agitated systems. *Powder Technol.* **1991**, *68*, 213–234. [CrossRef]
80. Sebastian Escotet-Espinoza, M.; Moghtadernejad, S.; Oka, S.; Wang, Y.; Roman-Ospino, A.; Schäfer, E.; Cappuyns, P.; Van Assche, I.; Futran, M.; Ierapetritou, M.; et al. Effect of tracer material properties on the residence time distribution (RTD) of continuous powder blending operations. Part I of II: Experimental evaluation. *Powder Technol.* **2019**, *342*, 744–763. [CrossRef]
81. Gao, Y.; Vanarase, A.; Muzzio, F.; Ierapetritou, M. Characterizing continuous powder mixing using residence time distribution. *Chem. Eng. Sci.* **2011**, *66*, 417–425. [CrossRef]
82. Vanarase, A.U.; Muzzio, F.J. Effect of operating conditions and design parameters in a continuous powder mixer. *Powder Technol.* **2011**, *208*, 26–36. [CrossRef]
83. Fan, L.; Chen, Y.m.; Lai, F. Recent developments in solids mixing. *Powder Technol.* **1990**, *61*, 255–287. [CrossRef]
84. Rosato, A.D.; Windows-Yule, C. *Segregation in Vibrated Granular Systems*; Academic Press: Cambridge, MA, USA, 2020; pp. 197–218.
85. Rennie, P.R.; Chen, X.D.; Hargreaves, C.; MacKereth, A.R. Study of the cohesion of dairy powders. *J. Food Eng.* **1999**, *39*, 277–284. [CrossRef]
86. De Villiers, M.M. Description of the kinetics of the deagglomeration of drug particle agglomerates during powder mixing. *Int. J. Pharm.* **1997**, *151*, 1–6. [CrossRef]
87. Mehrotra, A.; Muzzio, F.J.; Shinbrot, T. Spontaneous separation of charged grains. *Phys. Rev. Lett.* **2007**, *99*, 058001. [CrossRef] [PubMed]
88. Bekaert, B.; Grymonpré, W.; Novikova, A.; Vervaet, C.; Vanhoorne, V. Impact of blend properties and process variables on the blending performance. *Int. J. Pharm.* **2022**, *613*, 121421. [CrossRef] [PubMed]
89. Vanarase, A.U.; Osorio, J.G.; Muzzio, F.J. Effects of powder flow properties and shear environment on the performance of continuous mixing of pharmaceutical powders. *Powder Technol.* **2013**, *246*, 63–72. [CrossRef]
90. Gao, Y.; Ierapetritou, M.; Muzzio, F. Periodic section modeling of convective continuous powder mixing processes. *AIChE J.* **2012**, *58*, 69–78.
91. Portillo, P.M.; Ierapetritou, M.G.; Muzzio, F.J. Characterization of continuous convective powder mixing processes. *Powder Technol.* **2007**, *182*, 368–378. [CrossRef]
92. Werner, D.; Davison, H.; Robinson, E.; Sykes, J.; Seville, J.; Wellings, A.; Bhattacharya, S.; Monsalve, D.S.; Wheldon, T.K.; Windows-Yule, C. Effect of system composition on mixing in binary fluidised beds. *Chem. Eng. Sci.* **2023**, *271*, 118562. [CrossRef]
93. Goldhirsch, I. Introduction to granular temperature. *Powder Technol.* **2008**, *182*, 130–136. . [CrossRef]
94. Ogawa, S. Multitemperature theory of granular materials. In Proceedings of the US-Japan Seminar on Continuum Mechanical and Statistical Approaches in the Mechanics of Granular Materials, Sendai, Japan, 5–9 June 1978; pp. 208–217.
95. Gao, Y.; Muzzio, F.J.; Ierapetritou, M.G. Optimizing continuous powder mixing processes using periodic section modeling. *Chem. Eng. Sci.* **2012**, *80*, 70–80. [CrossRef]
96. Sarkar, A.; Wassgren, C.R. Simulation of a continuous granular mixer: Effect of operating conditions on flow and mixing. *Chem. Eng. Sci.* **2009**, *64*, 2672–2682. [CrossRef]
97. Tomita, Y.; Nagato, T.; Takeuchi, Y.; Takeuchi, H. Control of residence time of pharmaceutical powder in a continuous mixer with impeller and scraper. *Int. J. Pharm.* **2020**, *586*, 119520. [CrossRef]
98. Tran, D.; Komínová, P.; Kulaviak, L.; Zámostný, P. Evaluation of multifunctional magnesium aluminosilicate materials as novel family of glidants in solid dosage products. *Int. J. Pharm.* **2021**, *592*, 120054. [CrossRef] [PubMed]

99. Toson, P.; Siegmann, E.; Trogrlic, M.; Kureck, H.; Khinast, J.; Jajcevic, D.; Doshi, P.; Blackwood, D.; Bonnassieux, A.; Daugherity, P.D.; et al. Detailed modeling and process design of an advanced continuous powder mixer. *Int. J. Pharm.* **2018**, *552*, 288–300. [CrossRef] [PubMed]
100. Kale, D.P.; Zode, S.S.; Bansal, A.K. Challenges in translational development of pharmaceutical cocrystals. *J. Pharm. Sci.* **2017**, *106*, 457–470. [CrossRef] [PubMed]
101. Zakhvatayeva, A.; Zhong, W.; Makroo, H.A.; Hare, C.; Wu, C.Y. An experimental study of die filling of pharmaceutical powders using a rotary die filling system. *Int. J. Pharm.* **2018**, *553*, 84–96. [CrossRef] [PubMed]
102. Schomberg, A.K.; Kwade, A.; Finke, J.H. The challenge of die filling in rotary presses—A systematic study of material properties and process parameters. *Pharmaceutics* **2020**, *12*, 248. [CrossRef] [PubMed]
103. Qu, L.; Stewart, P.J.; Hapgood, K.P.; Lakio, S.; Morton, D.A.; Zhou, Q.T. Single-step coprocessing of cohesive powder via mechanical dry coating for direct tablet compression. *J. Pharm. Sci.* **2017**, *106*, 159–167. [CrossRef] [PubMed]
104. Baserinia, R.; Sinka, I.C. Powder die filling under gravity and suction fill mechanisms. *Int. J. Pharm.* **2019**, *563*, 135–155. [CrossRef] [PubMed]
105. Tang, X.; Zakhvatayeva, A.; Zhang, L.; Wu, Z.F.; Sun, P.; Wu, C.Y. Flow behaviour of pharmaceutical powders during rotary die filling with a paddle feeder. *Int. J. Pharm.* **2020**, *585*, 119547. [CrossRef]
106. Jackson, S.; Sinka, I.C.; Cocks, A.C. The effect of suction during die fill on a rotary tablet press. *Eur. J. Pharm. Biopharm.* **2007**, *65*, 253–256. [CrossRef]
107. Goh, H.P.; Heng, P.W.S.; Liew, C.V. Understanding effects of process parameters and forced feeding on die filling. *Eur. J. Pharm. Sci.* **2018**, *122*, 105–115. [CrossRef]
108. Han, X.; Jallo, L.; To, D.; Ghoroi, C.; Davé, R. Passivation of high-surface-energy sites of milled ibuprofen crystals via dry coating for reduced cohesion and improved flowability. *J. Pharm. Sci.* **2013**, *102*, 2282–2296. [CrossRef]
109. Zakhvatayeva, A.; Hare, C.; Wu, C.Y. Size-induced segregation during die filling. *Int. J. Pharm. X* **2019**, *1*, 100032. [CrossRef] [PubMed]
110. Wu, C.Y. DEM simulations of die filling during pharmaceutical tabletting. *Particuology* **2008**, *6*, 412–418. [CrossRef]
111. Mittal, B. *How to Develop Robust Solid Oral Dosage Forms: From Conception to Post-Approval*; Academic Press: Cambridge, MA, USA, 2016.
112. Bansal, A.K.; Balwani, G.; Sheokand, S. *Critical Material Attributes in Wet Granulation*; Academic Press: Cambridge, MA, USA, 2019; pp. 421–453.
113. Eidevåg, T.; Abrahamsson, P.; Eng, M.; Rasmuson, A. Modeling of dry snow adhesion during normal impact with surfaces. *Powder Technol.* **2020**, *361*, 1081–1092. [CrossRef]
114. Peeters, E.; Beer, T.D.; Vervaet, C.; Remon, J.p.; Peeters, E.; Beer, T.D.; Vervaet, C.; Remon, J.p. Reduction of tablet weight variability by optimizing paddle speed in the forced feeder of a high-speed rotary tablet press forced feeder of a high-speed rotary tablet press. *Drug Dev. Ind. Pharm.* **2015**, *41*, 530–539. [CrossRef] [PubMed]
115. Kushner, J.; Moore, F. Scale-up model describing the impact of lubrication on tablet tensile strength. *Int. J. Pharm.* **2010**, *399*, 19–30. [CrossRef]
116. Mosig, J.; Kleinebudde, P. Critical evaluation of root causes of the reduced compactability after roll compaction/dry granulation. *J. Pharm. Sci.* **2015**, *104*, 1108–1118. [CrossRef] [PubMed]
117. Almaya, A.; Aburub, A. Effect of Particle Size on Compaction of Materials with Different Deformation Mechanisms with and without Lubricants. *Aaps Pharmscitech* **2008**, *9*, 414–418. [CrossRef] [PubMed]
118. Guo, Y.; Wu, C.Y.; Thornton, C. The effects of air and particle density difference on segregation of powder mixtures during die filling. *Chem. Eng. Sci.* **2011**, *66*, 661–673. [CrossRef]
119. Schneider, L.C.; Sinka, I.C.; Cocks, A.C. Characterisation of the flow behaviour of pharmaceutical powders using a model die-shoe filling system. *Powder Technol.* **2007**, *173*, 59–71. [CrossRef]
120. Kukkar, V.; Anand, V.; Kataria, M.; Gera, M.; Choudhury, P.K. Mixing and formulation of low dose drugs: Underlying problems and solutions. *Thai J. Pharm. Sci* **2008**, *32*, 43–58.
121. Byrne, F.P.; Jin, S.; Paggiola, G.; Petchey, T.H.M.; Clark, J.H.; Farmer, T.J.; Hunt, A.J.; Robert McElroy, C.; Sherwood, J. Tools and techniques for solvent selection: Green solvent selection guides. *Sustain. Chem. Process.* **2016**, *4*, 7. [CrossRef]
122. Javadzadeh, Y.; Vazifehasl, Z.; Maleki Dizahj, S.; Mokhtarpor, M. Spherical crystallization of drugs. *Acta Pharm.* **2012**, *62*, 1–14. [CrossRef]
123. Pitt, K.; Peña, R.; Tew, J.D.; Pal, K.; Smith, R.; Nagy, Z.K.; Litster, J.D. Particle design via spherical agglomeration: A critical review of controlling parameters, rate processes and modelling. *Powder Technol.* **2018**, *326*, 327–343. [CrossRef]
124. Quodbach, J.; Kleinebudde, P. A critical review on tablet disintegration. *Pharm. Dev. Technol.* **2016**, *21*, 763–774. [CrossRef]
125. Nicolas, V.; Chambin, O.; Andrès, C.; Rochat-Gonthier, M.H.; Pourcelot, Y. Preformulation: Effect of moisture content on microcrystalline cellulose (Avicel PH-302) and its consequences on packing performances. *Drug Dev. Ind. Pharm.* **1999**, *25*, 1137–1142. [CrossRef]
126. Tye, C.K.; Sun, C.; Amidon, G.E. Evaluation of the effects of tableting speed on the relationships between compaction pressure, tablet tensile strength, and tablet solid fraction. *J. Pharm. Sci.* **2005**, *94*, 465–472. [CrossRef] [PubMed]
127. Osei-Yeboah, F.; Zhang, M.; Feng, Y.; Sun, C.C. A formulation strategy for solving the overgranulation problem in high shear wet granulation. *J. Pharm. Sci.* **2014**, *103*, 2434–2440. [CrossRef]

128. McKenna, A.; McCafferty, D.F. Effect of particle size on the compaction mechanism and tensile strength of tablets. *J. Pharm. Pharmacol.* **1982**, *34*, 347–351. [CrossRef] [PubMed]
129. Hallam, C.N.; Gabbott, I.P. Increasing tensile strength by reducing particle size for extrudate-based tablet formulations. *J. Drug Deliv. Sci. Technol.* **2019**, *52*, 825–830. [CrossRef]
130. Sun, C.; Grant, D.J. Influence of crystal structure on the tableting properties of sulfamerazine polymorphs. *Pharm. Res.* **2001**, *18*, 274–280. [CrossRef] [PubMed]
131. Eriksson, M.; Alderborn, G. The effect of particle fragmentation and deformation on the interparticulate bond formation process during powder compaction. *Pharm. Res.* **1995**, *12*, 1031–1039. [CrossRef] [PubMed]
132. Wünsch, I.; Finke, J.H.; John, E.; Juhnke, M.; Kwade, A. The influence of particle size on the application of compression and compaction models for tableting. *Int. J. Pharm.* **2021**, *599*, 120424. [CrossRef]
133. Šimek, M.; Grünwaldová, V.; Kratochvíl, B. Comparison of compression and material properties of differently shaped and sized paracetamols. *KONA Powder Part J.* **2017**, *2017*, 197–206. [CrossRef]
134. Choi, D.H.; Kim, N.A.; Chu, K.R.; Jung, Y.J.; Yoon, J.h.; Jeong, S.H. Material Properties and Compressibility Using Heckel and Kawakita Equation with Commonly Used Pharmaceutical Excipients. *J. Pharm. Investig.* **2010**, *40*, 237–244. [CrossRef]
135. Prasad, K.V.; Sheen, D.B.; Sherwood, J.N. Fracture property studies of paracetamol single crystals using microindentation techniques. *Pharm. Res.* **2001**, *18*, 867–872. [CrossRef]
136. Roberts, R.; Rowe, R.C. The compaction of pharmaceutical and other model materials—A pragmatic approach. *Chem. Eng. Sci.* **1987**, *42*, 903–911. [CrossRef]
137. Skelbæk-pedersen, A.L.; Vilhelmsen, T.K.; Wallaert, V. Investigation of the e ff ects of particle size on fragmentation during tableting. *Int. J. Pharm.* **2020**, *576*, 118985. [CrossRef]
138. Wong, L.W.; Pilpel, N. The effect of particle shape on the mechanical properties of powders. *Int. J. Pharm.* **1990**, *59*, 145–154. [CrossRef]
139. Upadhyay, P.; Khomane, K.S.; Kumar, L.; Bansal, A.K. Relationship between crystal structure and mechanical properties of ranitidine hydrochloride polymorphs. *CrystEngComm* **2013**, *15*, 3959–3964. [CrossRef]
140. Khomane, K.S.; Bansal, A.K. Effect of particle size on in-die and out-of-die compaction behavior of ranitidine hydrochloride polymorphs. *AAPS PharmSciTech* **2013**, *14*, 1169–1177. [CrossRef] [PubMed]
141. Packham, D.E. Surface energy , surface topography and adhesion. *Int. J. Adhes. Adhes.* **2003**, *23*, 437–448. [CrossRef]
142. Annamalai, M.; Gopinadhan, K.; Han, S.A.; Saha, S.; Park, H.J.; Cho, E.B.; Kumar, B.; Patra, A.; Kim, S.W.; Venkatesan, T. Surface energy and wettability of van der Waals structures. *Nanoscale* **2016**, *8*, 5764–5770. [CrossRef]
143. Li, Q.; Rudolph, V.; Weigl, B.; Earl, A. Interparticle van der Waals force in powder flowability and compactibility. *Int. J. Pharm.* **2004**, *280*, 77–93. [CrossRef]
144. Fichtner, F.; Rasmuson, A.; Alderborn, G. Particle size distribution and evolution in tablet structure during and after compaction. *Int. J. Pharm.* **2005**, *292*, 211–225. [CrossRef] [PubMed]
145. Chen, L.; Ding, X.; He, Z.; Huang, Z.; Kunnath, K.T.; Zheng, K.; Davé, R.N. Surface engineered excipients: I. improved functional properties of fine grade microcrystalline cellulose. *Int. J. Pharm.* **2018**, *536*, 127–137. [CrossRef] [PubMed]
146. Ho, R.; Hinder, S.J.; Watts, J.F.; Dilworth, S.E.; Williams, D.R.; Heng, J.Y. Determination of surface heterogeneity of d-mannitol by sessile drop contact angle and finite concentration inverse gas chromatography. *Int. J. Pharm.* **2010**, *387*, 79–86. [CrossRef] [PubMed]
147. Shinebaum, R. Investigating the Mechanical Properties of Pharmaceutical Excipients in Granule and Tablet Production. Ph.D. Thesis, Univerisity of Birmingham, Edgbaston, UK, 2021.
148. Morin, G.; Briens, L. The effect of lubricants on powder flowability for pharmaceutical application. *AAPS PharmSciTech* **2013**, *14*, 1158–1168. [CrossRef] [PubMed]
149. Lakio, S.; Vajna, B.; Farkas, I.; Salokangas, H.; Marosi, G.; Yliruusi, J. Challenges in detecting magnesium stearate distribution in tablets. *AAPS PharmSciTech* **2013**, *14*, 435–444. [CrossRef] [PubMed]
150. Abe, H.; Otsuka, M. Effects of lubricant-mixing time on prolongation of dissolution time and its prediction by measuring near infrared spectra from tablets. *Drug Dev. Ind. Pharm.* **2012**, *38*, 412–419. [CrossRef]
151. Kunnath, K.; Huang, Z.; Chen, L.; Zheng, K.; Davé, R. Improved properties of fine active pharmaceutical ingredient powder blends and tablets at high drug loading via dry particle coating. *Int. J. Pharm.* **2018**, *543*, 288–299. [CrossRef]
152. Hentzschel, C.M.; Alnaief, M.; Smirnova, I.; Sakmann, A.; Leopold, C.S. Tableting properties of silica aerogel and other silicates. *Drug Dev. Ind. Pharm.* **2012**, *38*, 462–467. [CrossRef]
153. Khunawattanakul, W.; Puttipipatkhachorn, S.; Rades, T.; Pongjanyakul, T. Novel chitosan-magnesium aluminum silicate nanocomposite film coatings for modified-release tablets. *Int. J. Pharm.* **2011**, *407*, 132–141. [CrossRef] [PubMed]
154. Holman, J.; Tantuccio, A.; Palmer, J.; van Doninck, T.; Meyer, R. A very boring 120 h:15 million tablets under a continuous state of control. *Powder Technol.* **2021**, *382*, 208–231. [CrossRef]
155. Tahara, K.; Kono, Y.; Myerson, A.S.; Takeuchi, H. Development of Continuous Spherical Crystallization to Prepare Fenofibrate Agglomerates with Impurity Complexation Using Mixed-Suspension, Mixed-Product Removal Crystallizer. *Cryst. Growth Des.* **2018**, *18*, 6448–6454. [CrossRef]
156. Chen, Y.; Yao, Z.; Tang, S.; Tong, H.; Yanagishima, T.; Tanaka, H.; Tan, P. Morphology selection kinetics of crystallization in a sphere. *Nat. Phys.* **2021**, *17*, 121–127. [CrossRef]

157. Van Snick, B.; Holman, J.; Cunningham, C.; Kumar, A.; Vercruysse, J.; De Beer, T.; Remon, J.P.; Vervaet, C. Continuous direct compression as manufacturing platform for sustained release tablets. *Int. J. Pharm.* **2017**, *519*, 390–407. [CrossRef]
158. Ierapetritou, M.; Muzzio, F.; Reklaitis, G. Perspectives on the continuous manufacturing of powder-based pharmaceutical processes. *AIChE J.* **2016**, *62*, 1846–1862. [CrossRef]

MDPI
St. Alban-Anlage 66
4052 Basel
Switzerland
www.mdpi.com

Pharmaceutics Editorial Office
E-mail: pharmaceutics@mdpi.com
www.mdpi.com/journal/pharmaceutics

www.ingramcontent.com/pod-product-compliance
Lightning Source LLC
Chambersburg PA
CBHW061830210926
51850CB00006B/850